New Directions on Model Predictive Control

New Directions on Model Predictive Control

Special Issue Editors

Jinfeng Liu
Helen E. Durand

MDPI • Basel • Beijing • Wuhan • Barcelona • Belgrade

Special Issue Editors
Jinfeng Liu
University of Alberta
Canada

Helen E. Durand
Wayne State University
USA

Editorial Office
MDPI
St. Alban-Anlage 66
4052 Basel, Switzerland

This is a reprint of articles from the Special Issue published online in the open access journal *Mathematics* (ISSN 2227-7390) in 2018 (available at: https://www.mdpi.com/journal/mathematics/special_issues/New_Directions_Model_Predictive_Control)

For citation purposes, cite each article independently as indicated on the article page online and as indicated below:

LastName, A.A.; LastName, B.B.; LastName, C.C. Article Title. *Journal Name* **Year**, *Article Number*, Page Range.

ISBN 978-3-03897-420-8 (Pbk)
ISBN 978-3-03897-421-5 (PDF)

Contents

About the Special Issue Editors

Jinfeng Liu, Associate Professor, received B.S. and M.S. degrees in Control Science and Engineering in 2003 and 2006, respectively, both from Zhejiang University, as well as a Ph.D. degree in Chemical Engineering from UCLA in 2011. In 2012, he joined the faculty of the Department of Chemical and Materials Engineering, University of Alberta in Canada. Dr. Liu's research interests include the general areas of process control theory and practice, with an emphasis on model predictive control, networked and distributed state estimation and control, and fault-tolerant process control and their applications to chemical processes, biomedical systems, and water conservation in irrigation.

Helen E. Durand, Assistant Professor, received her B.S. in Chemical Engineering from UCLA, and upon graduation joined the Materials and Processes Engineering Department as an engineer at Aerojet Rocketdyne for two and a half years. She earned her M.S. in Chemical Engineering from UCLA in 2014, and her Ph.D. in Chemical Engineering from UCLA in 2017. She is currently an Assistant Professor in the Department of Chemical Engineering and Materials Science at Wayne State University. Her research interests are in the general area of process systems engineering with a focus on process control and process operational safety.

Preface to "New Directions on Model Predictive Control"

Model predictive control (MPC) has been an important and successful advanced control technology in various industries, mainly due to its ability to effectively handle complex systems with hard control constraints. At each sampling time, MPC solves a constrained optimal control problem online, based on the most recent state or output feedback to obtain a finite sequence of control actions, and only applies the first portion. MPC presents a very flexible optimal control framework that is capable of handling a wide range of industrial issues while incorporating state or output feedback to aid in the robustness of the design.

Traditionally, centralized MPC with quadratic cost functions have dominated the focus of MPC research. Advances in computing, communication, and sensing technologies in the last decades have enabled us to look beyond the traditional MPC and have brought new challenges and opportunities in MPC research. Two important examples of this technology-driven development are distributed MPC (in which multiple local MPC controllers carry out their calculations in separate processors collaboratively) and economic MPC (in which a general economic cost function that typically is not quadratic is optimized). There are already many results focused on advances such as these in MPC. However, there are also still many important problems that require investigation within and beyond the developments to date. Along with the theoretical development in MPC, we are also witnessing the application of MPC to many non-traditional control problems.

This book consists of a compilation of works covering a number of application domains, such as hydraulic fracturing, continuous pharmaceutical manufacturing, and mineral column flotation, in addition to works covering theoretical and practical developments in topics such as economic and distributed MPC. The purpose of this book is to assemble a collection of current research in MPC that handles practically-motivated theoretical issues as well as recent MPC applications, with the aim of highlighting the significant potential benefits of new MPC theory and design.

We would like to thank those who have contributed to this book. We would also like to thank the many researchers and industrial practitioners who have contributed to the advancement of MPC over the last several decades. We would like to thank those who performed reviews of the manuscripts which comprise this book. The feedback of these reviewers and their time is invaluable. We would like to thank Dr. Jean Wu for her great support as the Managing Editor throughout the process of putting together the Special Issue for *Mathematics*, which this work represents. We would also like to thank our colleagues at the University of Alberta and at Wayne State University for their continuous support. Finally, our deepest gratitude is extended to our families and friends for their constant encouragement and support. Without them, this work would never be possible.

Jinfeng Liu, Helen E. Durand

Special Issue Editors

mathematics

MDPI

Article

Data Driven Economic Model Predictive Control

Masoud Kheradmandi and Prashant Mhaskar *

Department of Chemical Engineering, McMaster University, Hamilton, ON L8S 4L7, Canada; kheradm@mcmaster.ca

* Correspondence: mhaskar@mcmaster.ca; Tel.: +1-905-525-9140-23273

Received: 7 March 2018; Accepted: 22 March 2018; Published: 2 April 2018

Abstract: This manuscript addresses the problem of data driven model based economic model predictive control (MPC) design. To this end, first, a data-driven Lyapunov-based MPC is designed, and shown to be capable of stabilizing a system at an unstable equilibrium point. The data driven Lyapunov-based MPC utilizes a linear time invariant (LTI) model cognizant of the fact that the training data, owing to the unstable nature of the equilibrium point, has to be obtained from closed-loop operation or experiments. Simulation results are first presented demonstrating closed-loop stability under the proposed data-driven Lyapunov-based MPC. The underlying data-driven model is then utilized as the basis to design an economic MPC. The economic improvements yielded by the proposed method are illustrated through simulations on a nonlinear chemical process system example.

Keywords: Lyapunov-based model predictive control (MPC); subspace-based identification; closed-loop identification; model predictive control; economic model predictive control

1. Introduction

Control systems designed to manage chemical process operations often face numerous challenges such as inherent nonlinearity, process constraints and uncertainty. Model predictive control (MPC) is a well-established control method that can handle these challenges. In MPC, the control action is computed by solving an open-loop optimal control problem at each sampling instance over a time horizon, subject to the model that captures the dynamic response of the plant, and constraints [1]. In early MPC designs, the objective function was often utilized as a parameter to ensure closed-loop stability. In subsequent contributions, Lyapunov-based MPC was proposed where feasibility and stability from a well characterized region was built into the MPC [2,3].

With increasing recognition (and ability) of MPC designs to focus on economic objectives, the notion of Economic MPC (EMPC) was developed for linear and nonlinear systems [4–6], and several important issues (such as input rate-of-change constraint and uncertainty) addressed. The key idea with the EMPC designs is the fact that the controller is directly given the economic objective to work with, and the controller internally determines the process operation (including, if needed, a set point) [7].

Most of the existing MPC formulations, economic or otherwise, have been illustrated using first principles models. With growing availability of data, there exists the possibility of enhancing MPC implementation for situations where a first principles model may not be available, and simple 'step-test', transfer-function based model identification approaches may not suffice. One of the widely utilized approaches in the general direction of model identification are latent variable methods, where the correlation between subsequent measurements is used to model and predict the process evolution [8,9]. In one direction, Dynamic Mode Decomposition with control (DMDc) has been utilized to extract low-order models from high-dimensional, complex systems [10,11]. In another direction, subspace-based system identification methods have been adapted for the purpose of model identification, where state-space model from measured data are identified using projection methods [12–14]. To handle the resultant plant model mismatch with data-driven model based approaches, monitoring of the model validity becomes especially important.

One approach to monitor the process is to focus on control performance [15], where the control performance is monitored and compared against a benchmark control design. To focus more explicitly on the model behavior, in a recent result [16], an adaptive data-driven MPC was proposed to evaluate model prediction performance and trigger model identification in case of poor model prediction. In another direction, an EMPC using empirical model was proposed [17]. The approach relies on a linearization approach, resulting in closed-loop stability guarantees for regions where the plant-model mismatch is sufficiently small, and illustrate results on stabilization around nominally stable equilibrium points. In summary, data driven MPC or EMPC approaches, which utilize appropriate modeling techniques to identify data from closed-loop tests to handle operation around nominally unstable equilibrium points, remain to be addressed.

Motivated by the above considerations, in this work, we address the problem of data driven model based predictive control at an unstable equilibrium point. In order to identify a model around an unstable equilibrium point, the system is perturbed under closed-loop operation. Having identified a model, a Lyapunov-based MPC is designed to achieve local and practical stability. The Lyapunov-based design is then used as the basis for a data driven Lyapunov-based EMPC design to achieve economical goals while ensuring boundedness. The rest of the manuscript is organized as follows: first, the general mathematical description for the systems considered in this work and a representative formulation for Lyapunov-based model predictive control are presented. Then, the proposed approach for closed-loop model identification is explained. Subsequently, a Lyapunov-based MPC is designed and illustrated through a simulation example. Finally, an economic MPC is designed to consider economical objectives. The efficacy of the proposed method is illustrated through implementation on a nonlinear continuous stirred-tank reactor (CSTR) with input rate of change constraints. Finally, concluding remarks are presented.

2. Preliminaries

This section presents a brief description of the general class of processes that are considered in this manuscript, followed by closed-loop subspace identification and Lyapunov based MPC formulation.

2.1. System Description

We consider a multi-input multi-output (MIMO) controllable systems where $u \in \mathbb{R}^{n_u}$ denotes the vector of constrained manipulated variables, taking values in a nonempty convex subset $\mathcal{U} \subset \mathbb{R}^{n_u}$, where $\mathcal{U} = \{u \in \mathbb{R}^{n_u} \mid u_{\min} \leq u \leq u_{\max}\}$, $u_{\min} \in \mathbb{R}^{n_u}$ and $u_{\max} \in \mathbb{R}^{n_u}$ denote the lower and upper bounds of the input variables, and $y \in \mathbb{R}^{n_y}$ denotes the vector of measured output variables. In keeping with the discrete implementation of MPC, u is piecewise constant and defined over an arbitrary sampling instance k as:

$$u(t) = u(k), \quad k\Delta t \leq t < (k+1)\Delta t,$$

where Δt is the sampling time and x_k and y_k denote state and output at the kth sample time. The central problem that the present manuscript addresses is that of designing a data driven modeling and control design for economic MPC.

2.2. System Identification

In this section, a brief review of a conventional subspace-based state space system identification methods is presented [16,18,19]. These methods are used to identify the system matrices for a discrete-time linear time invariant (LTI) system of the following form:

$$x_{k+1} = Ax_k + Bu_k + w_k, \tag{1}$$

$$y_k = Cx_k + Du_k + v_k, \tag{2}$$

where $x \in \mathbb{R}^{n_x}$ and $y \in \mathbb{R}^{n_y}$ denote the vectors of state variables and measured outputs, and $w \in \mathbb{R}^{n_x}$ and $v \in \mathbb{R}^{n_y}$ are zero mean, white vectors of process noise and measurement noise with the following covariance matrices:

$$E[\begin{pmatrix} w_i \\ v_j \end{pmatrix} \begin{pmatrix} w_i^T & v_j^T \end{pmatrix}] = \begin{pmatrix} Q & S \\ S^T & R \end{pmatrix} \delta_{ij}, \tag{3}$$

where $Q \in \mathbb{R}^{n_x \times n_x}$, $S \in \mathbb{R}^{n_x \times n_y}$ and $R \in \mathbb{R}^{n_y \times n_y}$ are covariance matrices, and δ_{ij} is the Kronecker delta function. The subspace-based system identification techniques utilize Hankel matrices constructed by stacking the output measurements and manipulated variables as follows:

$$U_{1|i} = \begin{bmatrix} u_1 & u_2 & \dots & u_j \\ u_2 & u_3 & \dots & u_{j+1} \\ \dots & \dots & \dots & \dots \\ u_i & u_{i+1} & \dots & u_{i+j-1} \end{bmatrix}, \tag{4}$$

where i is a user-specified parameter that limits the maximum order of the system (n), and, j is determined by the number of sample times of data. By using Equation (4), the past and future Hankel matrices for input and output are defined:

$$U_p = U_{1|i}, \quad U_f = U_{1|i}, \quad Y_p = Y_{1|i}, \quad Y_f = Y_{1|i}. \tag{5}$$

Similar block-Hankel matrices are made for process and measurement noises $V_p, V_f \in \mathbb{R}^{in_y \times j}$ and $W_p, W_f \in \mathbb{R}^{in_x \times j}$ are defined in the similar way. The state sequences are defined as follows:

$$X_p = \begin{bmatrix} x_1 & x_2 & \dots & x_j \end{bmatrix}, \tag{6}$$

$$X_f = \begin{bmatrix} x_{i+1} & x_{i+2} & \dots & x_{i+j} \end{bmatrix}. \tag{7}$$

Furthermore, these matrices are used in the algorithm:

$$\Psi_p = \begin{bmatrix} Y_p \\ U_p \end{bmatrix}, \quad \Psi_f = \begin{bmatrix} Y_f \\ U_f \end{bmatrix}, \quad \Psi_{pr} = \begin{bmatrix} R_f \\ \Psi_p \end{bmatrix}. \tag{8}$$

By recursive substitution into the state space model equations Equations (1) and (2), it is straightforward to show:

$$Y_f = \Gamma_i X_f + \Phi_i^d U_f + \Phi_i^s W_f + V_f, \tag{9}$$

$$Y_p = \Gamma_i X_p + \Phi_i^d U_p + \Phi_i^s W_p + V_p, \tag{10}$$

$$X_f = A^i X_p + \Delta_i^d U_p + \Delta_i^s W_p, \tag{11}$$

where:

$$\Gamma_i = \begin{bmatrix} C \\ CA \\ CA^2 \\ \vdots \\ CA^{i-1} \end{bmatrix}, \quad \Phi_i^d = \begin{bmatrix} D & 0 & 0 & \dots & 0 \\ CB & D & 0 & \dots & 0 \\ CAB & CB & D & \dots & 0 \\ \dots & \dots & \dots & \dots & \dots \\ CA^{i-2}B & CA^{i-3}B & CA^{i-4}B & \dots & D \end{bmatrix}, \tag{12}$$

$$\Phi_i^s = \begin{bmatrix} 0 & 0 & 0 & \dots & 0 & 0 \\ C & 0 & 0 & \dots & 0 & 0 \\ CA & C & 0 & \dots & 0 & 0 \\ \dots & \dots & \dots & \dots & 0 & 0 \\ CA^{i-2} & CA^{i-3} & CA^{i-4} & \dots & C & 0 \end{bmatrix}, \quad (13)$$

$$\Delta_i^d = \begin{bmatrix} A^{i-1}B & A^{i-2}B & \dots & AB & B \end{bmatrix}, \quad \Delta_i^s = \begin{bmatrix} A^{i-1} & A^{i-2} & \dots & A & I \end{bmatrix}. \quad (14)$$

Equation (9) can be rewritten in the following form to have the input and output data at the left hand side of the equation [20]:

$$\begin{bmatrix} I & -\Phi_i^d \end{bmatrix} \begin{bmatrix} Y_f \\ U_f \end{bmatrix} = \Gamma_i X_f + \Phi_i^s W_f + V_f. \quad (15)$$

In open loop identification methods, in the next step, by orthogonal projecting of Equation (15) onto Ψ_p:

$$\begin{bmatrix} I & -\Phi_i^d \end{bmatrix} \Psi_f / \Psi_p = \Gamma_i X_f / \Psi_p. \quad (16)$$

Note that, the last two terms in RHS of Equation (15) are eliminated since the noise terms are independent, or othogonal to the future inputs. Equation (16) indicates that:

$$Column\_Space(W_f / W_p) = Column\_Space(({\Gamma_i^{\perp}}^T \begin{bmatrix} I & -H_i^d \end{bmatrix})^T). \quad (17)$$

Therefore, Γ_i and H_i^d can be calculated using Equation (17) by decomposition methods. These can in turn be utilized to determine the system matrices (some of these details are deferred to Section 3.1). For further discussion on system matrix extraction, the readers are referred to references [18,19].

2.3. Lyapunov-Based MPC

The Lyapunov-based MPC (LMPC) for linear system has the following form:

$$\min_{\tilde{u}_k,\dots,\tilde{u}_{k+P}} \sum_{j=1}^{N_y} ||\tilde{y}_{k+j} - y_{k+j}^{\mathrm{SP}}||_{Q_y}^2 + \sum_{j=1}^{N_u} ||\tilde{u}_{k+j} - \tilde{u}_{k+j-1}||_{R_{du}}^2, \quad (18)$$

subject to: (19)

$$\tilde{x}_{k+1} = A\tilde{x}_k + B\tilde{u}_k, \quad (20)$$

$$\tilde{y}_k = C\tilde{x}_k + D\tilde{u}_k, \quad (21)$$

$$\tilde{u} \in \mathcal{U}, \qquad \Delta\tilde{u} \in \mathcal{U}_\circ, \qquad \tilde{x}(k) = \hat{x}_l, \quad (22)$$

$$V(\tilde{x}_{k+1}) \leq \alpha V(\tilde{x}_k) \ \forall \ V(\tilde{x}_k) > \epsilon^*, \quad (23)$$

$$V(\tilde{x}_{k+1}) \leq \epsilon^* \ \forall \ V(\tilde{x}_k) \leq \epsilon^*, \quad (24)$$

where $\tilde{x}_{k+j}$, $\tilde{y}_{k+j}$, y_{k+j}^{SP} and $\tilde{u}_{k+j}$ denote predicted state and output, output set-point and calculated manipulated input variables j time steps ahead computed at time step k, and $\hat{x}_l$ is the current estimation of state, and $0 < \alpha < 1$ is a user defined parameter. The operator $||.||_Q^2$ denotes the weighted Euclidean norm defined for an arbitrary vector x and weighting matrix W as $||x||_W^2 = x^T W x$. Furthermore, $Q_y > 0$ and $R_{du} \geq 0$ denote the positive definite and positive semi-definite weighting matrices for penalizing deviations in the output predictions and for the rate of change of the manipulated inputs, respectively. Moreover, N_y and N_u denote the prediction and control horizons, respectively, and the input rate of change, given by $\Delta\tilde{u}_{k+j} = \tilde{u}_{k+j} - \tilde{u}_{k+j-1}$, takes values in a nonempty convex subset $\mathcal{U}_\circ \subset \mathbb{R}^m$, where $\mathcal{U}_\circ = \{\Delta u \in \mathbb{R}^{n_u} \mid \Delta u_{\min} \leq \Delta u \leq \Delta u_{\max}\}$. Note finally that, while the system dynamics are described in continuous time, the objective function and constraints are defined in discrete time to be consistent with the discrete implementation of the control action.

Equations (23) and (24) are representatives of Lyapunov-based stability constraint [21,22], where $V(x_k)$ is a suitable control Lyapunov function, and $\alpha, \epsilon^* > 0$ are user-specified parameters. In the presented formulation, $\epsilon^* > 0$ enables practical stabilization to account for the discrete nature of the control implementation.

Remark 1. *Existing Lyapunov-based MPC approaches exploit the fact that the feasibility (and stability) region can be pre-determined. The feasibility region, among other things, depends on the choice of the parameter α, the requested decay factor in the value of the Lyapunov function at each time step. If (reasonably) good first principles models are available, then these features of the MPC formulation provide excellent confidence over the operating region under closed-loop. In contrast, in the presence of significant plant-model mismatch (as is possibly the case with data driven models), the imposition of such decay constraints could result in unnecessary infeasibility issues. In designing the LMPC formulation with a data driven model, this possible lack of feasibility must be accounted for (as is done in Section 3.2).*

3. Integrating Lyapunov-Based MPC with Data Driven Models

In this section, we first utilize an identification approach necessary to identify good models for operation around an unstable equilibrium point. The data driven Lyapunov-based MPC design is presented next.

3.1. Closed-Loop Model Identification

Note that, when interested in identifying the system around an unstable equilibrium point, open-loop data would not suffice. To begin with, nominal open-loop operation around an unstable equilibrium point is not possible. If the nominal operation is under closed-loop, but the loop is opened to perform step tests, the system would move to the stable equilibrium point corresponding to the new input value, thereby not providing dynamic information around the desired operating point. The training data, therefore, has to be obtained using closed-loop step tests, and an appropriate closed-loop model identification method employed. Such a method is described next.

In employing closed-loop data, note that the assumption of future inputs being independent of future disturbances no longer holds, and, if not recognized, can cause biased results in system identification [18]. In order to handle this issue, the closed-loop identification approach in the projection utilizes a different variable Ψ_{pr} instead of Ψ_p. The new instrument variable, which satisfies the independence requirement, is used to project both sides of Equation (15) and the result is used to determine LTI model matrices. For further details, refer to [16,18,23].

By projecting Equation (15) onto Ψ_{pr} we get:

$$\begin{bmatrix} I & -\Phi_i^d \end{bmatrix} \Psi_f/\Psi_{pr} = \Gamma_i X_f/\Psi_{pr} + \Phi_i^s W_f/\Psi_{pr} + V_f/\Psi_{pr}. \tag{25}$$

Since the future process and measurement noises are independent of the past input/output and future setpoint in Equation (25), the noise terms cancel, resulting in:

$$\begin{bmatrix} I & -\Phi_i^d \end{bmatrix} \Psi_f/\Psi_{pr} = \Gamma_i X_f/\Psi_{pr}. \tag{26}$$

By multiplying Equation (26) by the extended orthogonal observability $\Gamma_i^{\perp}$, the state term is eliminated:

$$(\Gamma_i^{\perp})^T \begin{bmatrix} I & -\Phi_i^d \end{bmatrix} \Psi_f/\Psi_{pr} = 0. \tag{27}$$

Therefore, the column space of Ψ_f/Ψ_{pr} is orthogonal to the row space of $\begin{bmatrix}(\Gamma_i^{\perp})^T & -(\Gamma_i^{\perp})^T\Phi_i^d\end{bmatrix}$. By performing singular value decomposition (SVD) of Ψ_f/Ψ_{pr}:

$$\Psi_f/\Psi_{pr} = U\Sigma V = \begin{bmatrix}U_1 & U_2\end{bmatrix}\begin{bmatrix}\Sigma_1 & 0\\ 0 & 0\end{bmatrix}\begin{bmatrix}{V_1}^T\\ {V_2}^T\end{bmatrix}, \tag{28}$$

where Σ_1 contains dominant singular values of Ψ_f/Ψ_{pr} and, theoretically, it has the order $n_u i + n$ [18,23].

Therefore, the order of the system can be determined by the number of the dominant singular values of the Ψ_f/Ψ_{pr} [20]. The orthogonal column space of Ψ_f/Ψ_{pr} is U_2M, where $M \in \mathbb{R}^{(n_y-n)i\times(n_y-n)i}$ is any constant nonsingular matrix and is typically chosen as an identity matrix [18,23]. One approach to determine the LTI model is as follows [18]:

$$(\begin{bmatrix}\Gamma_i^{\perp} & -\Gamma_i^{\perp}\Phi_i^d\end{bmatrix})^T = U_2M. \tag{29}$$

From Equation (29), Γ_i and Φ_i^d can be estimated:

$$\begin{bmatrix}{\Gamma_i}^{\perp}\\ -(\Phi_i^d)^T{\Gamma_i}^{\perp}\end{bmatrix} = U_2, \tag{30}$$

which results in (using MATLAB (2017a, MathWorks, Natick, MA, USA) matrix index notation):

$$\begin{cases}\hat{\Gamma}_i = U_2(1:n_yi,:)^{\perp},\\ \hat{\Phi}_i^d = -\left(U_2(1:n_yi,:)^T\right)^{\dagger}U_2(n_yi+1:end,:)^T.\end{cases} \tag{31}$$

The past state sequence can be calculated as follows:

$$\hat{X}_i = \hat{\Gamma}_i^{\dagger}\begin{bmatrix}I & -\hat{\Phi}_i^d\end{bmatrix}\Psi_f/\Psi_{pr}. \tag{32}$$

The future state sequence can be calculated by changing data Hankel matrices as follows [18]:

$$R_f = R_{i+2|2i}, \tag{33}$$
$$U_p = U_{1|i+1}, \tag{34}$$
$$Y_p = Y_{1|i+1}, \tag{35}$$
$$U_f = U_{i+2|2i}, \tag{36}$$
$$Y_f = Y_{i+2|2i}, \tag{37}$$
$$\Rightarrow \hat{X}_{i+1} = \underline{\hat{\Gamma}}_i^{\dagger}\begin{bmatrix}I & -\underline{\hat{H}}_i^d\end{bmatrix}\Psi_f/\Psi_{pr}, \tag{38}$$

where $\underline{\hat{\Gamma}}_i$ is obtained by eliminating the last n_y rows of Γ_i, and $\underline{H}_i^d$ is obtained by eliminating the last n_y rows and the last n_u columns of H_i^d. Then, the model matrices can be estimated using least squares:

$$\begin{bmatrix}X_{i+1}\\ Y_{i|i}\end{bmatrix} = \begin{bmatrix}A & B\\ C & D\end{bmatrix}\begin{bmatrix}X_i\\ U_{i|i}\end{bmatrix} + \begin{bmatrix}W_{i|i}\\ V_{i|i}\end{bmatrix}. \tag{39}$$

Note that the difference between the proposed method in [18] and described method is that, in order to ensure that the observer is stable (eigenvalues of $A - KC$ are inside unit circle), instead of innovation form of LTI model, Equations (1) and (2) are used [16] to derive extended state space equations. The system matrices can be calculated as follows:

$$\begin{bmatrix} \hat{A} & \hat{B} \\ \hat{C} & \hat{D} \end{bmatrix} = \begin{bmatrix} X_{i+1} \\ Y_{i|i} \end{bmatrix} \begin{bmatrix} X_i \\ U_{i|i} \end{bmatrix}^{\dagger}. \tag{40}$$

With the proposed approach, process and measurement noise Hankel matrices can be calculated as the residual of the least square solution of Equation (39):

$$\begin{bmatrix} \hat{W}_{i|i} \\ \hat{V}_{i|i} \end{bmatrix} = \begin{bmatrix} X_{i+1} \\ Y_{i|i} \end{bmatrix} - \begin{bmatrix} \hat{A} & \hat{B} \\ \hat{C} & \hat{D} \end{bmatrix} \begin{bmatrix} X_i \\ U_{i|i} \end{bmatrix}. \tag{41}$$

Then, the covariances of plant noises can be estimated as follows:

$$\begin{bmatrix} \hat{Q} & \hat{S} \\ \hat{S}^T & \hat{R} \end{bmatrix} = E(\begin{bmatrix} \hat{W}_{i|i} \\ \hat{V}_{i|i} \end{bmatrix} \begin{bmatrix} \hat{W}_{i|i}^T & \hat{V}_{i|i}^T \end{bmatrix}). \tag{42}$$

Model identification using closed-loop data has a positive impact on the predictive capability of the model (see the simulation section for a comparison with a model identified using open-loop data).

3.2. Control Design and Implementation

Having identified an LTI model for the system (with its associated states), the MPC implementation first requires a determination of the state estimates. To this end, an appropriate state estimator needs to be utilized. In the present manuscript, a Luenberger observer is utilized for the purpose of illustration. Thus, at the time of control implementation, state estimates $\hat{x}_k$ are generated as follows:

$$\hat{x}_{k+1} = A\hat{x}_k + Bu_k + L(y_k - C\hat{x}_k), \tag{43}$$

where L is the observer gain and is computed using pole placement method, and y_k is the vector of measured variables (in deviation form, from the set point).

In order to stabilize the system at an unstable equilibrium point, a Lyapunov-based MPC is designed. The control calculation is achieved using a two-tier approach (to decouple the problem of stability enforcement and objective function tuning). The first layer calculates the minimum value of Lyapunov function that can be reached subject to the constraints. This tier is formulated as follows:

$$\begin{aligned} V_{min} = \min_{\tilde{u}_k^1} (V(\tilde{x}_{k+1})), \\ \text{subject to:} \\ \tilde{x}_{k+1} = A\tilde{x}_k + B\tilde{u}_k^1, \\ \tilde{y}_k = C\tilde{x}_k + D\tilde{u}_k^1, \\ \tilde{u}^1 \in \mathcal{U}, \quad \Delta\tilde{u}^1 \in \mathcal{U}_\circ, \quad \tilde{x}(k) = \hat{x}_l - x^{SP}, \end{aligned} \tag{44}$$

where $\tilde{x}$, $\tilde{y}$ are predicted state and output and $\tilde{u}^1$ is the candidate input computed in the first tier. x^{SP} is underlying state setpoint (in deviation form from the nominal equilibrium point), which here is the desired unstable equilibrium point (and therefore zero in terms of deviation variables). For setpoint tracking, this value can be calculated using the target calculation method; readers are referred to [24] for further details.

Note that the first tier has a prediction horizon of 1 because the objective is to only compute the immediate control action that would minimize the value of the Lyapunov function at the next time step. V is chosen as a quadratic Lyapunov function with the following form:

$$V(\tilde{x}) = \tilde{x}^T P \tilde{x}, \tag{45}$$

where P is a positive definite matrix computed by solving the Riccati equation with the LTI model matrices as follows:

$$A^T PA - P - A^T PB(B^T PB + R)^{-1} + Q = 0, \tag{46}$$

where $Q \in \mathbb{R}^{n_x \times n_x}$ and $R \in \mathbb{R}^{n_u \times n_u}$ are positive definite matrices. Then, in the second tier, this minimum value is used as a constraint (upper bound for Lyapunov function value at the next time step). The second tier is formulated as follows:

$$\begin{aligned} &\min_{\tilde{u}^2_k,\dots,\tilde{u}^2_{k+N_p}} \sum_{j=1}^{N_y} ||\tilde{y}_{k+j} - \tilde{y}^{SP}_{k+j}||^2_{Q_y} + ||\tilde{u}^2_{k+j} - \tilde{u}^2_{k+j-1}||^2_{R_{du}}, \\ &\text{subject to:} \\ &\quad \tilde{x}_{k+1} = A\tilde{x}_k + B\tilde{u}_k, \\ &\quad \tilde{y}_k = C\tilde{x}_k + D\tilde{u}_k, \\ &\quad \tilde{u}^2 \in \mathcal{U}, \qquad \Delta\tilde{u}^2 \in \mathcal{U}_\circ, \qquad \tilde{x}(k) = \hat{x}_l, \\ &\quad V(\tilde{x}_{k+1}) \leq V_{min} \; \forall \; V(\tilde{x}_k) > \epsilon^*, \\ &\quad V(\tilde{x}_{k+1}) \leq \epsilon^* \; \forall \; V(\tilde{x}_k) \leq \epsilon^* \end{aligned} \tag{47}$$

where N_p is the prediction horizon and $\tilde{u}^2$ denotes the control action computed by the second tier. In essence, in the second tier, the controller calculates a control action sequence that can take the process to the setpoint in an optimal fashion optimally while ensuring that the system reaches the minimum achievable Lyapunov function value at the next time step. Note that, in both of the tiers, the input sequence is a decision variable in the optimization problem, but only the first value of the input sequence of the second tier is implemented on the process. The solution of the first tier, however, is used to ensure and generate a feasible initial guess for the second tier. The two-tiered control structure is schematically presented in Figure 1.

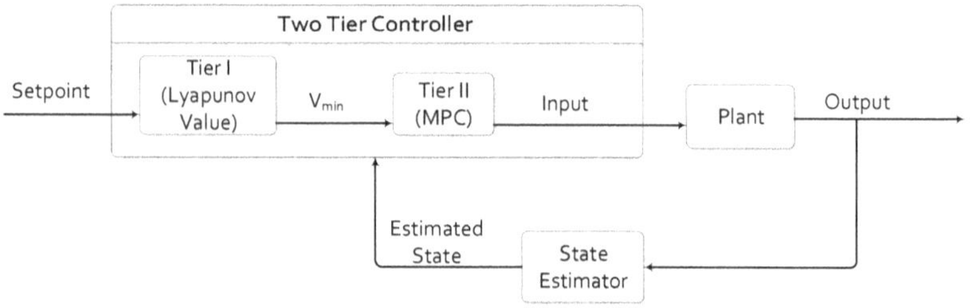

Figure 1. Two-tier control strategy.

Remark 2. *Note that Tiers 1 and 2 are executed in series and at the same time, and the implementation does not require a time scale separation. The overall optimization is split into two tiers to guarantee feasibility of the optimization problem. In particular, the first tier computes an input move with the objective function only focusing on minimizing the Lyapunov function value at the next time step. Notice that the constraints in the first tier are such that the optimization problem is guaranteed to be feasible. With this feasible solution, the second tier is used to determine the input trajectory that achieves the best performance, while requiring the Lyapunov* *function to decay. Again, since the second tier optimization problem uses the solution from Tier 1 to impose the stability constraint, feasibility of the second tier optimization problem, and, hence, of the MPC optimization problem, is guaranteed. In contrast, if one were to require the Lyapunov function to decay by an arbitrary chosen factor, determination of that factor in a way that guarantees feasibility of the optimization problem would be a non-trivial task.*

Remark 3. *It is important to recognize that, in the present formulation, feasibility of the optimization problem does not guarantee closed-loop stability. A superfluous (and incorrect) reason is as follows: the first tier computes the control action that minimizes the value of the Lyapunov function at the next step, but does not require that it be smaller than the previous time step, leading to potential destabilizing control action. The key point to realize here, however, is that if such a control action were to exist (that would lower the value of the Lyapunov function at the next time step), the optimization problem would determine that value by virtue of the Lyapunov function being the objective function, and lead to closed-loop stability. The reasons closed-loop stability may not be achieved are two: (1) the current state might be such that closed-loop stability is not achievable for the system dynamics and constraints; and (2) due to plant model mismatch, where the control action that causes the Lyapunov function to decay for the identified model does not do so for the system in question. The first reason points to a fundamental limitation due to the presence of input constraints, while the second is due to the lack of availability of the 'correct' system dynamics, and as such will be true in general for data driven MPC formulations. Note that inclusion of a noise/plant model mismatch term in the model may help with the predictive capability of the model, however, unless a bound on the uncertainty can be assumed, closed-loop stability can not be guaranteed.*

Remark 4. *Along similar lines, consider the scenario where, based on the model, and constraints, an input value exists for which* $V(x(k)) <= V(x(k-1))$ *is achievable. It can be readily shown that any solution computed by the first tier of the optimization problem would also result in* $V(x(k)) <= V(x(k-1))$ *by virtue of the objective function being the Lyapunov function at the next time step. Thus, in such a case, the explicit incorporation of the constraint* $V(x(k)) <= V(x(k-1))$ *(as is traditionally done in Lyapunov based MPC) does not help, and is not required. On the other hand, for the scenario where such an input does not exist, the inclusion of the constraint will cause the optimization problem to be infeasible. In contrast, in the proposed formulation, the MPC will compute a control action where the value of the Lyapunov function might be greater than the previous value, but greater by the smallest margin possible. The real impact of this phenomenon is in making the MPC formulation more pliable, especially when dealing with plant-model mismatch. In such scenarios, the proposed MPC continues to compute feasible (best possible, in terms of stabilizing behavior) solutions, and, should the process move into a region from where stabilization is possible, smoothly transits to computing stabilizing control action.*

Remark 5. *In the current manuscript, we focus on the cases where a first principal model is not available. If a good first principles model was available, it could be utilized directly in a nonlinear MPC design, or linearized if one were to implement a linear MPC. In the case of linearization, the applicability would be limited by the region over which the linearization holds. In contrast, note that the model utilized in the present manuscript does not result from a linearization of a nonlinear model. Instead, it is a linear model, possibly with a higher number of states than the original nonlinear model, albeit identified, and applicable, over a 'larger' region of operation, compared to a linearized model.*

Remark 6. *To account for possible plant-model mismatch, model validity can be monitored with model monitoring methods [16], resulting in appropriately triggering re-identification in case of poor model prediction. In another direction, in line with control performance monitoring approaches, the Lyapunov function value could be utilized. Thus, unacceptable increases in Lyapunov function value could be utilized as a means of triggering re-identification.*

Remark 7. *As mentioned previously, in order to create rich training data around unstable operating points, closed-loop data must be generated. In turn, since open-loop methods result in biased estimation [25,26] in model identification, a suitable closed-loop identification method is utilized, and adapted to ensure that the model accurately captures the key dynamics.*

4. Simulation Results

We next illustrate the proposed approach using a nonlinear CSTR example [27]. To this end, consider a CSTR where a first-order, exothermic and irreversible reaction of the form $A \xrightarrow{k} B$ takes place. The mass and energy conservation laws results in the following mathematical model:

$$\begin{aligned} \dot{C}_A &= \frac{F}{V}(C_{A0} - C_A) - k_0 e^{\frac{-E}{RT_R}} C_A, \\ \dot{T}_R &= \frac{F}{V}(T_{A0} - T_R) + \frac{(-\Delta H)}{\rho c_p} k_0 e^{\frac{-E}{RT_R}} C_A + \frac{Q}{\rho c_p V}. \end{aligned} \tag{48}$$

The description of the process variables and the values of the system parameters are presented in Table 1. The control objective is to stabilize the system at an unstable equilibrium point using inlet concentration, C_{A_0}, and the rate of heat input, Q, while the manipulated inputs are constrained to be within the limits $|C_{A_0}| \leq 1$ kmol/m^3 and $|Q| \leq 9 \times 10^3$ KJ/min, and the input rate is constrained as $|\Delta C_{A_0}| \leq 0.1$ Kmol/m^3 and $|\Delta Q| \leq 9 \times 200$ KJ/min. We assume that both of the states are measured. The system has an unstable equilibrium point at $C_A = 0.573$ Kmol/m^3 and $T = 395.3$ K. The goal is to stabilize the system at this equilibrium point. To this end, first an LTI model is identified using closed-loop data; then, an MPC is designed to stabilize the system at the unstable equilibrium point.

Table 1. Variable and parameter description and values for the continuous stirred-tank reactor (CSTR) example.

Variable	Description	Unit	Value
$C_{A,S}$	Nominal Value of Concentration	$\frac{kmol}{m^3}$	0.573
$T_{R,S}$	Nominal Value of Reactor Temperature	K	395
F	Flow Rate	$\frac{m^3}{min}$	0.2
V	Volume of the Reactor	$\frac{m^3}{min}$	0.2
$C_{A0,S}$	Nominal Inlet Concentration	$\frac{kmol}{m^3}$	0.787
k_0	Pre-Exponential Constant	$-$	72×10^9
E	Activation Energy	$\frac{kJ}{mol}$	8.314×10^4
R	Ideal Gas Constant	$\frac{kJ}{KmolK}$	8.314
T_{A0}	Inlet Temperature	K	352.6
ΔH	Enthalpy of the Reaction	$\frac{kJ}{Kmol}$	4.78×10^4
ρ	Fluid Density	$\frac{kg}{m^3}$	10^3
c_p	Heat Capacity	$\frac{kj}{kg.K}$	0.239

For system identification of the CSTR model, proportional–integral (PI) controllers (pairing C_A with $C_{A,in}$ and T with Q) are implemented in the process. In particular, pseudo-random binary signals are used as set-points for PI controllers. The identified LTI model order is selected as $n = 4$ and $i = 12$, in order to achieve the best fit in model prediction (using cross-validation). Note that these four states are the states of the identified LTI model. When dealing with setpoint tracking, these states can be augmented with additional states and utilized as part of an offset-free MPC design. Model validation results under a different set of set-point changes from training data are presented in Figures 2 and 3. The identified system is unstable with absolute eigenvalues $\begin{bmatrix} 0.9311 & 0.9311 & 0.9998 & 1.0002 \end{bmatrix}$, which has an eigenvalue outside unit circle. The unstable nature of the identified model is consistent with the operation of the system around the unstable equilibrium point.

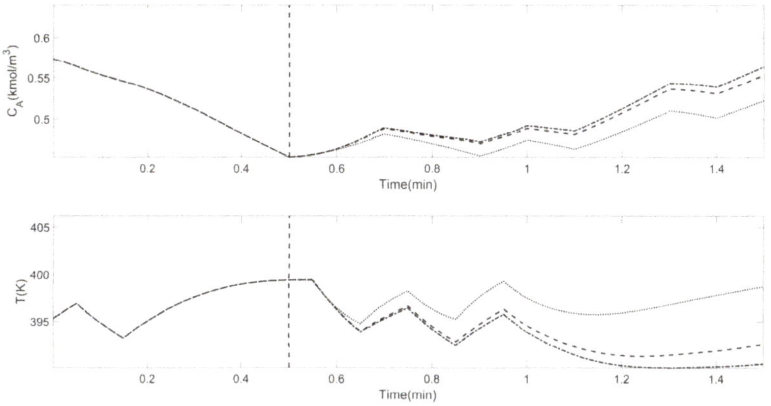

Figure 2. Data driven model validation results: measured outputs (dash-dotted line), state and output estimates using the (linear time invariant) LTI model model from closed-loop data and identification (dashed line), state and output estimates using the LTI model from open-loop data and identification (dotted line), observer stopping point (vertical dashed line).

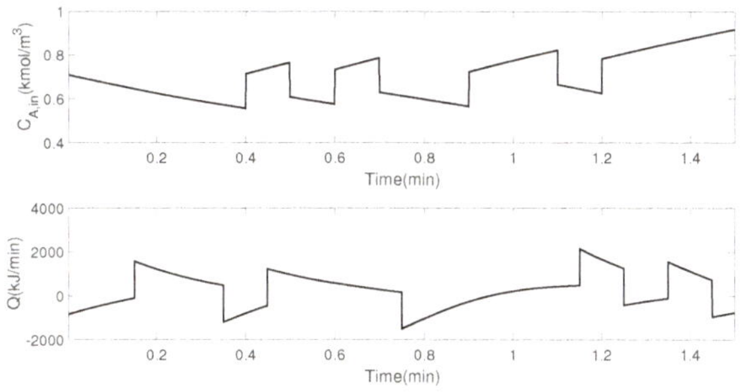

Figure 3. Model validation data: manipulated inputs under a proportional–integral (PI) controller.

For the model validation, initially, a steady state Kalman filter (gain calculated by the identification method) is utilized to update state estimate until $t = 0.8$ min and after convergence of the states (gaged via convergence of the outputs), the model and the input trajectory (without the state estimator) are used to predict the future output. Figure 2 illustrates the results of the model validation, and compares against a model obtained from open-loop step pseudo-random binary sequence (PRBS) on the input. As expected, the model identified using closed-loop data predicts better.

Next, closed-loop simulation results for proposed controller and conventional MPC (i.e., MPC without Lyapunov constraint) with horizons 1 and 10 are presented in Figures 4–7. The controllers parameters are presented in Table 2. As can be seen, the LMPC has the best performance in stabilizing the system at the unstable equilibrium point. The MPC with a horizon of 1 is not capable of stabilizing the system, and the controller with a horizon of 10 reaches the set-point later compared to the LMPC. In addition, the evolution of the subspace states indicates better performance under the proposed LMPC.

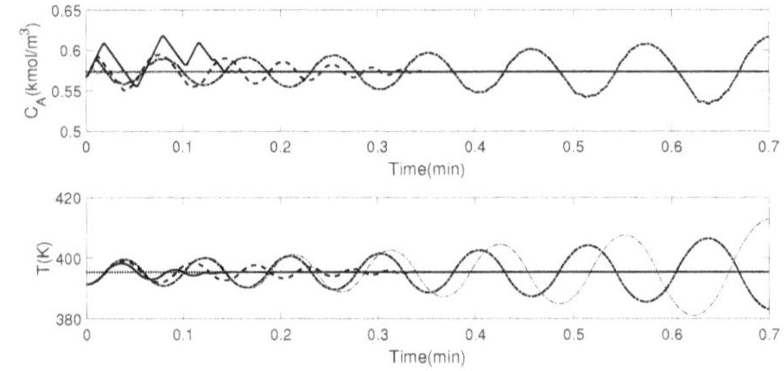

Figure 4. Closed-loop profiles of the measured variables obtained from the proposed Lyapunov-based MPC (continuous line), MPC with horizon 1 (dash-dotted line), MPC with horizon 10 (dashed line), and MPC with horizon 1 and open-loop identification (narrow dash-dotted line) and set-point (dashed line).

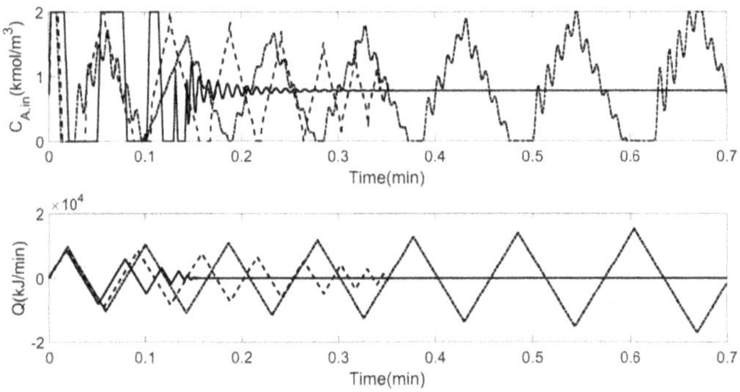

Figure 5. Closed-loop profiles of the manipulated variables obtained from the proposed LMPC (continuous line), MPC with horizon 1 (dash-dotted line), MPC with horizon 1 and open-loop identification (narrow dash-dotted line) and MPC with horizon 10 (dashed line).

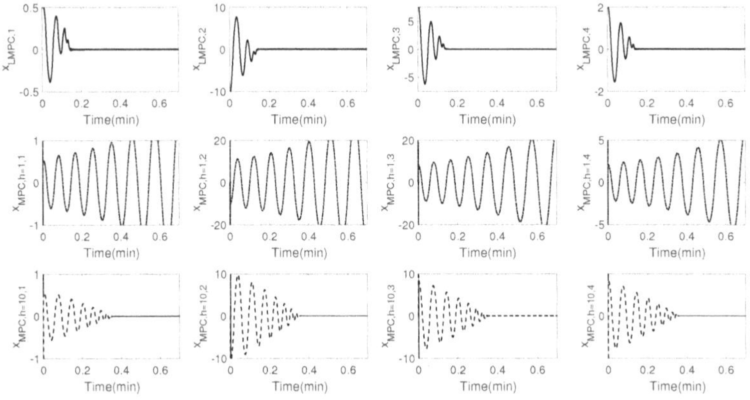

Figure 6. Closed-loop profiles of the LTI model states obtained from the proposed LMPC (continuous line), MPC with horizon 1 (dash-dotted line) and MPC with horizon 10 (dashed line).

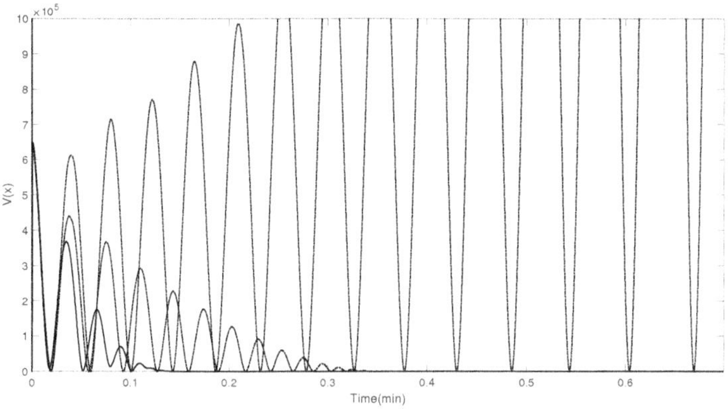

Figure 7. Closed-loop Lyapunov function profiles obtained from the proposed LMPC (continuous line), MPC with horizon 1 (dash-dotted line) and MPC with horizon 10 (dashed line).

Table 2. List of controllers parameters for the CSTR reactor.

Variable	Value
Δt	0.2 min
Q_x	$\begin{bmatrix} 1 & 0 \\ 0 & 1 \end{bmatrix}$
$Q_{x,MPC}$	$10 \times diag([1/C_{A,s}, 1/T_{R,S}])$
$R_{\Delta u,MPC}$	$diag([1/C_{A0,max}, 1/Q_{max}])$
Q_K	$diag([10^3, 10^3])$
R_K	$diag([10^{-3}, 10^{-3}])$
τ_{min}	0
τ_{max}	5
ε_i	$10^{-3} \times x_{i,Sp}$
Δu_{min}	$\begin{bmatrix} -0.1 & -200 \end{bmatrix}$
Δu_{max}	$\begin{bmatrix} 0.1 & 200 \end{bmatrix}$
ϵ^*	1
$V(x)$	$(x - x_{sp})^T(x - x_{sp})$
c_y	$\begin{bmatrix} 10^8 & 0 \end{bmatrix}^T$
c_u	$\begin{bmatrix} 0 & 0.1 \end{bmatrix}^T$
ρ	7.83×10^5

5. Data-Driven EMPC Design and Illustration

Having illustrated the ability of the LMPC to achieve stabilization, it is next utilized to achieve economical objectives while ensuring stability. The Lyapunov based EMPC formulation is as follows:

$$\max_{\tilde{u}_k,\ldots,\tilde{u}_{k+P}} \sum_{j=1}^{N_y} c_y^T \tilde{y}_{k+j} - c_u^T \tilde{u}_{k+j},$$

subject to:

$$\begin{aligned} &\tilde{x}_{k+1} = A\tilde{x}_k + B\tilde{u}_k, \\ &\tilde{y}_k = C\tilde{x}_k + D\tilde{u}_k, \\ &\tilde{u} \in \mathcal{U}, \qquad \Delta\tilde{u} \in \mathcal{U}_\circ, \qquad \tilde{x}(k) = \hat{x}_l, \\ &V(\tilde{x}_{k+j}) \leq \rho \text{ for } j = 1, \ldots, P, \end{aligned} \tag{49}$$

where the value of ρ dictates the neighborhood that the process states are allowed to evolve within. c_y and c_u indicate output and input cost vectors. Other variables have the same definition as Equation (47).

Remark 8. *In recent contributions [17,28], a Lyapunov-Based EMPC is proposed that utilizes data-driven methods to identify an empirical model for the system where the number of empirical model states is equal to the order of the plant model. In contrast, in the present work, the order of the model is selected based on the ability of the model to fit and predict dynamic behavior over a suitable range of operation, in turn allowing for an EMPC design that can reliably operate over a larger region.*

Remark 9. *The EMPC formulation in the present manuscript utilizes a linear form of the cost function for the purpose of illustration. The proposed approach is not limited by this particular choice. Any other form of the cost function, including those where the costs could be time dependent, could be readily utilized within the proposed formulation. In such scenarios, the presence of the stability constraints provide the safeguards that allow the EMPC to move the process as needed to achieve economical goals.*

Remark 10. *The use of linear models in the control design opens up the possibility of utilizing MPC formulations [3,29] that enable stabilization from the entire null controllable region (the region from which stabilization is achievable subject to input constraints). The use of the NCR can, in turn, be utilized to maximize the region over which the EMPC can be implemented, thereby maximizing the potential economic benefit. Such an implementation, however, needs to account for potential plant model mismatch owing to the use of the linear model, and remains the subject of future work.*

Next, the proposed Lyapunov-based EMPC (LEMPC) is implemented on the CSTR simulation example and compared to the LMPC implementation. The closed-loop results are presented in Figures 8–11. Exploiting the flexibility of operation within a neighborhood of the origin, the LEMPC drives the system to a point on the border of that neighborhood, which happens to be the optimal operating point, instead of the nominal operating point. Figure 12 shows the comparison of the LEMPC and LMPC. As expected, the LEMPC achieves improved economic returns compared to the conventional MPC.

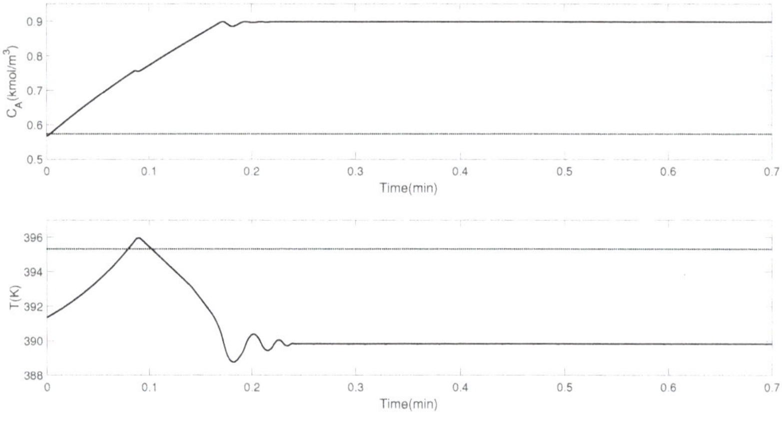

Figure 8. Closed-loop profiles of the measured variables obtained from the proposed Lyapunov-based economic MPC (continuous line) and the nominal equilibrium point (dashed line).

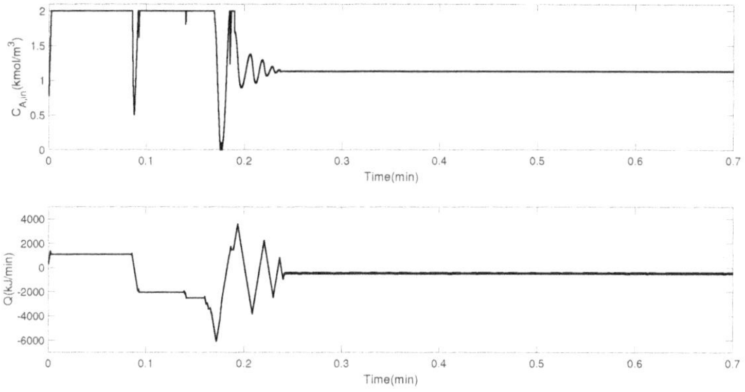

Figure 9. Closed-loop profiles of the manipulated variables obtained from the proposed LEMPC (continuous line).

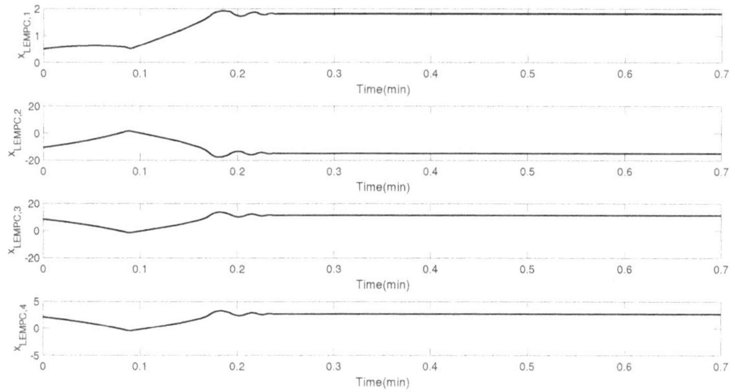

Figure 10. Closed-loop profiles of the identified model states obtained from the proposed LEMPC (continuous line).

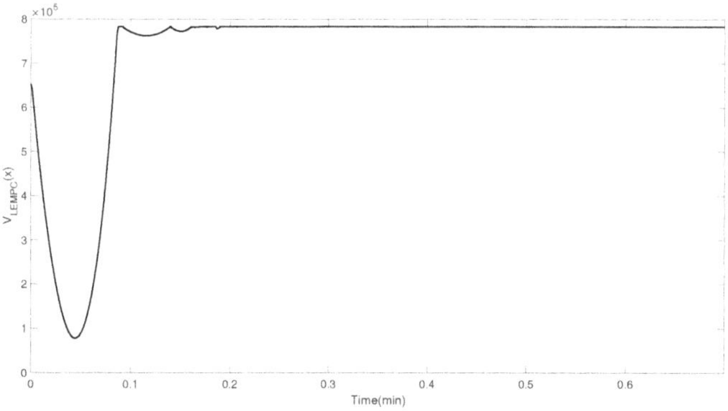

Figure 11. Closed-loop Lyapunov function profiles obtained from the proposed LEMPC (continuous line). Note that the LEMPC drives the system to a point within the acceptable neighborhood of the origin.

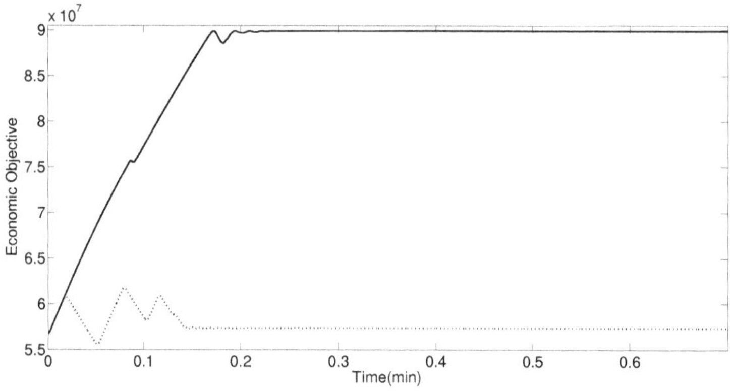

Figure 12. A comparison of the economic cost between the LEMPC (continuous line) and LMPC (dotted line).

6. Conclusions

In this study, a novel data-driven MPC is developed that enables stabilization at nominally unstable equilibrium points. This LMPC is then utilized within an economic MPC formulation to yield a data driven EMPC formulation. The proposed approach is described and compared against a representative MPC, and shown to be able to provide improved closed-loop performance.

Acknowledgments: Financial support from the McMaster Advanced Control Consortium (MACC) is gratefully acknowledged.

Author Contributions: Masoud Kheradmandi as the lead author was the primary contributor, contributed to conceiving and designing of the framework, performed all the simulations and wrote the first draft of the manuscript, and made susbequent revisions. The advisor Prashant Mhaskar contributed to conceiving and designing of the framework, analyzing the data and revising the paper.

Conflicts of Interest: The authors declare no conflict of interest.

References

1. Rawlings, J.B.; Mayne, D.Q. *Model Predictive Control: Theory and Design*; Nob Hill Publishing: Madison, Wisconsin, 2009.
2. Mhaskar, P.; El-Farra, N.H.; Christofides, P.D. Predictive control of switched nonlinear systems with scheduled mode transitions. *IEEE Trans. Autom. Control* **2005**, *50*, 1670–1680.
3. Mahmood, M.; Mhaskar, P. Constrained control Lyapunov function based model predictive control design. *Int. J. Robust Nonlinear Control* **2014**, *24*, 374–388.
4. Angeli, D.; Amrit, R.; Rawlings, J.B. On average performance and stability of economic model predictive control. *IEEE Trans. Autom. Control* **2012**, *57*, 1615–1626.
5. Bayer, F.A.; Lorenzen, M.; Müller, M.A.; Allgöwer, F. Improving Performance in Robust Economic MPC Using Stochastic Information. *IFAC-PapersOnLine* **2015**, *48*, 410–415.
6. Liu, S.; Liu, J. Economic model predictive control with extended horizon. *Automatica* **2016**, *73*, 180–192.
7. Müller, M.A.; Angeli, D.; Allgöwer, F. Economic model predictive control with self-tuning terminal cost. *Eur. J. Control* **2013**, *19*, 408–416.
8. Golshan, M.; MacGregor, J.F.; Bruwer, M.J.; Mhaskar, P. Latent Variable Model Predictive Control (LV-MPC) for trajectory tracking in batch processes. *J. Process Control* **2010**, *20*, 538–550.
9. MacGregor, J.; Bruwer, M.; Miletic, I.; Cardin, M.; Liu, Z. Latent variable models and big data in the process industries. *IFAC-PapersOnLine* **2015**, *48*, 520–524.
10. Narasingam, A.; Siddhamshetty, P.; Kwon, J.S.I. Handling Spatial Heterogeneity in Reservoir Parameters Using Proper Orthogonal Decomposition Based Ensemble Kalman Filter for Model-Based Feedback Control of Hydraulic Fracturing. *Ind. Eng. Chem. Res.* **2018**, doi:10.1021/acs.iecr.7b04927.

11. Narasingam, A.; Kwon, J.S.I. Development of local dynamic mode decomposition with control: Application to model predictive control of hydraulic fracturing. *Comput. Chem. Eng.* **2017**, *106*, 501–511.
12. Huang, B.; Kadali, R. *Dynamic Modeling, Predictive Control and Performance Monitoring: A Data-Driven Subspace Approach*; Springer; Berlin/Heidelberg, Germany, 2008.
13. Li, W.; Han, Z.; Shah, S.L. Subspace identification for FDI in systems with non-uniformly sampled multirate data. *Automatica* **2006**, *42*, 619–627.
14. Hajizadeh, I.; Rashid, M.; Turksoy, K.; Samadi, S.; Feng, J.; Sevil, M.; Hobbs, N.; Lazaro, C.; Maloney, Z.; Littlejohn, E.; et al. Multivariable Recursive Subspace Identification with Application to Artificial Pancreas Systems. *IFAC-PapersOnLine* **2017**, *50*, 886–891.
15. Shah, S.L.; Patwardhan, R.; Huang, B. Multivariate controller performance analysis: methods, applications and challenges. In *AICHE Symposium Series*; American Institute of Chemical Engineers: New York, NY, USA, 1998; Volume 2002, pp. 190–207.
16. Kheradmandi, M.; Mhaskar, P. Model predictive control with closed-loop re-identification. *Comput. Chem. Eng.* **2017**, *109*, 249–260.
17. Alanqar, A.; Ellis, M.; Christofides, P.D. Economic model predictive control of nonlinear process systems using empirical models. *AIChE J.* **2015**, *61*, 816–830.
18. Huang, B.; Ding, S.X.; Qin, S.J. Closed-loop subspace identification: An orthogonal projection approach. *J. Process Control* **2005**, *15*, 53–66.
19. Qin, S.J. An overview of subspace identification. *Comput. Chem. Eng.* **2006**, *30*, 1502–1513.
20. Wang, J.; Qin, S.J. A new subspace identification approach based on principal component analysis. *J. Process Control* **2002**, *12*, 841–855.
21. Mhaskar, P.; El-Farra, N.H.; Christofides, P.D. Stabilization of nonlinear systems with state and control constraints using Lyapunov-based predictive control. *Syst. Control Lett.* **2006**, *55*, 650–659.
22. Mayne, D.Q.; Rawlings, J.B.; Rao, C.V.; Scokaert, P.O. Constrained model predictive control: Stability and optimality. *Automatica* **2000**, *36*, 789–814.
23. Qin, S.J.; Ljung, L. Closed-loop subspace identification with innovation estimation. *IFAC Proc. Vol.* **2003**, *36*, 861–866.
24. Pannocchia, G.; Rawlings, J.B. Disturbance models for offset-free model-predictive control. *AIChE J.* **2003**, *49*, 426–437.
25. Ljung, L. System identification. In *Signal Analysis and Prediction*; Springer: Boston, MA, USA, 1998.
26. Forssell, U.; Ljung, L. Closed-loop identification revisited. *Automatica* **1999**, *35*, 1215–1241.
27. Wallace, M.; Pon Kumar, S.S.; Mhaskar, P. Offset-free model predictive control with explicit performance specification. *Ind. Eng. Chem. Res.* **2016**, *55*, 995–1003.
28. Alanqar, A.; Durand, H.; Christofides, P.D. Fault-Tolerant Economic Model Predictive Control Using Error-Triggered Online Model Identification. *Ind. Eng. Chem. Res.* **2017**, *56*, 5652–5667.
29. Mahmood, M.; Mhaskar, P. Enhanced Stability Regions for Model Predictive Control of Nonlinear Process Systems. *AIChE J.* **2008**, *54*, 1487–1498.

mathematics

MDPI

Article

A Novel Distributed Economic Model Predictive Control Approach for Building Air-Conditioning Systems in Microgrids

Xinan Zhang †, Ruigang Wang and Jie Bao *

School of Chemical Engineering, University of New South Wales, Sydney, NSW 2052, Australia; zhangxn@ntu.edu.sg (X.Z.); ruigang.wang@unsw.edu.au (R.W.)

* Correspondence: j.bao@unsw.edu.au

† Current address: School of Electrical & Electronic Engineering, Nanyang Technological University, 50 Nanyang Avenue, Singapore.

Received: 26 March 2018; Accepted: 13 April 2018; Published: 17 April 2018

Abstract: With the penetration of grid-connected renewable energy generation, microgrids are facing stability and power quality problems caused by renewable intermittency. To alleviate such problems, demand side management (DSM) of responsive loads, such as building air-conditioning system (BACS), has been proposed and studied. In recent years, numerous control approaches have been published for proper management of single BACS. The majority of these approaches focus on either the control of BACS for attenuating power fluctuations in the grid or the operating cost minimization on behalf of the residents. These two control objectives are paramount for BACS control in microgrids and can be conflicting. As such, they should be considered together in control design. As individual buildings may have different owners/residents, it is natural to control different BACSs in an autonomous and self-interested manner to minimize the operational costs for the owners/residents. Unfortunately, such "selfish" operation can result in abrupt and large power fluctuations at the point of common coupling (PCC) of the microgrid due to lack of coordination. Consequently, the original objective of mitigating power fluctuations generated by renewable intermittency cannot be achieved. To minimize the operating costs of individual BACSs and simultaneously ensure desirable overall power flow at PCC, this paper proposes a novel distributed control framework based on the dissipativity theory. The proposed method achieves the objective of renewable intermittency mitigation through proper coordination of distributed BACS controllers and is scalable and computationally efficient. Simulation studies are carried out to illustrate the efficacy of the proposed control framework.

Keywords: model predictive control (MPC); dissipativity; building air-conditioning system (BACS); microgrids

1. Introduction

In the past decade, electricity generation by using renewable energy resources, such as solar energy, becomes increasingly popular due to its capability of saving fossil fuels and reducing emissions. As a result, the number of grid-connected solar generation (SG) plants rises rapidly. One of the drawbacks of SG is the fluctuations in its output power caused by renewable intermittency. In practice, large and rapid output power oscillations are often experienced in SG plants, which may lead to bus voltage instability or even blackouts in the electric grid [1–3]. Such stability problems become far more significant in the microgrids, where high penetration of SG plants can be expected [4]. One of the widely accepted solutions to the aforementioned stability problems is to increase the operating reserve of electric grid. This includes the installation of extra generators, deployment of a large amount of battery energy storage systems (BESSs), and demand side management (DSM), etc. Obviously, large scale implementation of either extra generators or BESSs will introduce very high costs. In comparison, DSM utilizes the load shifting potential of end-users and does not require additional infrastructure. Thus, DSM is considered to be one of the most cost-effective methods for providing operating reserve to the grid [5,6].

Conceptually, DSM indicates the management of electrical loads to diminish undesirable fluctuations in power flow while satisfying customer requirements. Some electrical loads are controllable, including washing machines, air-conditioners and ventilation systems, etc. Among all these loads, air-conditioners typically consume a significant portion of energy. This is especially true for modern buildings that employ central air-conditioning systems. Statistically, nearly 40% of the world's end-use electric energy is consumed by buildings and more than 50% of the building energy is used for ventilation and air-conditioning [7–9]. This shows that control of building air-conditioning systems (BACS) can be crucial for maintaining power balance in the grid, and the thermal capacity of buildings can be used as an effective tool for smoothing the power flow and shaving peak power in microgrids.

In recent years, model predictive control (MPC) has been investigated in the optimal management and operation of energy systems (including BACSs) [10–17]. For example, Maasoumy et al. [14] proposed an MPC approach to regulate building heating, ventilation and air-conditioning (HVAC) systems to offer an ancillary service to automatic generation control (AGC). The proposed method contributes to improving the accuracy of AGC in power systems. However, the operating costs of building HVAC systems and the scenarios of large scale power systems with SGs are not considered. In [15–17], researchers proposed other model based control methods to manipulate the aggregated demand of BACSs to compensate for the power fluctuations caused by SG units. These methods essentially increase the operating reserve of electric grids. Nevertheless, the associated operating costs of BACS are still not considered. In fact, the operating costs are one of the main concerns of building residents and must be taken into account in BACS control. From the perspective of building residents, the main control objective of BACSs should be the minimization of operating cost so that their electricity bills can be reduced. In [18], a hierarchical economic MPC framework based on the time-scale difference between HVAC and building thermal energy storage was developed to improve the total operation cost. In this work, distributed control is adopted since different buildings are usually subject to different energy demands and ownerships.

To study general situations in microgrids, dynamic electricity prices are employed in this paper for energy trading of distributed buildings with SGs and BACSs. Theoretically, dynamic prices are based on the current and predicted power supply/demand information that is available in the commonly used day-ahead market. Certainly, such dynamic "prices" can be either actual electricity prices for a microgrid with financially independent buildings (where the prices impact owners' economic costs) or virtual prices for a microgrid owned by one organization such as a university or company campus (where the "prices" are used as a token for the coordination of energy consumption of different buildings). With respect to the dynamic prices, each individual BACS controller can minimize its operating cost economically through demand management. This mechanism encourages BACS controllers to shave peak power demand in the high price region and shift it to a low price region. However, if not appropriately coordinated, a positive feedback loop might be formed. In this case, the prediction of an increasing electricity price stimulates building controllers to purchase and use more energy at the current step to save the predicted future SG outputs for possible energy selling at higher prices. This subsequently results in a further boosted electricity price. The presence of such a mechanism can significantly deteriorate the collective power flow profile of all participated buildings in the microgrid. Under some extreme conditions, voltage instability might also be incurred. Therefore, the formulation of the aforementioned positive feedback loop must be avoided through appropriate coordination of different BACS controllers. Proper coordination of microgrid users is necessary to attenuate excessive energy trading behaviors effectively. Advanced control methods, such as process control techniques, can be applied in this scenario to improve system-wide stability and performance [10,13,19,20]. Although some articles are published on the coordination of distributed controllers for microgrid applications [21–23], their nature of achieving economic optimum at an aggregation level makes them computationally complicated and not scalable for large scale systems, such as microgrids. Thus, distributed control without online centralized optimization is a better solution to the aforementioned problem.

In this paper, a novel distributed economic MPC approach for BACS in microgrids is developed, based on the dissipativity theory. This approach allows individual BACS MPCs to minimize their own operational costs while attenuating the fluctuations of the total power demand and ensure microgrid level stability. To allow large scale implementation, the basic idea is to constrain the behaviors of BACS controllers with additional conditions to achieve the microgrid level performance and stability. In this work, such conditions are developed based on concept of dissipativity. Being an input and output property of dynamical systems [24,25], dissipativity was found useful for stability design for feedback systems (e.g., [26,27]) and was recently applied to develop plantwide interaction analysis and distributed control approaches (e.g., [28–31]). In this paper, microgrid-wide performance and stability are represented as microgrid dissipativity conditions that in turn are translated into the constraints that each BACS controller has to satisfy. To reduce the conservativeness of the dissipativity conditions, dynamic supply rates in Quadratic Difference Form (QdF) [32] are adopted in this approach, similar to [33]. Each BACS controller can optimize its own "selfish" economic objective based on the local information (e.g., the indoor temperature) and the electricity price, subject to the above discussed constraint, without iterative optimization or negotiations.

The remaining part of this paper is organized as follows: Section 2 introduces thermal modeling of buildings with AC and SG for microgrid applications. In Section 3, the effect of dynamic electricity price on energy trading in microgrids is discussed and an illustrative price scheme is provided. The dissipativity theory is briefly reviewed in Section 4. Subsequently, microgrid-wide dissipativity is analyzed and the proposed control framework is presented. Based on the general case of dynamic electricity price, simulations are carried out with the results presented in Section 5 to show the effectiveness of the proposed approach on improving collective performance of BACSs in microgrids. Finally, a conclusion is drawn in Section 6.

2. Buildings with Air-Conditioning and Solar Generation in Microgrids

2.1. Building Thermal Modeling

To effectively reduce the energy consumption of BACS without sacrificing the comfort of residents, thermal dynamics of buildings have to be studied. Therefore, a suitable building thermal model is necessary. In literature, a number of models are proposed to quantitatively evaluate building thermal dynamics [34–38]. Typically, three exogenous disturbances, including ambient temperature, solar irradiation and heat generated by internal electrical appliances, are adopted in these models and ground temperature is neglected [35]. The complexity of these models are basically dependent on their accuracy and the number of thermal zones considered in a building. It is noted that linear state space models are commonly used and found to be accurate enough [36–38]. In addition, the complexity of such models rises drastically with the increase of the number of building thermal zones [39]. Therefore, in real-time control applications, a reduced order model is usually desirable. Without loss of generality, a lumped parameter building thermal model [38] is employed in this paper as follows:

$$
\begin{aligned}
\frac{dT_d}{dt} &= \frac{1}{C_d R_a}(T_a - T_d) + \frac{1}{C_d R_d}(T_w - T_d) + \frac{1}{C_d R_w}(T_e - T_d) + \frac{1}{C_d}\left[(P_{heat} + P_{cool} + \rho P_{ld}) + \Phi_s A \mu_1\right], \\
\frac{dT_w}{dt} &= \frac{1}{C_w R_d}(T_d - T_w) + \frac{1}{C_w}\Phi_s A \mu_2, \\
\frac{dT_e}{dt} &= \frac{1}{C_e R_w}(T_d - T_e) + \frac{1}{C_e R_{am}}(T_a - T_e),
\end{aligned}
\tag{1}
$$

where the definitions and units of notations used in Equation (1) are given in Table 1. Noticeably, expressions $\Phi_s A\mu_1$ and $\Phi_s A\mu_2$ represent the solar radiation transferred through an external building envelop to heat up indoor air and interior walls, respectively. Considering the effect of shading, the usage of heat insulation materials in the envelop of most buildings and the heat transfer rate between air and wall, the values of μ_1 and μ_2 are chosen to be 0.02 and 0.0075 in this paper. It should be pointed out that the control framework to be presented in Section 4 can be applied with any building thermal model, including those more detailed models that consider the dynamics of each individual thermal zone.

Table 1. Variables in the thermal model.

Variable	Definition	Unit
T_d	Temperature of building indoor air	°C
T_w	Temperature of building interior walls	°C
T_e	Temperature of building envelop	°C
T_a	Temperature of ambient environment	°C
Φ_s	Solar radiation	kW/m^2
P_{heat}	Heating power from air-conditioning system (positive)	kW
P_{cool}	Cooling power from air-conditioning system (negative)	kW
P_{ld}	Power consumption of indoor appliances (excluding AC system)	kW
ρ	Fraction of heat generated from the operation of indoor appliances	
R_a	Thermal resistance between indoor air and ambient environment	°C/kW
R_d	Thermal resistance between interior walls and indoor air	°C/kW
R_w	Thermal resistance between indoor air and building envelop	°C/kW
R_{am}	Thermal resistance between building envelop and ambient environment	°C/kW
C_d	Heat capacity of indoor air	kW/°C
C_w	Heat capacity of interior walls	kW/°C
C_e	Heat capacity of building envelop	kW/°C
A	Effective area of building envelop	m^2
μ_1	Coefficient of solar radiation transferred through building envelop to heat up indoor air	
μ_2	Coefficient of solar radiation transferred through building envelop to heat up interior walls	

2.2. *Building with Air-Conditioning and Solar Generation*

As mentioned before, modern buildings are usually equipped with automatically controlled air-conditioners (AC), which includes on/off type and inverter based variable frequency type. In most cases, the latter shows superior performance over the former even though it is more expensive [40]. According to some previous studies, the variable frequency air-conditioner (VFAC) can achieve significant energy savings and simultaneously provide better comfort level [40–42] compared to its on/off counterpart. This means the increased capital cost of VFAC can be readily paid back through the reduction of electricity bills. Consequently, nowadays, VFACs are widely used to replace the conventional on/off models [43]. In view of such circumstances, the VFAC, which is capable of continuously adjusting its output heating/cooling power, is assumed to be the default air-conditioner for buildings in this paper.

In the microgrids, it is common to install solar photovoltaic (PV) systems on the roof of buildings to reduce buildings' energy dependency on the grid. In this way, many buildings can have their own clean energy sources to support loads or even sell surplus electric energy back to the grid. In order to describe a general situation, SG systems are assumed to be present in the buildings of microgrids in this work. Conceptually, the studied building system with AC and SG can be briefly depicted by Figure 1. Denote P_b and P_s as the power purchased from and sold to the microgrid by the building, respectively. Then, the power balance equation can be expressed as follows:

$$P_b + P_s + P_{pv} = P_{heat} + P_{cool} + P_{ld}, \tag{2}$$

where P_{pv} and P_{ld} are defined as the power generated by SG plant and the power consumed by the other electrical loads (excluding AC system), respectively.

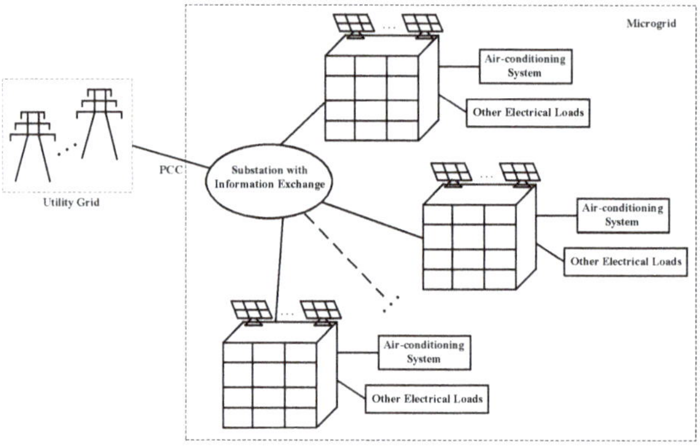

Figure 1. Block diagram of buildings with air-conditioner (AC) and solar generation (SG) in microgrids.

The output power of SG can be estimated/predicted from forecasted solar radiation and ambient temperature together with the technical specifications of the PV panels and the incremental conductance based maximum power point tracking algorithm [44].

2.3. State-Space Representation

The aim of controller is to maintenance the building's indoor temperature with a certain range while dynamically adjusting the power demands within the microgrid to achieve better economy. Thus, the state variables x, output (or controlled) variables y, manipulated variables u and disturbance variables d can be chosen as follows:

$$\begin{aligned} x &= \begin{bmatrix} T_d & T_w & T_e \end{bmatrix}^T, \; y = \begin{bmatrix} T_d & P_b & P_s \end{bmatrix}^T, \\ u &= \begin{bmatrix} P_{heat} & P_{cool} & P_b \end{bmatrix}^T, \; d = \begin{bmatrix} P_{ld} & P_{pv} & T_a & \Phi_s \end{bmatrix}^T, \end{aligned} \tag{3}$$

respectively. The discrete-time state-space model of Equations (1)–(2) can be expressed in the following compact form:

$$\begin{aligned} x(k+1) &= Ax(k) + Bu(k) + Ed(k), \\ y(k) &= Cx(k) + Du(k) + Fd(k), \end{aligned} \tag{4}$$

where

$$\begin{aligned} A &= I_3 + T_s \begin{bmatrix} -\frac{1}{C_d}\left(\frac{1}{R_a} + \frac{1}{R_d} + \frac{1}{R_w}\right) & \frac{1}{R_d C_d} & \frac{1}{R_w C_d} \\ \frac{1}{R_d C_w} & -\frac{1}{R_d C_w} & 0 \\ \frac{1}{R_w C_e} & 0 & -\frac{1}{C_e}\left(\frac{1}{R_w} + \frac{1}{R_{am}}\right) \end{bmatrix}, \; B = \frac{T_s}{C_d}\begin{bmatrix} 1 & 1 & 0 \\ 0 & 0 & 0 \\ 0 & 0 & 0 \end{bmatrix}, \\ C &= \begin{bmatrix} 1 & 0 & 0 \\ 0 & 0 & 0 \\ 0 & 0 & 0 \end{bmatrix}, \; D = \begin{bmatrix} 0 & 0 & 0 \\ 0 & 0 & 1 \\ 1 & 1 & -1 \end{bmatrix}, \; E = T_s \begin{bmatrix} \frac{\rho}{C_d} & 0 & \frac{1}{R_a C_d} & \frac{\mu_1 A}{C_d} \\ 0 & 0 & 0 & \frac{\mu_2 A}{C_w} \\ 0 & 0 & \frac{1}{R_{am} C_e} & 0 \end{bmatrix}, \; F = \begin{bmatrix} 0 & 0 & 0 & 0 \\ 0 & 0 & 0 & 0 \\ 1 & -1 & 0 & 0 \end{bmatrix}, \end{aligned} \tag{5}$$

with T_s as the controller sampling period and I_3 as a 3×3 identity matrix.

3. Electricity Price Policy for Energy Trading in Microgrids

There are research efforts on minimizing the operating costs of BACS based on the predictions of electricity price while respecting thermal comfort constraints [45–48]. Nonetheless, the attentions of these proposals are only based on static electricity price. Instead, the general case of dynamic electricity pricing, which varies with respect to real-time power supply and demand conditions, is not investigated. In practice, dynamic electricity prices are widely proposed for microgrid and smart grid applications [22,49–51] because they are effective tools for achieving power and financial balances. Furthermore, the existing proposals also neglect the effect of cost minimization by distributed BACS controllers on the overall power flow profile of microgrid. Actually, this effect is very important since the detrimental intermittent power fluctuations caused by renewable intermittency can be aggregated rather than mitigated if there is no coordination among distributed BACS controllers. The reason for such phenomenon is that buildings in one geographical area, such as a microgrid, are typically subject to the same electricity price and very similar weather conditions. As a consequence, simultaneous cost minimization of BACSs can lead to similar control actions, producing similar building power flow profiles even though there are some differences in the output of SGs installed on buildings.

It is acknowledged that a constant electricity price does not give customers incentives to change their load patterns, which can lead to supply issues in power systems during peak demand hours [9,52]. In addition, it implies that people who use electricity during off-peak hours are essentially subsidizing peak hour users [53]. Undoubtedly, such a price policy is undesirable for both the grid operator and the users. To effectively shave the peak of power demand, time-of-use (TOU) price policy is proposed and employed [47,54,55]. For example, a typical TOU price α_{UG} adopted by the state of New South Wales in Australia [54,55] is illustrated in Figure 2, from which it is seen that TOU electricity price varies with time periodically with large price difference between peak and non-peak hours. In this way, the shifting of electrical loads from peak hours to non-peak (off-peak or shoulder) hours can help reduce users' electricity cost to a large extent. Therefore, TOU price policy motivates users to shift their electricity usage to low price regions.

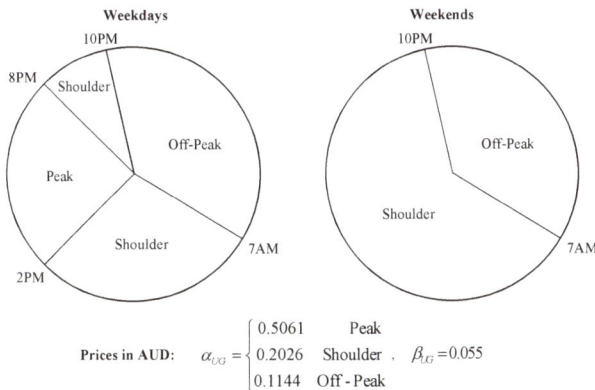

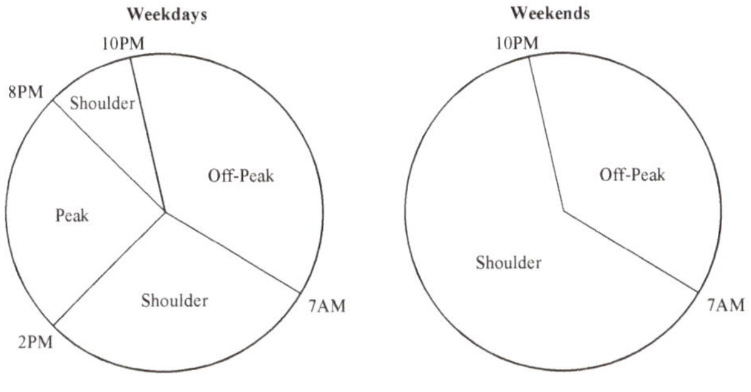

Prices in AUD: $$\alpha_{UG} = \begin{cases} 0.5061 & \text{Peak} \\ 0.2026 & \text{Shoulder} \\ 0.1144 & \text{Off-Peak} \end{cases}, \quad \beta_{UG} = 0.055$$

Figure 2. Illustrative example of time-of-use (TOU) electricity price and solar feed-in tariff.

Indeed, there are drawbacks of the existing TOU price policies. Firstly, researchers have pointed out that the ratio of peak to off-peak TOU prices has to be significant. Otherwise, the profile of user power demand will not change effectively [56]. Nevertheless, large price differences in different time intervals can result in higher overall cost for some users, and, thus, may not be preferable. Secondly, since the TOU price policy is static, dynamic supply and demand information cannot be reflected in the electricity price. Consequently, it is impossible for the electricity market to use price as an effective tool to affect power balance in the grid. In practice, dynamic electricity price has great potential to influence both the supply and demand to alleviate power imbalances in areas with high penetration of SG plants [49], such as microgrids. Theoretically, electric energy can be traded among users in microgrids with reasonable prices and this is beneficial for both grid operator and users. For the grid operator, electricity price can be regulated based on real-time supply and demand information to mitigate microgrid-wide power imbalance. Alternatively, for users in the microgrid, trading with the other users instead of the grid gives them the opportunity of getting a better electricity price. This contributes to reducing their overall cost. Therefore, dynamic electricity prices are promising for microgrid applications.

Of course, to implement dynamic electricity prices in the microgrid, energy trading between microgrid and utility grid (UG) is inevitable because power imbalances in microgrid have to be compensated by UG. As a result, the existing TOU price profile employed by UG must be considered. In general, the dynamic electricity prices should possess three features. First of all, they ought to be functions of the real-time power supply and demand in microgrid. Secondly, the users who buy energy should be charged at a price not higher than that of the UG and the users who sell energy should be paid at a price not lower than that of the UG. This feature motivates users to participate in energy trading activities. Thirdly, despite their nature of benefiting users, the dynamic electricity prices must guarantee that the microgrid can be financially self-sustained in its transactions with UG. In other words, the financial gain of microgrid through energy selling must be able to cover the financial loss of microgrid through energy buying. In theory, many pricing policies can satisfy the above features. One of the examples is

$$\begin{cases} \alpha_{mg} = [1 + (1-\gamma) P_s^{\Sigma} / P_b^{\Sigma}] \alpha_{UG}, \ \beta_{mg} = \gamma \alpha_{UG}, & \text{if } P_b^{\Sigma} + P_s^{\Sigma} > 0, \\ \alpha_{mg} = \beta_{mg} = \beta_{UG}, & \text{if } P_b^{\Sigma} + P_s^{\Sigma} \leq 0, \end{cases} \tag{6}$$

where α_{mg} and β_{mg} represent the electricity price charged on energy buyers and paid to energy sellers in microgrid, respectively, and $P_{b,\Sigma}$ (≥ 0) and $P_{s,\Sigma}$ (≤ 0) are the total electric power purchased from and sold to the microgrid by internal users, i.e.,

$$P_b^{\Sigma} = \sum_{i=1}^{M} P_b^i, \quad P_s^{\Sigma} = \sum_{i=1}^{M} P_s^i, \tag{7}$$

with P_b^i and P_s^i as the trading behaviors of individual users and M as the number of users. Notations α_{UG} and β_{UG} denote the electricity price charged on and paid to microgrid by UG when there are transactions between them. The values of α_{UG} and β_{UG} can be determined from Figure 2. Symbol γ is a user-defined constant in the range $0 < \gamma < 1$.

From Equation (6), it can be seen that

$$\begin{cases} \alpha_{mg} P_b^{\Sigma} + \beta_{mg} P_s^{\Sigma} = (P_b^{\Sigma} + P_s^{\Sigma}) \alpha_{UG} & \text{if } P_b^{\Sigma} + P_s^{\Sigma} > 0, \\ \alpha_{mg} P_b^{\Sigma} + \beta_{mg} P_s^{\Sigma} = (P_b^{\Sigma} + P_s^{\Sigma}) \beta_{UG} & \text{if } P_b^{\Sigma} + P_s^{\Sigma} \leq 0. \end{cases} \tag{8}$$

In both equations, the left-hand side represents financial gain/loss of microgrid's trading with internal users and the right-hand side indicates financial gain/loss of microgrid's trading with UG. Obviously, equivalence of the two sides implies zero financial gain/loss of microgrid. Consequently, by employing such a pricing policy, the microgrid serves as a non-profit information platform that facilitates energy trading among internal users. In practical applications, if certain operation and maintenance costs are associated with this information platform, a monthly service charge can be imposed on users. However, this service charge should be very low due to the large amount of users in microgrid and the marginal cost of running an information platform.

4. Dissipativity Based Distributed Control Framework

The distributed control diagram of building thermal systems in the microgrid is depicted by Figure 3, where B_i and C_i represent the i-th building and its corresponding controller. During each sampling period, individual controllers receive price information (both current and predicted purchasing/selling prices) from the energy trading unit, retrieve load profiles from historical data and measure building temperatures. Then, it calculates the control inputs by minimizing its economical cost and sending the power demand/supply information to the energy trading unit. The further price profile is generated by this centralized component based on the price scheme in Equation (6) and redistributed to individual controllers at the next time step.

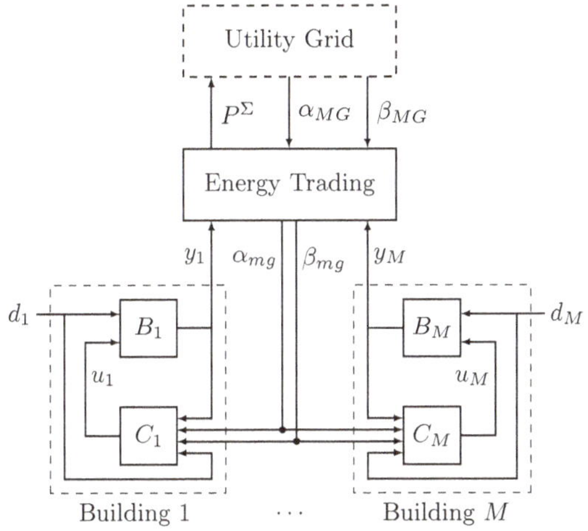

Figure 3. Block diagram of signal flows of a controlled building thermal system.

As shown in our previous work [57], without proper coordination, the "selfish" nature of each economic MPC controller could generate excessive energy trading behaviors, which may cause undesirable oscillations. In this section, the dissipativity theory, which characterizes system behaviors from input–output trajectories, is employed to resolve this issue. The whole system in Figure 3 is treated as a network of dynamical interacted subsystems. Firstly, the dissipativity analysis is performed on each components, e.g., building thermal model, controller and energy trading unit. Then, a microgrid-wide supply rate is obtained by the linear combination of individual subsystems' supply rates. Finally, a microgrid-wide dissipativity synthesis is performed offline and the online implementation involves solving distributed economic model predictive control (DEMPC) problems subject to additional dissipativity based coordination constraints.

4.1. Dissipativity and Dissipative Conditions

Consider a general discrete-time system expressed as follows:

$$x(k+1) = f(x(k), u(k)), \quad y(k) = h(x(k), u(k)), \tag{9}$$

where $x \in \mathbb{R}^n$, $u \in \mathbb{R}^p$ and $y \in \mathbb{R}^q$ are defined as the state, input and output variables, respectively. Notation k denotes the k-th sampling instant. This system is said to be dissipative if there exists a positive semidefinite function $\psi(x)$ defined on the state, called storage function, and a function $\phi(y, u)$

defined on the input and output, which is known as supply rate, such that the following inequality holds [24]

$$\psi(x(k+1)) - \psi(x(k)) \leq \phi(y(k), u(k)), \ \forall k \geq 0. \tag{10}$$

Commonly, the following (Q, S, R)-type supply rate is used

$$\phi(y(k), u(k)) = \begin{bmatrix} y(k) \\ u(k) \end{bmatrix}^T \begin{bmatrix} Q_\phi & S_\phi \\ S_\phi^T & R_\phi \end{bmatrix} \begin{bmatrix} y(k) \\ u(k) \end{bmatrix}, \tag{11}$$

where $Q_\phi = Q_\phi^T \in \mathbb{R}^{q\times q}$, $R_\phi = R_\phi^T \in \mathbb{R}^{p\times p}$ and $S_\phi \in \mathbb{R}^{q\times p}$ are parametric matrices. In general, the information of system gain is contained in Q_ϕ, R_ϕ and the phase relation is indicated by S_ϕ. For example, $Q_\phi = -I_q, S_\phi = 0, R_\phi = \rho^2 I_p$ implies a system with bounded $\mathcal{L}_2$ gain (with an $\mathcal{H}_\infty$ system norm of ρ).

Since the (Q, S, R)-type supply rate only contains input and output information at the current sampling instant, i.e., $u(k)$ and $y(k)$, it can be very conservative. The conservativeness of the dissipativity conditions can be reduced by introducing the concept of dynamic supply rate, e.g., in the quadratic differential form (QDF) [32], or the QdF for discrete time systems [58]. Such a dynamic supply rate is a function of the input and output trajectories and, consequently, captures more behavioral features of the system. An exemplary illustration of an n-th order QdF supply rate is

$$\Phi(\hat{y}, \hat{u}) := \begin{bmatrix} \hat{y}(k) \\ \hat{u}(k) \end{bmatrix}^T \begin{bmatrix} Q_\phi & S_\phi \\ S_\phi^T & R_\phi \end{bmatrix} \begin{bmatrix} \hat{y}(k) \\ \hat{u}(k) \end{bmatrix}, \tag{12}$$

where $Q_\phi = Q_\phi^T \in \mathbb{R}^{nq\times nq}$, $S_\phi \in \mathbb{R}^{nq\times np}$, $R_\phi = R_\phi^T \in \mathbb{R}^{np\times np}$, and $\hat{u}(k), \hat{y}(k)$ are the input and output trajectories defined by:

$$\hat{y}(k) = \mathrm{col}(y(k), \ldots, y(k+n-1)), \quad \hat{u}(k) = \mathrm{col}(u(k), \ldots, u(k+n-1)) \tag{13}$$

where $\mathrm{col}(y_1, \ldots, y_m) = \begin{bmatrix} y_1^T & \cdots & y_m^T \end{bmatrix}^T$. By using the above notation, a QdF storage function $\Psi(\hat{y}, \hat{u})$ can be derived as follows:

$$\Psi(\hat{y}, \hat{u}) := \begin{bmatrix} \hat{y}(k) \\ \hat{u}(k) \end{bmatrix}^T \begin{bmatrix} \mathrm{diag}(X_\psi, 0_{q\times q}) & \mathrm{diag}(Y_\psi, 0_{q\times p}) \\ \mathrm{diag}(Y_\psi^T, 0_{p\times q}) & \mathrm{diag}(Z_\psi, 0_{p\times p}) \end{bmatrix} \begin{bmatrix} \hat{y}(k) \\ \hat{u}(k) \end{bmatrix}, \tag{14}$$

where $X_\psi = X_\psi^T \in \mathbb{R}^{(n-1)q\times(n-1)q}$, $Y_\psi \in \mathbb{R}^{(n-1)q\times(n-1)p}$, $Z_\psi = Z_\psi^T \in \mathbb{R}^{(n-1)p\times(n-1)p}$. It is noted that zero vectors are concatenated with X_ψ , Y_ψ and Z_ψ to extend the dimension of parametric matrix. With slight abuse of notations, we use Φ and Ψ to denote the coefficient matrix of supply rate and storage function, i.e.,

$$\Phi = \begin{bmatrix} Q_\phi & S_\phi \\ S_\phi^T & R_\phi \end{bmatrix}, \quad \Psi = \begin{bmatrix} \mathrm{diag}(X_\psi, 0_{q\times q}) & \mathrm{diag}(Y_\psi, 0_{q\times p}) \\ \mathrm{diag}(Y_\psi^T, 0_{p\times q}) & \mathrm{diag}(Z_\psi, 0_{p\times p}) \end{bmatrix}. \tag{15}$$

Subsequently, the QdF-type dissipativity [33] can be defined as follows.

Definition 1. *System (9) is said to be dissipative with respect to the n-th order QdF-type supply rate $\Phi(\hat{y}, \hat{u})$ if there exists n-th order QdF-type storage function $\Psi(\hat{y}, \hat{u}) \geq 0$ satisfying the following dissipation inequality:*

$$\nabla\Psi(\hat{y}, \hat{u}) = \Psi(\hat{y}(k+1), \hat{u}(k+1)) - \Psi(\hat{y}(k), \hat{u}(k)) \leq \Phi(\hat{y}(k), \hat{u}(k)), \quad \forall k \geq 0. \tag{16}$$

It is noted that the change of storage function $\nabla\Psi(\hat{y}, \hat{u})$ is expressed as

$$\nabla\Psi = \begin{bmatrix} \mathrm{diag}(0_{q\times q}, X_\psi) - \mathrm{diag}(X_\psi, 0_{q\times q}) & \mathrm{diag}(0_{q\times p}, Y_\psi) - \mathrm{diag}(Y_\psi, 0_{q\times p}) \\ \mathrm{diag}(0_{p\times q}, Y_\psi^T) - \mathrm{diag}(Y_\psi^T, 0_{p\times q}) & \mathrm{diag}(0_{p\times p}, Z_\psi) - \mathrm{diag}(Z_\psi, 0_{p\times p}) \end{bmatrix}. \tag{17}$$

4.2. Dissipativity Analysis of an Individual Building in the Microgrid

Assume that there are M buildings participating in the energy trading in microgrid and the i-th ($i = 1, \ldots, M$) building system is expressed as follows:

$$\begin{aligned} x_i(k+1) &= A_i x_i(k) + B_i u_i(k) + E_i d_i(k), \\ y_i(k) &= C_i x_i(k) + D_i u_i(k) + F_i d_i(k), \end{aligned} \tag{18}$$

where variables x_i, u_i, d_i, y_i and matrices $A_i, B_i, C_i, D_i, E_i, F_i$ are defined in a similar way as those in (3)–(5).

The dissipativity property of individual building thermal model can be obtained as follows.

Proposition 1. *System (18) is dissipative with supply rate $\Phi_i(\hat{y}_i, \hat{u}_i, \hat{d}_i)$, if there exists a storage function $\Psi_i(\hat{y}_i, \hat{u}_i, \hat{d}_i)$ satisfying the following linear matrix inequalities (LMIs)*

$$\Psi_i \geq 0, \quad \begin{bmatrix} \hat{C}_i & \hat{D}_i & \hat{F}_i \\ 0 & I & 0 \\ 0 & 0 & I \end{bmatrix}^T (\Phi_i - \nabla\Psi_i) \begin{bmatrix} \hat{C}_i & \hat{D}_i & \hat{F}_i \\ 0 & I & 0 \\ 0 & 0 & I \end{bmatrix} \geq 0, \tag{19}$$

where

$$\hat{C}_i = \begin{bmatrix} C_i \\ C_i A_i \\ \vdots \\ C_i A_i^n \end{bmatrix}, \hat{D}_i = \begin{bmatrix} D_i & 0 & \cdots & 0 \\ C_i B_i & D_i & \cdots & 0 \\ \vdots & \vdots & \ddots & 0 \\ C_i A_i^{n-1} B_i & C_i A_i^{n-2} B_i & \cdots & D_i \end{bmatrix}, \hat{F}_i = \begin{bmatrix} F_i & 0 & \cdots & 0 \\ C_i E_i & F_i & \cdots & 0 \\ \vdots & \vdots & \ddots & 0 \\ C_i A_i^{n-1} E_i & C_i A_i^{n-2} E_i & \cdots & F_i \end{bmatrix}. \tag{20}$$

Proof. From (18), we have $\hat{y}_i(k) = \hat{C}_i x_(k) + \hat{D}_i \hat{u}_i(k) + \hat{F}_i \hat{d}_i(k)$. By submitting it into the dissipation inequality (16), then (19) follows as $x(k), \hat{u}_i(k), \hat{d}_i(k)$ are independent. □

4.3. Dissipativity Based DEMPC

In this work, a DEMPC approach is developed to control each BACS as MPC implement cost functions that directly reflect the actual operational costs of air-conditioners and can deal with constraints easily. For the i-th building with AC and SG, the economic optimal control problem can be expressed as follows:

$$\begin{aligned} \min_{\mathbf{u}^i} \quad & \sum_{j=0}^{N-1} \alpha_{mg}(k+j) P_b^i(k+j) + \beta_{mg}(k+j) P_s^i(k+j), \\ \text{s.t.} \quad & \tilde{x}^i(k+j+1) = A\tilde{x}^i(k+j) + Bu^i(k+j) + E\tilde{d}^i(k+j), \ \hat{x}^i(k) = x^i(k) \\ & \tilde{y}^i(k+j) = C\tilde{x}^i(k+j) + Du^i(k+j) + F\tilde{d}^i(k+j) \\ & \tilde{y}^i(k+j) \in [\underline{T}_i, \overline{T}_i] \times [0, \overline{P}_t] \times [-\overline{P}_t, 0] \\ & \tilde{u}^i(k+j) \in [0, \overline{P}_{AC}] \times [-\overline{P}_{AC}, 0] \times [0, \overline{P}_t], \end{aligned} \tag{21}$$

where $\mathbf{u}^i = \{u^i(k), \ldots, u^i(k+N-1)\}$ is the vector of decision variable and N is the prediction horizon. In addition, the constraint inequalities indicate limits imposed by user comfort temperature zone $[\underline{T}_i, \overline{T}_i]$, power distribution line limit $\overline{P}_t$ and rating of air-conditioner $\overline{P}_{AC}$.

Similar to [57], an additional dissipativity based constraint is added to individual DEMPC controllers to achieve microgrid-wide stability and performance. Since MPC is a static control law without any storage function, it could be very conservative to impose the dissipation inequality (16) to the DEMPC formulation (21). To solve this problem, the concept of dissipative trajectory, which is the integral version of (16), is adopted in this work:

Definition 2 ([31]). *An MPC controller with the supply rate $\Phi_c(\hat{y}(k), \hat{u}(k))$ is said to trace a dissipative trajectory if the following condition is satisfied:*

$$W_k = \sum_{j=0}^{k} \Phi_c(\hat{y}(j), \hat{u}(j)) \geq 0, \quad \forall k \geq 0. \tag{22}$$

To ensure that the DEMPC controller in (21) is dissipative with respect to supply rate $\Phi_{c,i}(\hat{y}_i, \hat{u}_i, \hat{d}_i)$, the following constraint

$$W_{k-1} + \hat{y}_{c,i}^T Q_{\phi_{c,i}} \hat{y}_{c,i} + \begin{bmatrix} \hat{y}_{c,i}(k) \\ \hline \hat{u}_{c,i}(k) \\ \hat{d}_{c,i}(k) \\ \hat{v}_{c,i}(k) \end{bmatrix}^T \begin{bmatrix} 0 & S_{\phi_{c,i}} \\ S_{\phi_{c,i}}^T & R_{\phi_{c,i}} \end{bmatrix} \begin{bmatrix} \hat{y}_{c,i}(k) \\ \hline \hat{u}_{c,i}(k) \\ \hat{d}_{c,i}(k) \\ \hat{v}_{c,i}(k) \end{bmatrix} \geq 0, \tag{23}$$

where $y_{c,i} = u_i$, $u_{c,i} = y_i$, $d_{c,i} = d_i$ and $v_{c,i} = \begin{bmatrix} \alpha_{mg} & \beta_{mg} \end{bmatrix}^T$ are imposed to the optimization problem (21). To ensure its recursive feasibility, the controller's supply matrix $\Phi_{c,i}$ needs to satisfy [33] the following conditions:

$$-Q_{\phi_{c,i}} \geq 0, \quad \begin{bmatrix} 0 & S_{\phi_{c,i}} \\ S_{\phi_{c,i}}^T & R_{\phi_{c,i}} \end{bmatrix} \geq 0. \tag{24}$$

4.4. Analysis of Dissipativity of Price Controller in Microgrid

The energy trading unit is a memoryless rational function of total power supply (P_s^Σ) and demand (P_b^Σ). The dissipation inequality of the CPC can be expressed as follows:

$$\begin{bmatrix} \alpha_{mg} \\ \beta_{mg} \\ P_b^\Sigma \\ P_s^\Sigma \end{bmatrix}^T \phi_e \begin{bmatrix} \alpha_{mg} \\ \beta_{mg} \\ P_b^\Sigma \\ P_s^\Sigma \end{bmatrix} \geq 0, \tag{25}$$

where ϕ_e is the quadratic supply rate (QSR) matrix. The problem in (25) can be solved efficiently by the sum-of-squares (SOS) programming method.

Here is a brief introduction to the basic concept of SOS programming. Let $\mathbb{R}[x]$ be the set of all polynomials in x with real coefficients and

$$\Sigma[x] := \{p \in \mathbb{R}[x] \mid p = p_1^2 + p_2^2 + \cdots + p_n^2,\ p_1, \ldots, p_n \in \mathbb{R}[x]\} \tag{26}$$

be the subset of $\mathbb{R}[x]$ containing the SOS polynomials. Finding a sum of squares polynomial $p(x)$ is equivalent to determination of the existence of a positive semidefinite matrix Q such that

$$p(x) = m^T(x) Q m(x) \tag{27}$$

where $m(x)$ is a vector of monomials. The SOS decomposition (27) can be efficiently and reliably achieved through semidefinite programming (SDP) [59]. In this paper, open source MATLAB toolbox YALMIP (version R20170921, Linkoping University, Linköping, Sweden) [60] and SDP solver SeDuMi (version 1.05R5, maintained by CORAL Lab, Department of Industrial and Systems Engineering, Lehigh University, Bethlehem, PA, USA) [61] are used for finding the Q matrix.

By substituting the price scheme (6) into (25), we can have the following SOS programming problem:

$$\begin{bmatrix} \alpha_{UG}[P_b^{\Sigma} + (1-\gamma)P_s^{\Sigma}] \\ \gamma\alpha_{UG}P_b^{\Sigma} \\ (P_b^{\Sigma})^2 \\ P_s^{\Sigma}P_b^{\Sigma} \end{bmatrix}^T \phi_e \begin{bmatrix} \alpha_{UG}[P_b^{\Sigma} + (1-\gamma)P_s^{\Sigma}] \\ \gamma\alpha_{UG}P_b^{\Sigma} \\ (P_b^{\Sigma})^2 \\ P_s^{\Sigma}P_b^{\Sigma} \end{bmatrix} \geq 0. \tag{28}$$

The following n-th order QdF supply rate (augment of QSR supply rate) can be written as

$$\Phi_E = \Pi_m^T \operatorname{diag}(\phi_e, \ldots, \phi_e)\Pi_m, \tag{29}$$

where the permutation matrix Π_m is defined by

$$\begin{bmatrix} \alpha_{mg}(k) \\ \beta_{mg}(k) \\ P_b^{\Sigma}(k) \\ P_s^{\Sigma}(k) \\ \vdots \\ \alpha_{mg}(k+n-1) \\ \beta_{mg}(k+n-1) \\ P_b^{\Sigma}(k+n-1) \\ P_s^{\Sigma}(k+n-1) \end{bmatrix} = \Pi_m \begin{bmatrix} \hat{\alpha}_{mg}(k) \\ \hat{\beta}_{mg}(k) \\ \hat{P}_b^{\Sigma}(k) \\ \hat{P}_s^{\Sigma}(k) \end{bmatrix}. \tag{30}$$

4.5. Microgrid-Wide Dissipativity Synthesis

Let the independent variables for the networked system be partitioned into the following sets:

$$\begin{aligned} y_g &= \operatorname{col}(P_b^1, P_s^1, \ldots, P_b^M, P_s^M), \\ d_g &= \operatorname{col}(d_1, \ldots, d_M), \\ w_g &= \operatorname{col}(P_{heat}^1, P_{cool}^1, T_d^1, \ldots, P_{heat}^M, P_{cool}^M, T_d^M, \alpha_{mg}, \beta_{mg}). \end{aligned} \tag{31}$$

Then, the microgrid-wide supply rate, which is the linear combination of the supply rates of individual subsystems, distributed controllers and energy trading unit can be represented as

$$\Phi_g(\hat{y}_g, \hat{d}_g, \hat{w}_g) = \begin{bmatrix} \hat{y}_g \\ \hat{d}_g \\ \hat{w}_g \end{bmatrix}^T \Phi_g \begin{bmatrix} \hat{y}_g \\ \hat{d}_g \\ \hat{w}_g \end{bmatrix}, \tag{32}$$

where $\Phi_g = \Pi^T \operatorname{diag}(\Phi_1, \ldots, \Phi_M, \Phi_{c,1}, \ldots, \Phi_{c,M}, \Phi_E)\Pi$ and $\Pi = \begin{bmatrix} \Pi_1^T & \Pi_2^T & \Pi_3^T \end{bmatrix}^T$ with permutation matrices Π_1, Π_2, Π_3 satisfying

$$\begin{bmatrix} \hat{y}_1 \\ \hat{u}_1 \\ \hat{d}_1 \\ \vdots \\ \hat{y}_M \\ \hat{u}_M \\ \hat{d}_M \end{bmatrix} = \Pi_1 \begin{bmatrix} \hat{y}_g \\ \hat{d}_g \\ \hat{w}_g \end{bmatrix}, \quad \begin{bmatrix} \hat{y}_{c,1} \\ \hat{u}_{c,1} \\ \hat{d}_{c,1} \\ \hat{v}_{c,1} \\ \vdots \\ \hat{y}_{c,M} \\ \hat{u}_{c,M} \\ \hat{d}_{c,M} \\ \hat{v}_{c,M} \end{bmatrix} = \Pi_2 \begin{bmatrix} \hat{y}_g \\ \hat{d}_g \\ \hat{w}_g \end{bmatrix}, \quad \begin{bmatrix} \hat{y}_{c,1} \\ \hat{u}_{c,1} \\ \hat{d}_{c,1} \\ \hat{v}_{c,1} \\ \vdots \\ \hat{y}_{c,M} \\ \hat{u}_{c,M} \\ \hat{d}_{c,M} \\ \hat{v}_{c,M} \end{bmatrix} = \Pi_3 \begin{bmatrix} \hat{y}_g \\ \hat{d}_g \\ \hat{w}_g \end{bmatrix}. \tag{33}$$

By imposing different constraints on the above microgrid-wide supply rate, we can achieve different performances of the collective behavior of all buildings. In practice, the UG operator mainly pays attention to the power flow at point of common coupling (PCC) depicted in Figure 1 because it directly affects the stability and power quality of UG. Instead, the power flows within the microgrid are usually not of concern on condition that the limits of power distribution lines are taken care of by distributed controllers.

In the context of microgrid with a high penetration of SG plants, desirable performances at PCC include: (1) reduced peak-to-peak amplitude of power flow profile; and (2) attenuated amplitude of rapid power flow fluctuations. To satisfy these two requirements, a frequency weighted $\mathcal{H}_\infty$ microgrid-wide performance (34) is employed in this paper for the collective behavior of all buildings

$$\frac{||W(z)y_g||_2}{||d_g||_2} \leq 1, \tag{34}$$

where y_g and d_g are defined in (31) and $W(z)$ is a frequency dependent weighting function utilized to penalize the mid to high frequency fluctuations of total power flow of all buildings and the excessive energy trading by individual buildings. An example of $W(z)$ is

$$W(z) = w(z)\Omega, \quad \Omega = \begin{bmatrix} 1 & 1 & 1 & \dots & 1 \\ 0 & \xi & 0 & \dots & 0 \\ 0 & 0 & \xi & \dots & 0 \\ \vdots & \vdots & \vdots & \ddots & \vdots \\ 0 & 0 & 0 & \dots & \xi \end{bmatrix} \otimes \begin{bmatrix} 1 & 1 \end{bmatrix}, \tag{35}$$

where the Kronecker operator $\otimes$ is defined by $A \otimes B = \begin{bmatrix} a_{11}B & \cdots & a_{1n}B \\ \vdots & \ddots & \vdots \\ a_{n1}B & \cdots & a_{nn}B \end{bmatrix}$. The weighting function $w(z)$ can be chosen as follows to attenuate the high frequency power fluctuations:

$$w(z) = K\frac{2T_2(1-z^{-1}) + T_s(1+z^{-1})}{2T_1(1-z^{-1}) + T_s(1+z^{-1})}, \tag{36}$$

with coefficients T_s, $T_i (i = 1, 2)$ and K representing controller sampling period, time constants and attenuation gain, respectively. In addition, Ω is a linear transformation matrix that puts weightings on both the overall power flow at PCC and the net power flow of each individual building as interpreted by

$$\Omega y_g = \begin{bmatrix} P_b^\Sigma + P_s^\Sigma \\ P_b^2 + P_s^2 \\ \vdots \\ P_b^M + P_s^M \end{bmatrix}. \tag{37}$$

In this paper, the weightings in (35) are normalized with unity weighting assigned to the overall power flow at PCC (i.e., $P_b^\Sigma + P_s^\Sigma$) and a small positive weighting of $\xi < 1$ assigned to the net power flow of each building. Physically, this means that the penalty on the amplitude of high frequency power fluctuations is mainly imposed on the overall power flow, while the excessive energy trading behavior of each building is also constrained.

The condition on the microgrid $\mathcal{H}_\infty$ performance (34) can be reinterpreted into the microgrid supply rate condition to ensure the minimum performance level of microgrid observed at PCC [33,57]. To illustrate this, the microgrid-wide supply rate is partitioned as follows:

$$\Phi_g(\hat{y}_g, \hat{d}_g, \hat{w}_g) = \begin{bmatrix} \hat{y}_g \\ \hat{d}_g \\ \hat{w}_g \end{bmatrix}^T \begin{bmatrix} \Lambda_{yy} & \Lambda_{yd} & \Lambda_{yw} \\ \Lambda_{yd}^T & \Lambda_{dd} & \Lambda_{dw} \\ \Lambda_{yw}^T & \Lambda_{dw}^T & \Lambda_{ww} \end{bmatrix} \begin{bmatrix} \hat{y}_g \\ \hat{d}_g \\ \hat{w}_g \end{bmatrix}. \tag{38}$$

According to [33], the $\mathcal{L}_2$-gain condition in (34) can be converted into the following LMIs:

$$\begin{aligned} \Lambda_{yd} = 0,\ \Lambda_{yw} = 0,\ \Lambda_{dw} = 0, \\ \Lambda_{yy} \leq -N^T N,\ \Lambda_{dd} \geq D^T D,\ \Lambda_{ww} \leq 0, \end{aligned} \tag{39}$$

where $N = \text{diag}\{K(T_s - 2T_1), K(T_s + 2T_1)\} \otimes \Omega$ and $D = \text{diag}\{T_s - 2T_2, T_s + 2T_2\} \otimes I_M$.

4.6. Distributed Control Design and Implmentation

The proposed dissipativity based DEMPC involves two steps:

- **Off-line dissipativity synthesis**: The dissipativity property for a given system is not unique. A system can have different supply rates that represent different aspects of the process dynamics (e.g., the gain and phase conditions). Therefore, dissipativity conditions for all subsystems including individual buildings, BACS controllers and the pricing controller that *allow* the required microgrid-wide stability and performance condition in (34) need to be found during the offline design step. This is done by solving LMIs in (19) for dissipativity conditions for buildings (corresponding to the building model in (18)), feasibility conditions for individual EMPC controllers in (24), the dissipativity condition for the pricing controller in (28), and the dissipativity condition representing the microgrid-wide stability and performance in (39) *simultaneously*. The outcome of this step is the dissipativity conditions (more specifically, the supply rates $Q_{\phi_{c,i}}$, $S_{\phi_{c,i}}$ and $R_{\phi_{c,i}}$ for the i-th controller) that individual EMPC controllers need to satisfy.
- **Online implementation**: solve the DEMPC optimization problem in (21) subject to an additional dissipativity based coordination constraint in (23).

5. Simulation Results

To demonstrate the efficacy of the proposed control framework, a simulation model of microgrid consisting of eight buildings with AC and SG is developed. These buildings can be divided into four groups with each two buildings in one group sharing the same thermal parameters. Simulation studies are carried out based on the illustrative dynamic price scheme presented in Section 3. The data of power consumption by electrical appliances in buildings are downloaded from Australian Energy Market Operator (AEMO) [62] and the information of weather conditions, including ambient temperature and solar radiation, is obtained from the weather station of Murdoch University [63]. For the reference, profiles of the aforementioned data for one building are plotted in Figure 4. Furthermore, the values of thermal parameters for different types of buildings (with their definitions given in Table 1) are given in Table 2. In addition, an exemplary weighting function $W(z)$ is designed by selecting $K = 2$, $T_1 = 2\pi \times 3$ rad/h, $T_2 = \frac{2\pi}{48}$ rad/h, and $\xi = 0.1$. The controller sampling period is selected to be 10 min and the online computation time for individual DEMPC controller is less than 10 s, which is negligible. The small control latency is due to the non-iterative feature of the proposed approach.

Table 2. Values of thermal parameters for four types of buildings.

Buildings	R_a	R_d	R_w	R_{am}	C_d	C_w	C_e	μ_1	μ_2	A	ρ
Type 1	3.26	0.21	0.132	0.0389	76.02	874.94	2767.1	0.02	0.0075	10500	0.1
Type 2	4.22	0.22	0.142	0.133	37.04	337	1465.28	0.02	0.0075	3600	0.1
Type 3	3.95	0.23	0.146	0.12	36.72	300	1302.5	0.02	0.0075	3200	0.1
Type 4	3.54	0.2	0.144	0.142	32.85	312	1310	0.02	0.0075	3200	0.1

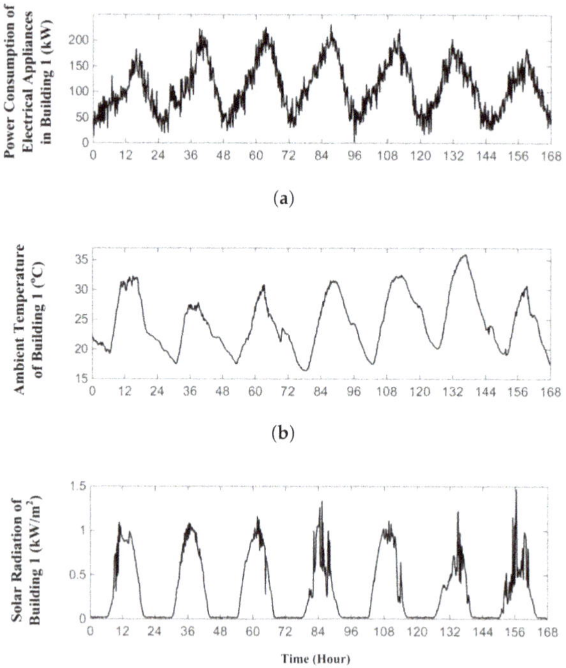

(**a**)

(**b**)

(**c**)

Figure 4. Representative profiles of: (**a**) power consumption of electrical appliances (excluding building air-conditioning systems (BACS)) in building 7 (kW); (**b**) ambient temperature of building 7 (°C); (**c**) solar radiation of buiding 1 (kW/m^2).

Firstly, the dynamic responses of different buildings are simulated by using distributed MPC as BACS controller without dissipativity based coordination (23). The results are given in Figures 5a and 6a, from which high amplitude fluctuations are seen in both the dynamic electricity prices and the overall power flow of microgrid. This is caused by the "selfish" optimizations by distributed BACS controllers, which formulate a positive feedback loop in energy trading as analyzed in the Introduction.

After implementing the dissipativity based coordination (23), which imposes constraints on the overall power flow of all participating buildings in the microgrid, comparative simulations are run under the same conditions of Figures 5a and 6a. The corresponding results are shown in Figures 5b and 6b. By comparing these four figures, it is seen that high amplitude fluctuations in electricity prices and the overall power flow are effectively attenuated, which implies successful mitigation of excessive energy trading of buildings. Moreover, from the third subplots of Figure 6, it is observable that the peak-to-peak amplitude of overall power flow is reduced by approximately 50% with the application of dissipativity based coordination. This means that coordination of distributed BACS controllers can reduce microgrid's energy dependency on UG.

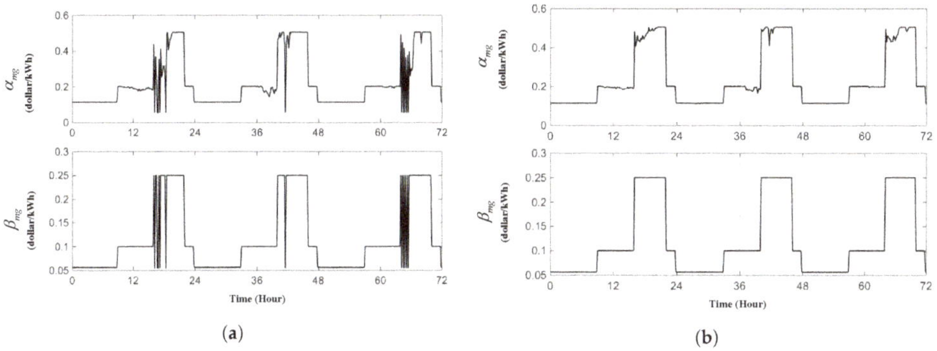

Figure 5. Dynamic electricity price for energy trading: (**a**) with; (**b**) without dissipativity based coordination.

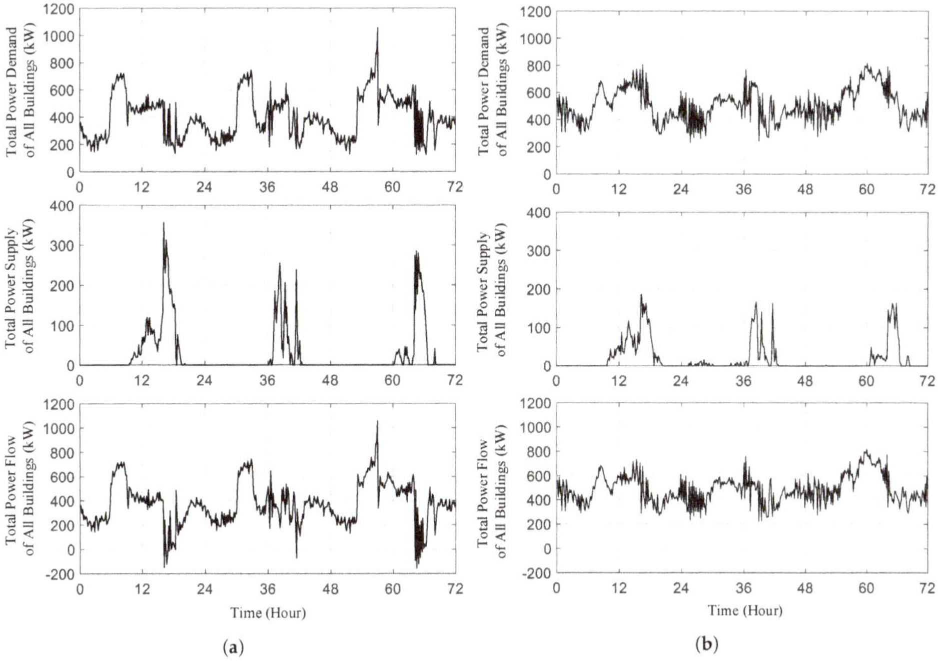

Figure 6. Overall power flow profiles of all buildings: (**a**) without dissipativity based coordination; (**b**) with dissipativity based coordination.

To investigate the impact of dissipativity based coordination on the response of individual buildings, corresponding simulation results are presented in Figure 7, from which it can be seen that the indoor temperature is kept within the required comfort zone and fluctuations in the total power flow and the consumption by AC of an individual building is also attenuated.

To further demonstrate the effectiveness of the proposed control framework, the responses of electricity prices and overall power flow profiles are simulated during the weekend. The corresponding results are given in Figures 8 and 9, from which improvements on the damping of rapid fluctuations and peak-to-peak amplitude of power flow can also be seen in the responses with dissipativity based coordination.

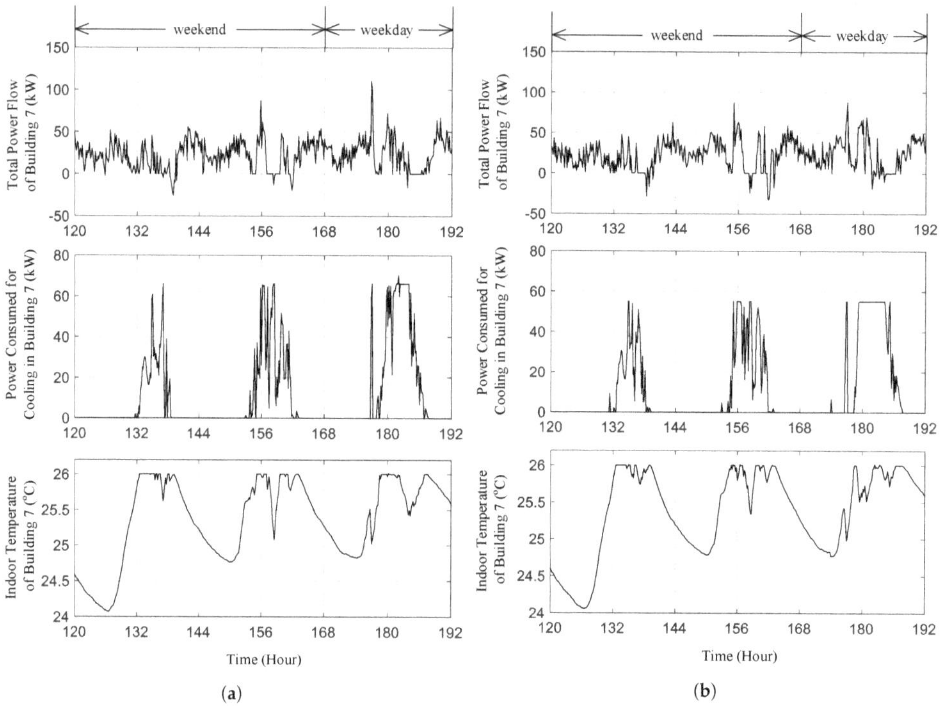

Figure 7. Response of building 7: (**a**) without dissipativity based coordination; (**b**) with dissipativity based coordination.

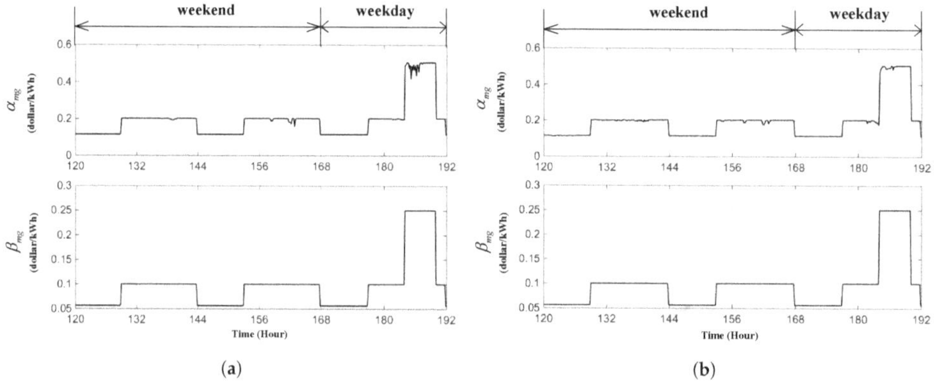

Figure 8. Weekend dynamic electricity prices: (**a**) without dissipativity based coordination; (**b**) with dissipativity based coordination.

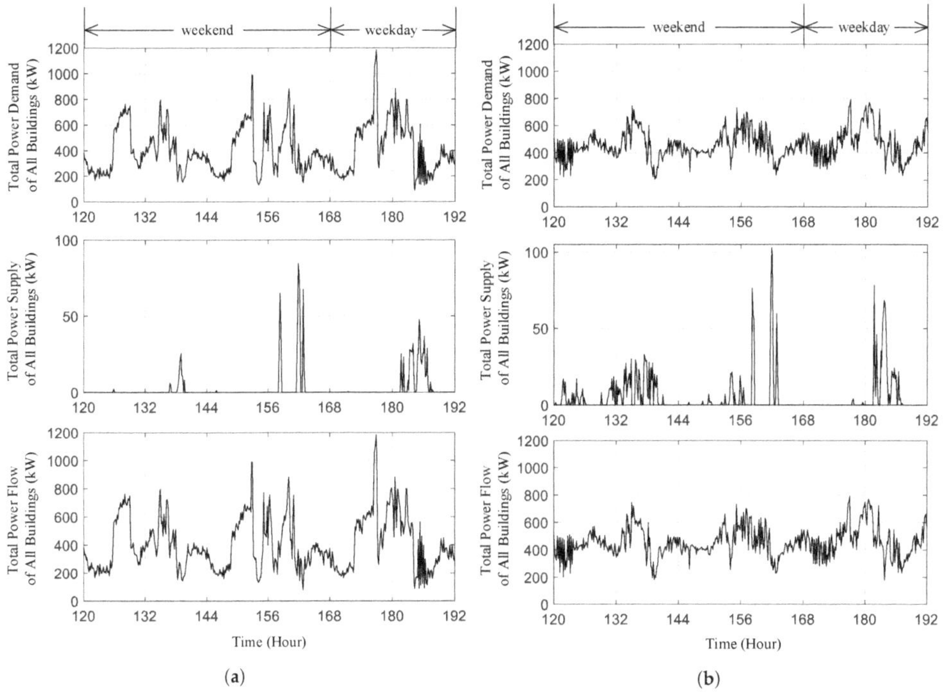

Figure 9. Weekend overall power flow profiles of microgrid: (**a**) without dissipativity based coordination; (**b**) with dissipativity based coordination.

6. Conclusions

This paper proposes a novel distributed control framework for the management of buildings with air-conditioners and SG in the context of microgrid. It allows the freedom of individually distributed building air-conditioner controllers in order to minimize their own operating costs, while achieving appropriate coordination among them to produce desirable overall power flow profile at the PCC. The effectiveness of the proposed control framework is demonstrated through simulation studies in comparison with conventional building air-conditioning control without coordination.

Acknowledgments: This work was supported by the Australian Research Council (Discovery Projects DP150103100).

Author Contributions: Xinan Zhang developed the main results, performed the simulation studies and prepared the initial draft of the paper. Ruigang Wang contributed to the distributed control theoretical developments. Jie Bao developed the dissipativity based distributed control idea, oversaw all aspects of the research and revised this manuscript.

Conflicts of Interest: The authors declare that they have no conflict of interest regarding the publication of the research article.

References

1. Tonkoski, R.; Turcotte, D.; El-Fouly, T.H. Impact of high PV penetration on voltage profiles in residential neighborhoods. *IEEE Trans. Sustain. Energy* **2012**, *3*, 518–527.
2. Eftekharnejad, S.; Vittal, V.; Heydt, G.T.; Keel, B.; Loehr, J. Impact of increased penetration of photovoltaic generation on power systems. *IEEE Trans. Power Syst.* **2013**, *28*, 893–901.
3. Rahouma, A.; El-Azab, R.; Salib, A.; Amin, A.M. Frequency response of a large-scale grid-connected solar photovoltaic plant. In Proceedings of the SoutheastCon 2015, Fort Lauderdale, FL, USA, 9–12 April 2015; pp. 1–7.

4. Wang, Y.; Zhang, P.; Li, W.; Xiao, W.; Abdollahi, A. Online overvoltage prevention control of photovoltaic generators in microgrids. *IEEE Trans. Smart Grid* **2012**, *3*, 2071–2078.
5. Jiang, B.; Muzhikyan, A.; Farid, A.M.; Youcef-Toumi, K. Demand side management in power grid enterprise control: A comparison of industrial & social welfare approaches. *Appl. Energy* **2017**, *187*, 833–846.
6. Ramchurn, S.D.; Vytelingum, P.; Rogers, A.; Jennings, N. Agent-based control for decentralised demand side management in the smart grid. In Proceedings of the 10th International Conference on Autonomous Agents and Multiagent Systems, Taipei, Taiwan, 2–6 May 2011; Volume 1, pp. 5–12.
7. Bellia, L.; Capozzoli, A.; Mazzei, P.; Minichiello, F. A comparison of HVAC systems for artwork conservation. *Int. J. Refrig.* **2007**, *30*, 1439–1451.
8. Lauro, F.; Moretti, F.; Capozzoli, A.; Panzieri, S. Model Predictive Control for Building Active Demand Response Systems. *Energy Procedia* **2015**, *83*, 494–503.
9. Xue, X.; Wang, S.; Yan, C.; Cui, B. A fast chiller power demand response control strategy for buildings connected to smart grid. *Appl. Energy* **2015**, *137*, 77–87.
10. Qi, W.; Liu, J.; Christofides, P.D. A distributed control framework for smart grid development: Energy/water system optimal operation and electric grid integration. *J. Process Control* **2011**, *21*, 1504–1516.
11. Qi, W.; Liu, J.; Chen, X.; Christofides, P.D. Supervisory predictive control of standalone wind/solar energy generation systems. *IEEE Trans. Control Syst. Technol.* **2011**, *19*, 199–207.
12. Qi, W.; Liu, J.; Christofides, P.D. Supervisory predictive control for long-term scheduling of an integrated wind/solar energy generation and water desalination system. *IEEE Trans. Control Syst. Technol.* **2012**, *20*, 504–512.
13. Qi, W.; Liu, J.; Christofides, P.D. Distributed supervisory predictive control of distributed wind and solar energy systems. *IEEE Trans. Control Syst. Technol.* **2013**, *21*, 504–512.
14. Maasoumy, M.; Sanandaji, B.M.; Sangiovanni-Vincentelli, A.; Poolla, K. Model predictive control of regulation services from commercial buildings to the smart grid. In Proceedings of the 2014 American Control Conference, Portland, OR, USA, 4–6 June 2014; pp. 2226–2233.
15. Zhang, W.; Lian, J.; Chang, C.Y.; Kalsi, K. Aggregated modeling and control of air-conditioning loads for demand response. *IEEE Trans. Power Syst.* **2013**, *28*, 4655–4664.
16. Bashash, S.; Fathy, H.K. Modeling and control of aggregate air-conditioning loads for robust renewable power management. *IEEE Trans. Control Syst. Technol.* **2013**, *21*, 1318–1327.
17. Liu, M.; Shi, Y. Model predictive control of aggregated heterogeneous second-order thermostatically controlled loads for ancillary services. *IEEE Trans. Power Syst.* **2016**, *31*, 1963–1971.
18. Touretzky, C.R.; Baldea, M. Integrating scheduling and control for economic MPC of buildings with energy storage. *J. Process Control* **2014**.
19. Soroush, M.; Chmielewski, D.J. Process systems opportunities in power generation, storage and distribution. *Comput. Chem. Eng.* **2013**, *51*, 86–95.
20. Chen, X.; Heidarinejad, M.; Liu, J.; Christofides, P.D. Distributed economic MPC: Application to a nonlinear chemical process network. *J. Process Control* **2012**, *22*, 689–699.
21. Vytelingum, P.; Voice, T.D.; Ramchurn, S.D.; Rogers, A.; Jennings, N.R. Agent-based micro-storage management for the smart grid. In Proceedings of the 9th International Conference on Autonomous Agents and Multiagent Systems, Toronto, ON, Canada, 10–14 May 2010; pp. 39–46.
22. Maity, I.; Rao, S. Simulation and pricing mechanism analysis of a solar-powered electrical microgrid. *IEEE Syst. J.* **2010**, *4*, 275–284.
23. Stephens, E.R.; Smith, D.B.; Mahanti, A. Game theoretic model predictive control for distributed energy demand-side management. *IEEE Trans. Smart Grid* **2015**, *6*, 1394–1402.
24. Willems, J.C. Dissipative dynamical systems part I: General theory. *Arch. Ration. Mech. Anal.* **1972**, *45*, 321–351.
25. Weiland, S.; Willems, J.C. Dissipative dynamical systems in a behavioral context. *Math. Models Methods Appl. Sci.* **1991**, *1*, 1–25.
26. Hill, D.; Moylan, P. Stability Results for Nonlinear Feedback Systems. *Automatica* **1977**, *13*, 377–382.
27. Moylan, P.; Hill, D. Stability criteria for large-scale systems. *IEEE Trans. Autom. Control* **1978**, *23*, 143–149.
28. Xu, S.; Bao, J. Distributed control of plantwide chemical processes. *J. Process Control* **2009**, *19*, 1671–1687.
29. Xu, S.; Bao, J. Control of chemical processes via output feedback controller networks. *Ind. Eng. Chem. Res.* **2010**, *49*, 7421–7445.

30. Tippett, M.J.; Bao, J. Dissipativity based distributed control synthesis. *J. Process Control* **2013**, *23*, 755–766.
31. Tippett, M.J.; Bao, J. Control of plant-wide systems using dynamic supply rates. *Automatica* **2014**, *50*, 44–52.
32. Willems, J.; Trentelman, H. On quadratic differential forms. *SIAM J. Control Optim.* **1998**, *36*, 1703–1749.
33. Tippett, M.J.; Bao, J. Distributed model predictive control based on dissipativity. *AIChE J.* **2013**, *59*, 787–804.
34. Yu, Z.; Jia, L.; Murphy-Hoye, M.C.; Pratt, A.; Tong, L. Modeling and stochastic control for home energy management. *IEEE Trans. Smart Grid* **2013**, *4*, 2244–2255.
35. Hazyuk, I.; Ghiaus, C.; Penhouet, D. Optimal temperature control of intermittently heated buildings using Model Predictive Control: Part I–Building modeling. *Build. Environ.* **2012**, *51*, 379–387.
36. Ghosh, S.; Reece, S.; Rogers, A.; Roberts, S.; Malibari, A.; Jennings, N.R. Modeling the Thermal Dynamics of Buildings: A Latent-Force-Model-Based Approach. *ACM Trans. Intell. Syst. Technol.* **2015**, *6*, 7.
37. Ma, Y.; Anderson, G.; Borrelli, F. A distributed predictive control approach to building temperature regulation. In Proceedings of the 2011 American Control Conference, San Francisco, CA, USA, 29 June–1 July 2011; pp. 2089–2094.
38. Thavlov, A.; Bindner, H.W. Thermal models for intelligent heating of buildings. In Proceedings of the International Conference on Applied Energy, Suzhou, China, 5–8 July 2012.
39. Haghighi, M.M. Controlling Energy-Efficient Buildings in the Context of Smart Grid: A Cyber Physical System Approach. Ph.D. Thesis, EECS Department, University of California, Berkeley, CA, USA, 2013.
40. Nasution, H.; Hassan, M.N.W. Potential electricity savings by variable speed control of compressor for air-conditioning systems. *Clean Technol. Environ. Policy* **2006**, *8*, 105–111.
41. Hunt, W.; Amarnath, A. Cooling Efficiency Comparison between Residential Variable Capacity and Single Speed Heat Pump. *ASHRAE Trans.* **2013**, *119*, Q1.
42. Funami, K.; Nishi, H. Evaluation of power consumption and comfort using inverter control of air-conditioning. In Proceedings of the 37th Annual Conference on IEEE Industrial Electronics Society, Melbourne, VIC, Australia, 7–10 November 2011; pp. 3236–3241.
43. Energy of the Environment Air Conditioners. 2016. Available online: http://www.energyrating.gov.au/products/space-heating-and-cooling/air-conditioners (Australian Government Equipment Energy Efficiency Program) (accessed on 24 April 2016).
44. Lokanadham, M.; Bhaskar, K.V. Incremental conductance based maximum power point tracking (MPPT) for photovoltaic system. *Int. J. Eng. Res. Appl.* **2012**, *2*, 1420–1424.
45. Ma, J.; Qin, S.J.; Salsbury, T. Experimental study of economic model predictive control in building energy systems. In Proceedings of the 2013 American Control Conference, Washington, DC, USA, 17–19 June 2013; pp. 3753–3758.
46. Putta, V.; Zhu, G.; Kim, D.; Hu, J.; Braun, J. Comparative evaluation of model predictive control strategies for a building HVAC system. In Proceedings of the 2013 American Control Conference, Washington, DC, USA, 17–19 June 2013; pp. 3455–3460.
47. Ma, J.; Qin, S.J.; Salsbury, T. Application of economic MPC to the energy and demand minimization of a commercial building. *J. Process Control* **2014**, *24*, 1282–1291.
48. Brundage, M.P.; Chang, Q.; Li, Y.; Xiao, G.; Arinez, J. Energy efficiency management of an integrated serial production line and HVAC system. *IEEE Trans. Autom. Sci. Eng.* **2014**, *11*, 789–797.
49. Braun, P.; Grüne, L.; Kellett, C.M.; Weller, S.R.; Worthmann, K. A real-time pricing scheme for residential energy systems using a market maker. In Proceedings of the 2015 5th Australian Control Conference (AUCC), Gold Coast, Australia, 5–6 November 2015; pp. 259–262.
50. Ito, K. Nonlinear Pricing in Energy and Environmental Markets. Ph.D. Thesis, Agricultural & Resource Economics, University of California, Berkeley, CA, USA, 2011.
51. Okawa, Y.; Namerikawa, T. Distributed dynamic pricing based on demand-supply balance and voltage phase difference in power grid. *Control Theory Technol.* **2015**, *13*, 90–100.
52. Yang, L.; Dong, C.; Wan, C.J.; Ng, C.T. Electricity time-of-use tariff with consumer behavior consideration. *Int. J. Prod. Econ.* **2013**, *146*, 402–410.
53. Rowlands, I.H.; Furst, I.M. The cost impacts of a mandatory move to time-of-use pricing on residential customers: An Ontario (Canada) case-study. *Energy Effic.* **2011**, *4*, 571–585.
54. Matters, E. Feed-in Tariff for Grid-Connected Solar Power Systems. 2011. Available online: http://www.energymatters.com.au/rebates-incentives/feedintariff (accessed on 15 September 2015).

55. Matters, E. Battery Storage And Solar Feed in Tariffs—State Of Play. 2015. Available online: http://www.energymatters.com.au/renewable-news/solar-fit-batteries-em5074/ (accessed on 15 September 2015).
56. Muzmar, M.; Abdullah, M.; Hassan, M.; Hussin, F. Time of Use pricing for residential customers case of Malaysia. In Proceedings of the 2015 IEEE Student Conference on Research and Development (SCOReD), Kuala Lumpur, Malaysia, 13–14 December 2015; pp. 589–593.
57. Zhang, X.; Bao, J.; Wang, R.; Zheng, C.; Skyllas-Kazacos, M. Dissipativity based distributed economic model predictive control for residential microgrids with renewable energy generation and battery energy storage. *Renew. Energy* **2017**, *100*, 18–34.
58. Kojima, C.; Takaba, K. A generalized Lyapunov stability theorem for discrete-time systems based on quadratic difference forms. In Proceedings of the 44th IEEE Conference on Decision and Control, Seville, Spain, 15 December 2005; pp. 2911–2916.
59. Parrilo, P.A. Structured Semidefinite Programs and Semialgebraic Geometry Methods in Robustness and Optimization. Ph.D. Thesis, California Institute of Technology, Pasadena, CA, USA, 2000.
60. Löfberg, J. YALMIP: A toolbox for modeling and optimization in MATLAB. In Proceedings of the CACSD Conference, Taipei, Taiwan, 2–4 September 2004; pp. 284–289.
61. Sturm, J.F. Using SeDuMi 1.02, a MATLAB toolbox for optimization over symmetric cones. *Optim. Methods Softw.* **1999**, *11*, 625–653.
62. Fan, S.; Hyndman, R.J. Short-term load forecasting based on a semi-parametric additive model. *IEEE Trans. Power Syst.* **2012**, *27*, 134–141.
63. Station, M.U.W. 2017. Available online: http://wwwmet.murdoch.edu.au/downloads (accessed on 24 April 2016).

mathematics

MDPI

Article

Economic Model Predictive Control with Zone Tracking

Su Liu and Jinfeng Liu *

Department of Chemical & Materials Engineering, University of Alberta, Edmonton, AB T6G 2V4, Canada; su7@ualberta.ca

* Correspondence: jinfeng@ualberta.ca; Tel.: +1-780-492-1317

Received: 1 April 2018; Accepted: 23 April 2018; Published: 25 April 2018

Abstract: In this work, we propose a framework for economic model predictive control (EMPC) with zone tracking. A zone tracking stage cost is incorporated into the existing EMPC framework to form a multi-objective optimization problem. We provide sufficient conditions for asymptotic stability of the optimal steady state and characterize the exact penalty for the zone tracking cost which prioritizes zone tracking objective over economic objective. Moreover, an algorithm to modify the target zone based on the economic performance and reachability of the optimal steady state is proposed. The modified target zone effectively decouples the dynamic zone tracking and economic objectives and simplifies parameter tuning.

Keywords: predictive control; process optimization; soft constraint; zone control

1. Introduction

Process control of chemical plants needs to address a number of objectives including safety, environmental regulations, product quality, energy efficiency, profitability, etc. Based on the emphasis of the control objectives in practice, three elements or three facets of control exist [1]: regulatory control, constraint control and maneuvering control. Regulatory control refers to the conventional setpoint tracking control which minimizes the variance of controlled variables to the setpoint. Constraint control, or zone control, prevents the system from violating its boundary and steers the system back into the zone whenever constraint violation happens. No control action is required if the system is (predicted to be) in the target zone. Maneuvering control moves the system from the current operating point to a new operating point, typically due to economic considerations or change of operating conditions. Generally speaking, a well designed control system necessarily integrates all three control types, although emphasis on the three facets of control may vary from application to application.

Model predictive control (MPC) has been the most widely applied advanced control technique. The ability to handle constraints and to incorporate economic considerations makes MPC an ideal platform for integrating different control objectives. The literature is rich with theories for conventional setpoint tracking MPC ([2]). The past decade has seen an increasing academic interest in economic model predictive control (EMPC) ([3–6]) which integrates economic objectives into regulatory setpoint tracking control. On the contrary, zone control has received less attention. In the MPC framework, zone control is usually dealt with by the so-called soft constraint technique ([7–11]). As its name suggests, soft constraint is often dismissed as a trick to avoid feasibility issue with hard constraint, and is discussed separately from set-point tracking and economic objectives. To the best of the authors' knowledge, only a few MPC frameworks explicitly handle zone tracking objectives ([12–14]). A practical challenge for the design and implementation of an integrated control system is the difficulty in parameter tunning. How to tradeoff different control objectives via the tuning parameters to get the desired closed-loop performance is crucial to the successful implementation of any control framework.

In this work, we propose an EMPC framework with integrated zone control. A zone tracking stage cost which penalizes a weighted l_1 norm and squared l_2 norm distance to the target zone is incorporated into the EMPC framework to form a multi-objective optimization problem. We provide sufficient conditions for asymptotic stability of the optimal steady state and characterize the exact penalty for the zone tracking cost which prioritizes the zone tracking objective over the economic objective. Moreover, we propose an algorithm to modify the target zone based on the economic performance and reachability of the optimal steady state in the target zone. The modified target zone is constructed as an invariant subset of the original target zone in which closed-loop transient economic performance is guaranteed. EMPC with the modified target zone effectively decouples the zone tracking and economic objectives and enjoys a simplified and more transparent parameter tuning procedure. Finally, two numerical examples are investigated which reveal the intrinsic difficulties in parameter tunning for EMPC with zone tracking and demonstrate the efficacy of the proposed approach.

2. Problem Setup

2.1. Notation

Throughout this work, $\|x\|_p$ denotes the l_p norm of the vector x such that $\|x\|_p = \left(\sum |x^{(i)}|^p\right)^{1/p}$. The operator $|\cdot|$ denotes the l_2 norm or Euclidean norm of a scalar or a vector. The symbol $\mathbb{I}_M^N$ denotes the set of integers $\{M, M+1, ..., N\}$. The symbol $\mathbb{I}_{\geq 0}$ denotes the set of non-negative integers $\{0, 1, 2, ...\}$. The symbol $proj_{\mathbb{X}}(\mathbb{O})$ denotes projection of the set $\mathbb{O}$ onto its subspace $\mathbb{X}$. A function $l(x) : \mathbb{X} \to \mathbb{R}$ is said to be positive definite with respect to a set $\mathbb{X}_t \subset \mathbb{X}$, if $l(x) = 0$ for all $x \in \mathbb{X}_t$ and $l(x) > 0$ otherwise.

2.2. System Description and Control Objective

We consider the following nonlinear discrete time system:

$$x(n+1) = f(x(n), u(n)) \tag{1}$$

where $x(n) \in \mathbb{X} \subset \mathbb{R}^{n_x}$, $u(n) \in \mathbb{U} \subset \mathbb{R}^{n_u}$, $n \in \mathbb{I}_{\geq 0}$, denote the state and input at time n, respectively. The vector function $f(\cdot) : \mathbb{R}^{n_x} \times \mathbb{R}^{n_u} \to \mathbb{R}^{n_x}$ is continuous. The system is subject to coupled state and input constraint:

$$(x(n), u(n)) \in \mathbb{Z} \subseteq \mathbb{X} \times \mathbb{U}, \qquad n \in \mathbb{I}_{\geq 0}$$

where $\mathbb{X}, \mathbb{U}, \mathbb{Z}$ are all compact sets. The primary control objective is to steer and maintain the system in a compact set $\mathbb{Z}_t \subset \mathbb{Z}$. The distance to the target zone is measured by the function $\ell_z(x(n), u(n)) : \mathbb{Z} \to \mathbb{R}$ which is positive definite with respect to the target zone $\mathbb{Z}_t$. There is also a secondary economic objective to minimize the operational cost characterized by the function $\ell_e(x, u) : \mathbb{Z} \to \mathbb{R}$. Both $\ell_z(\cdot)$ and $\ell_e(\cdot)$ are continuous functions. Since zone tracking objectives are usually associated with important process specifications concerning safety or product quality, an ideal control strategy should satisfy zone objectives whenever possible and allow zone tracking violation only for a short period of time. This leads to the following formal statement of the control objective as an infinite-horizon optimization problem:

$$\begin{aligned} \min \quad & \sum_{n=0}^{\infty} \ell_e(x(n), u(n)) + \sum_{n=0}^{K} \ell_z(x(n), u(n)) + K \\ \text{s.t.} \quad & (x(n), u(n)) \in \mathbb{Z}, \quad n \in \mathbb{I}_{\geq 0} \\ & (x(n), u(n)) \in \mathbb{Z}_t, \quad n \geq K+1 \end{aligned} \tag{2}$$

The above optimization problem optimizes the economic performance over an infinite horizon and minimizes the zone tracking error for the first K time steps when the system is outside the target zone $\mathbb{Z}_t$. When the system is steered into the target zone, the zone tracking error $\ell_z(\cdot)$ is zero. The duration of zone target violation K is also an explicit objective which ensures that the zone tracking objective is satisfied in finite time. The optimization of Equation (2) is essentially a multi-objective optimization problem in which the magnitude, duration of zone target violation, as well as the economic performance are traded off.

For simplicity of exposition, we assume that the optimization of Equation (2) is well defined. This implies that the system can be steered to the target zone $\mathbb{Z}_t$ in finite time without violating the constraint. Moreover, the infinite sum $\sum_{n=0}^{\infty} \ell_e(x(n), u(n))$ is bounded. This condition is satisfied for strictly dissipative systems if we assume without loss of generality that the economic performance at the optimal steady state is zero [6].

Let (x_s, u_s) denote the economically optimal steady state in the target zone $\mathbb{Z}_t$. That is:

$$\begin{aligned} (x_s, u_s) = \;& \arg\min_{x,u} \; \ell_e(x, u) \\ \text{s.t.} \;\; & x = f(x, u) \\ & (x, u) \in \mathbb{Z}_t \end{aligned} \tag{3}$$

We assume that (x_s, u_s) uniquely solves the above steady-state optimization problem. Note that the economic cost function $\ell_e(\cdot)$ is not necessarily positive definite with respect to (x_s, u_s).

3. EMPC with Zone Tracking

In this section, we propose a general framework for EMPC with zone tracking to tackle the infinite-horizon optimization problem in Equation (2). The EMPC is formulated as a finite horizon optimization problem which is solved repeatedly in a receding horizon fashion to approximate the optimal solution to Equation (2). We provide sufficient conditions for asymptotic stability of the optimal steady state and discuss how zone tracking can be prioritized using exact penalty.

3.1. EMPC Formulation

At a sampling time n, the EMPC is formulated as the following finite-horizon optimization problem:

$$\min_{u_0, \cdots, u_{N-1}} \quad \textstyle\sum_{i=0}^{N-1} \ell_e(x_i, u_i) + \ell_z(x_i, u_i) \tag{4}$$

$$\text{s.t.} \quad x_{i+1} = f(x_i, u_i), \;\; i \in \mathbb{I}_0^{N-1} \tag{5}$$

$$x_0 = x(n) \tag{6}$$

$$(x_i, u_i) \in \mathbb{Z}, \;\; i \in \mathbb{I}_0^{N-1} \tag{7}$$

$$x_N = x_s \tag{8}$$

In the above optimization, the objective function Equation (4) minimizes the zone tracking error and economic performance over a finite horizon of N steps, Equation (5) is the system model, Equation (6) specifies the initial condition, Equation (7) sets the state and input constraints. The point-wise terminal constraint Equation (8) requires the terminal state to arrive at the optimal steady state x_s after N steps. This implicitly imposes constraint on the duration of zone objective violation (K in Equation (2)).

Let $u^*(i|n), i \in \mathbb{I}_0^{N-1}$ denote the optimal solution. The input injected to the system at time n is: $u(n) = u^*(0|n)$. At the next sampling time $n+1$, the optimization of Equation (4) is re-evaluated, generating an implicit feedback control law $u(n) = \kappa_N(x(n))$. We denote the feasibility region of the optimization problem of Equation (4) by $\mathbb{X}_N$. Due to the terminal constraint Equation (8), $\mathbb{X}_N$ is forward invariant under the EMPC design. In other words, the EMPC design is recursively feasible. The zone tracking penalty $\ell_z(x,u)$ is defined by the following function:

$$\begin{aligned} \ell_z(x,u) = \min_{x^z,u^z} \quad & c_1(\|x-x^z\|_1 + \|u-u^z\|_1) + c_2(\|x-x^z\|_2^2 + \|u-u^z\|_2^2) \\ \text{s.t.} \quad & (x^z,u^z) \in \mathbb{Z}_t \end{aligned} \tag{9}$$

where c_1 and c_2 are positive scalars. The zone tracking penalty defined in Equation (9) is formulated implicitly as an optimization problem which minimizes weighted l_1 norm and squared l_2 norm distance from the point (x,u) to the target zone $\mathbb{Z}_t$. The distance to the target zone is evaluated by introducing artificial variables (x_z,u_z) which are bounded in the target zone $\mathbb{Z}_t$. It is easy to verify that $\ell_z(x,u)$ defined in Equation (9) is positive definite with respect to the target zone $\mathbb{Z}_t$. When the zone tracking penalty of Equation (9) is used, Equation (4) is equivalent to the following optimization problem:

$$\begin{aligned} \min_{u_i,x_i^z,u_i^z} \quad & \sum_{i=0}^{N-1} \ell_e(x_i,u_i) + c_1(\|x_i-x_i^z\|_1 + \|u_i-u_i^z\|_1) + c_2(\|x_i-x_i^z\|_2^2 + \|u_i-u_i^z\|_2^2) \\ \text{s.t.} \quad & x_{i+1} = f(x_i,u_i),\ i \in \mathbb{I}_0^{N-1} \\ & x_0 = x(n) \\ & (x_i,u_i) \in \mathbb{Z},\ i \in \mathbb{I}_0^{N-1} \\ & x_N = x_s \\ & (x_i^z,u_i^z) \in \mathbb{Z}_t,\ i \in \mathbb{I}_0^{N-1} \end{aligned} \tag{10}$$

Remark 1. *The incorporation of the l_1 norm penalty allows for the so-called exact penalty which will be discussed in Section 3.3. From a multi-objective optimization point of view, the combined use of the linear (l_1 norm) penalty and quadratic (squared l_2 norm) penalty offers a way to trade off the magnitude and duration of zone tracking violation [7]. Larger linear penalty may, though not necessarily, result in more drastic control move with smaller duration of zone tracking violation. On the contrary, quadratic penalty generally leads to mild control action with smaller zone tracking violation but larger duration of violation. These results will be demonstrated in the simulation.*

Remark 2. *A well-known technique to address the feasibility issues in MPC is to employ the so-called soft constraint where slack variables are introduced to relax hard (state) constraint. Interested readers may refer to [7,9] which provide comprehensive discussions on constraint relaxation of MPC using soft constraint. Let the target zone $\mathbb{Z}_t$ be characterized by $\mathbb{Z}_t := \{(x,u) : g(x,u) \leq 0\}$ where $g(x,u) : \mathbb{R}^{n_x} \times \mathbb{R}^{n_u} \to \mathbb{R}^{n_y}$. When soft constraint is used, the penalty for constraint violation has the following form:*

$$\begin{aligned} \ell_s(x,u) = \min_{s} \quad & c_1\|s\|_1 + c_2\|s\|_2^2 \\ \text{s.t.} \quad & g(x,u) \leq s \end{aligned}$$

The function $\ell_s(x,u)$ defined above is also positive definite with respect to the target zone $\mathbb{Z}_t$. Note that the zone tracking penalty $\ell_s(x,u)$ is different from $\ell_z(x,u)$ in Equation (9). In the above example, implementing the soft-constraint penalty $\ell_s(x,u)$ requires n_y artificial variables whereas for $\ell_z(x,u)$ the number is $n_x + n_u$. The soft-constraint penalty $\ell_s(x,u)$ may be better described as an output zone tracking penalty (considering $y = g(x,u)$ as the system output) whereas $\ell_z(x,u)$ in Equation (9) is a zone tracking penalty for system state and input. The pros and cons of state zone tracking against output zone tracking calls for further investigation. One advantage of using the state zone tracking penalty $\ell_z(x,u)$ is that it allows set-theoretic method ([15]) in the EMPC design. In Section 4 an algorithm based on set-theoretic method will be proposed to modify the target zone.

3.2. Stability Analysis

In the following, we establish sufficient conditions for asymptotic stability of the optimal steady state x_s.

Definition 1. *(Strictly dissipative systems) The system of Equation (1) is strictly dissipative with respect to the supply rate* $s: \mathbb{X} \times \mathbb{U} \to \mathbb{R}$ *if there exists a continuous storage function* $\lambda(\cdot): \mathbb{X} \to \mathbb{R}$ *and a* $\mathcal{K}_\infty$ *function* $\alpha_1(\cdot)$ *such that the following holds for all* $x \in \mathbb{X}$ *and* $u \in \mathbb{U}$:

$$\lambda(f(x,u)) - \lambda(x) \leq s(x,u) - \alpha_l(|x - x_s|)$$

Assumption 1. *(Strict dissipativity, [16]) The system of Equation (1) is strictly dissipative with respect to the supply rate* $s(x,u) = \ell_e(x,u) - \ell_e(x_s,u_s)$

Assumption 2. *(weak controllability, ([17])) There exists a* $\mathcal{K}_\infty$ *function* $\gamma(\cdot)$ *such that for all* $x \in \mathbb{X}_N$, *there exists a feasible solution to Equation (4) such that:*

$$\sum_{i=0}^{N-1} |u_i - u_s| \leq \gamma(|x - x_s|)$$

Theorem 1. *If Assumptions 1 and 2 hold, then the optimal steady state* x_s *is asymptotically stable under the EMPC of Equation (4) with a region of attraction* $\mathbb{X}_N$.

Proof. To proceed, we define the rotated cost $\bar{\ell}_e(x,u)$ as:

$$\bar{\ell}_e(x,u) = \ell_e(x,u) - \ell_e(x_s,u_s) + \lambda(x) - \lambda(f(x,u)) \tag{11}$$

From Assumption 1, the rotated cost satisfies

$$\bar{\ell}_e(x,u) \geq \alpha_l(|x - x_s|) \tag{12}$$

Substitute Equation (11) into the Equation (4), the optimization problem of Equation (4) can be equivalently written as follows:

$$\begin{aligned}
\min_{u_0,\cdots,u_{N-1}} \quad & \sum_{i=0}^{N-1} \bar{\ell}_e(x_i,u_i) + \ell_z(x_i,u_i) - \lambda(x_0) + \lambda(x_N) + N\ell_e(x_s,u_s) \\
\text{s.t.} \quad & x_{i+1} = f(x_i,u_i), \ i \in \mathbb{I}_0^{N-1} \\
& x_0 = x(n) \\
& (x_i,u_i) \in \mathbb{Z}, \ i \in \mathbb{I}_0^{N-1} \\
& x_N = x_s
\end{aligned}$$

The last three terms in the objective function of the above optimization are all constant and can be dropped (because of constraints Equations (6) and (8)). The above optimization is then reduced to:

$$\begin{aligned}
V(x(n)) = \min_{u_0,\cdots,u_{N-1}} \quad & \sum_{i=0}^{N-1} L(x_i,u_i) \\
\text{s.t.} \quad & x_{i+1} = f(x_i,u_i), \ i \in \mathbb{I}_0^{N-1} \\
& x_0 = x(n) \\
& (x_i,u_i) \in \mathbb{Z}, \ i \in \mathbb{I}_0^{N-1} \\
& x_N = x_s
\end{aligned}$$

where $L(x,u) = \bar{\ell}_e(x,u) + \ell_z(x,u)$. Since $L(\cdot)$ is non-negative and taking into Equation (12), we have:

$$V(x(n)) \geq L(x_0, u_0) = \bar{\ell}_e(x_0, u_0) + \ell_z(x_0, u_0) \geq \bar{\ell}_e(x_0, u_0) \geq \alpha_l(|x(n) - x_s|) \tag{13}$$

Moreover, Assumption 2 implies the existence of a $\mathcal{K}_\infty$ function $\beta(\cdot)$ such that (see [17] Appendix):

$$V(x(n)) \leq \beta(|x(n) - x_s|) \tag{14}$$

Finally, the value function $V(x(n))$ is strictly non-increasing and satisfies:

$$V(x(n+1)) - V(x(n)) \leq -L(x(n), u(n)) \leq -\alpha_l(|x(n) - x_s|) \tag{15}$$

Equations (13)–(15) makes the value function $V(x(n))$ a Lyapunov function with respect to the optimal steady state x_s. Therefore the optimal steady state x_s is asymptotically stable. □

Remark 3. *If the optimal steady state lies in the interior of the target zone, then asymptotic stability of the optimal steady state implies finite-time convergence into the target zone. However, if the optimal steady state is on the boundary of the target zone, asymptotic stability of the optimal steady state does not imply finite-time convergence to the target zone. One way to still achieve nominal finite-time convergence to the target zone is to implement the whole predicted input sequence instead of only the first element. That is,* $u(n+i) = u^*(i|n)$ *for* $i \in \mathbb{I}_0^{N-1}$*. At the sampling time* $n+N$*, re-evaluate the optimization of Equation (4) based on the state measurement* $x(n+N)$*. Due to the terminal constraint, the nominal system will reach the optimal steady state in N steps. Another way is to employ sufficiently large* l_1 *norm zone tracking penalty (large* c_1*) which is known to result in deadbeat control policy ([18]).*

3.3. Prioritized Zone Tracking

In practice, zone control objectives are usually associated with important or urgent objectives such as operation safety and product specification. Thus a natural question to ask is how to pick the zone tracking penalty such that the zone tracking is prioritized over the economic objective? In other words, how to ensure that the system stays in the target zone whenever possible? The answer has to do with the so-called exact penalty function. To proceed with the discussion, we introduce the concept of the dual norm:

Definition 2. *(dual norm) Consider the p norm of a vector u,* $\|u\|_p$*,* $p \in \mathbb{I}_{\geq 0}$*. We refer to* $\|u\|_q$ *as the dual norm of* $\|u\|_p$ *where* $\|u\|_q$ *is defined as follows:*

$$\|u\|_q := \max_{\|v\| \leq 1} u^T v$$

It can be verified that $\|\cdot\|_1$ is the dual norm of $\|\cdot\|_\infty$ and vice versa. Consider the following constrained optimization problem:

$$\begin{aligned} \min_{u} \quad & V(u) \\ \text{s.t.} \quad & g(u) \leq 0 \end{aligned} \tag{16}$$

The above hard-constraint optimization problem can be recast as the following soft-constraint optimization problem:

$$\begin{aligned} \min_{u,\epsilon} \quad & V(u) + c\,\|\epsilon\|_p \\ \text{s.t.} \quad & g(u) \leq \epsilon \end{aligned} \tag{17}$$

where c is a positive scalar. It is conceivable that if the penalty c is sufficiently large, then the solution to the Equation (17) will be identical to Equation (16). The following well-known result specifies how large is sufficiently large for c:

Theorem 2. *Consider the optimization of Equations (16) and (17). Assume that the solutions to both problems satisfy second-order sufficient conditions, and let λ denote the vector of the Lagrange multipliers of Equation (16). If $c > \|\lambda\|_q$ where $\|\cdot\|_q$ is the dual norm of $\|\cdot\|_p$, then the solutions to Equations (16) and (17) are identical.*

Proof. The proof can be found in [19] Theorem 14.3.1. □

Any penalty function satisfying the conditions in Theorem 2 is called the exact penalty function. When exact penalty is used, the soft-constraint optimization problem has the same solution as the original problem. This means that if we treat the feasibility region, $\{x \mid g(x) \leq 0\}$, as the target zone and the penalty, $c\,\|\epsilon\|_p$, as the zone tracking penalty, the exact penalty prioritizes the zone tracking objective over the economic objective.

Remark 4. *The exact penalty function problem for Equation (16) is usually written in the following form:*

$$\min_u V(u) + c\,\|\max\{0, g(u)\}\|_p$$

The above formulation is equivalent to the problem of Equation (17). With slight abuse of language, we refer to the soft-constraint formulation in Equation (17) with $c > \|\lambda\|_q$ as the exact penalty. Essentially what allows the exact penalty to be exact or to prioritize the constraint objective is the non-smoothness of the penalty function at $\epsilon = 0$. In the case of equality constraint, the same result applies since one can also treat $h(u) = 0$ as a set of inequality constraints $h(u) \leq 0$ and $-h(u) \leq 0$. In this case, the exact penalty function problem becomes: $\min_u V(u) + c\,\|h(u)\|_p$

In the light of Theorem 2, we can construct the following optimization problem in which zone tracking is made the hard constraint:

$$\begin{aligned} \min_{u_i, x_i^z, u_i^z} \quad & \sum_{i=0}^{N-1} \ell_e(x_i, u_i) \\ \text{s.t.} \quad & x_{i+1} = f(x_i, u_i), \ i \in \mathbb{I}_0^{N-1} \\ & x_0 = x(n) \\ & (x_i, u_i) \in \mathbb{Z}, \ i \in \mathbb{I}_0^{N-1} \\ & x_N = x_s \\ & (x_i^z, u_i^z) \in \mathbb{Z}_t, \ i \in \mathbb{I}_0^{N-1} \\ & x_i = x_i^z, \ i \in \mathbb{I}_0^{N-1} \\ & u_i = u_i^z, \ i \in \mathbb{I}_0^{N-1} \end{aligned} \tag{18}$$

To make the statement of the final result more compact, we employ the definition of the N-step reachable set [20]:

Definition 3. *(N-step reachable set) We use $\mathbb{X}_N(\mathbb{Z}_t, x_s)$ to denote the set of states that can be steered to $\mathbb{X}_f$ in N steps while satisfying the state and input constraints $(x, u) \in \mathbb{Z}_t$. That is,*

$$\mathbb{X}_N(\mathbb{Z}_t, x_s) = \left\{x(0) \mid \exists\,(x(n), u(n)) \in \mathbb{Z}_t, n \in \mathbb{I}_0^{N-1}, x(N) = x_s\right\}$$

Based on the above definition, the EMPC is capable of maintaining the system in the target zone only if $x(0) \in \mathbb{X}_N(\mathbb{Z}_t, x_s)$. Now we are ready to state and prove the final result:

Theorem 3. *Let λ denote the Lagrange multiplier of the optimization of Equation (18). If $x(n) \in \mathbb{X}_N(\mathbb{Z}_t, x_s)$ and $c_1 > \|\lambda\|_\infty$, then the solutions to the MPC of Equations (10) and (18) are identical.*

Proof. We provide a sketch of the proof. Consider the exact penalty problem of Equation (18) with $p = 1$ and $q = \infty$ in Theorem 2. Since $c_1 > \|\lambda\|_\infty$, it can be verified that the optimization of Equation (18) is equivalent to the following optimization problem:

$$\begin{aligned} \min_{u_i, x_i^z, u_i^z} \quad & \sum_{i=0}^{N-1} \ell_e(x_i, u_i) + c_1(\|x_i - x_i^z\|_1 + \|u_i - u_i^z\|_1) \\ \text{s.t.} \quad & (5) - (8) \\ & (x_i^z, u_i^z) \in \mathbb{Z}_t, \ \ i \in \mathbb{I}_0^{N-1} \end{aligned}$$

Since the optimal solution satisfies $x_i = x_i^z$ and $u_i = u_i^z$ for $i \in \mathbb{I}_0^{N-1}$, adding the quadratic terms $c_2(\|x_i - x_i^z\|_2^2 + \|u_i - u_i^z\|_2^2)$ to the stage cost does not change the optimal solution to the above optimization problem. This implies that the solutions to the MPC of Equation (10) and Equation (18) are identical. □

Theorem 3 implies that if the l_1 norm penalty c_1 is sufficiently large, then the zone tracking objective is prioritized over the economic objective. Note that the constraints (18c)–(18e) can be combined by canceling the slack variables x_i^z and u_i^z into the compact form:

$$(x_i, u_i) \in \mathbb{Z}_t, i \in \mathbb{I}_0^{N-1}$$

Thus the EMPC of Equation (18) yields the same solution to the following EMPC constrained by the target zone:

$$\begin{aligned} \min_{u_0, \cdots, u_{N-1}} \quad & \textstyle\sum_{i=0}^{N-1} \ell_e(x_i, u_i) \\ \text{s.t.} \quad & x_{i+1} = f(x_i, u_i), \ \ i \in \mathbb{I}_0^{N-1} \\ & x_0 = x(n) \\ & (x_i, u_i) \in \mathbb{Z}_t, \ \ i \in \mathbb{I}_0^{N-1} \\ & x_N = x_s \end{aligned} \tag{19}$$

The constraint (6) is removed because $\mathbb{Z}_t \subset \mathbb{Z}$. Therefore, sufficiently large l_1 norm penalty term $c_1(\|x_i - x_i^z\|_1 + \|u_i - u_i^z\|_1)$ in effect can convert the zone tracking objective into hard constraints $(x_i, u_i) \in \mathbb{Z}_t$ whenever possible.

Remark 5. *Note that exact penalty cannot be achieved by the quadratic term $c_2(\|x - x^z\|_2^2 + \|u - u^z\|_2^2)$ unless c_2 can be made infinitely large [9]. Note also that to ensure $x(n) \in \mathbb{Z}_t$ for all $n \in \mathbb{I}_{\geq 0}$ and $x(0) \in \mathbb{X}_N(\mathbb{Z}_t, x_s)$, the condition $c_1 > \|\lambda\|_\infty$ needs to be satisfied for Lagrange multipliers λ associated with all $x(n) \in \mathbb{X}_N(\mathbb{Z}_t, x_s)$. Finding the exact lower bound for such c_1 is in general difficult. The task is possible for linear systems with quadratic tracking costs and polyhedral constraints ([8]).*

4. Modified Target Zone

The EMPC with zone tracking framework discussed in Section 3 is applicable to a broad class of control systems with multiple objectives involving economic optimization, zone tracking or setpoint tracking. A challenging problem in practice that largely affects the performance of MPC is parameter tuning. Specifically, for the EMPC of Equation (10), how does one pick c_1 and c_2 such that the closed-loop system achieves the desired tradeoff between magnitude and duration of zone tracking violation as well as economic performance? As a rule of thumb, if the zone objective is more important than economic objective, one should pick sufficiently large c_1 such that it makes zone tracking the exact penalty, and tune c_2 relative to c_1 to tradeoff magnitude and duration of constraint violation. While this method suits most scenarios, for certain systems which we will show in the simulation, naive choice of a large zone tracking penalty may lead to arbitrarily poor transient economic performance.

Parameter tuning can be a challenging task for a number of reasons: (i) the number of tuning parameters can be large depending on the number of system states and inputs. In the EMPC formulation Equation (10), we have lumped all tuning parameters into two parameters, c_1 and c_2, for simplicity of exposition. Note that in principle each element of the system state and input may be assigned a weight, for quadratic penalty a weighting matrix may be used; (ii) there is a lack of transparency in the relationship between closed-loop performance and tuning parameters; (iii) the difficulty may be intrinsic to the problem. That is, suitable tuning parameters may vary under different conditions or states. These will be illustrated in the simulation example.

Motivated by the above considerations, we propose an algorithm to modify the target zone. The idea is to construct a modified target zone which is an invariant subset of the original target zone in which closed-loop transient economic performance is guaranteed. In this way, EMPC tracking the modified target zone will have guaranteed economic performance once the system is in the modified target zone. Moreover, under exact zone tracking penalty, the closed-loop system will not leave the target zone once it enters. EMPC tracking the modified target zone allows the user to tradeoff magnitude and duration of zone tracking violation by tuning c_1 and c_2 without worrying about poor transient economic performance. The proposed algorithm is outlined below:

Algorithm 1: Modified target zone.

1. Choose some $M \in \mathbb{I}_1^N$ and $\alpha \geq 0$
2. Set $\mathbb{Z}_0 = (x_s, u_s)$
3. for $i = 0 : M-1$

   Calculate $\mathbb{Z}_{i+1}$ with Equation. (20)

   end
4. The modified target zone is $\mathbb{Z}_t' = \mathbb{Z}_M$

$$\mathbb{Z}_{i+1} = \left\{ (x,u) \;\middle|\; \begin{array}{c} f(x,u) \in proj_{\mathbb{X}}(\mathbb{Z}_i) \\ \ell_e(x,u) \leq \ell_e(x_s,u_s) + \alpha \\ (x,u) \in \mathbb{Z}_t \end{array} \right\} \tag{20}$$

The modified target zone $\mathbb{Z}_t'$, obtained by Algorithm 1 is a zone in which the system can be steered to the optimal steady state in M steps in the target zone $\mathbb{Z}_t$ while the economic performance of each step is upper bounded by $\ell_e(x_s, u_s) + \alpha$. Once the modified target zone is obtained, the EMPC of Equation (10) may be implemented with the original target zone $\mathbb{Z}_t$ replaced by the modified target zone $\mathbb{Z}_t'$. The properties of the EMPC that tracks the modified target zone $\mathbb{Z}_t'$ are summarized in the following theorem:

Theorem 4. *Consider the system of Equation (1) under the EMPC of Equation (10) with the target zone* $\mathbb{Z}_t$ *replaced with the modified target zone* $\mathbb{Z}_t'$ *and* $x(0) \in \mathbb{X}_N(\mathbb{Z}_t, x_s)$.

(i) *If* c_1 *is an exact zone tracking penalty for* $\mathbb{Z}_t'$ *for all* $x(n) \in \mathbb{X}_N(\mathbb{Z}_t, x_s)$, *then the modified target zone is forward invariant under the closed-loop system. That is,*

$$(x(n), u(n)) \in \mathbb{Z}_t' \implies (x(n+1), u(n+1)) \in \mathbb{Z}_t'$$

(ii) *If in addition Assumptions 1 and 2 hold, the transient economic performance in the modified target zone $\mathbb{Z}_t'$ is upper bounded such that for any time instant K where $(x(K), u(K)) \in \mathbb{Z}_t'$, the following holds:*

$$\sum_{n=K}^{\infty} \left(\ell_e(x(n), u(n)) - \ell_e(x_s, u_s)\right) \le M \cdot \alpha$$

Proof.

(i) First, we prove that the sets $\mathbb{Z}_i$, $i \in \mathbb{I}_0^M$, defined in Equation (20) are nested such that $\mathbb{Z}_i \subseteq \mathbb{Z}_{i+1}$ for $i \in \mathbb{I}_0^{M-1}$. We can start from $M = 2$. It is easy to verify that $(x_s, u_s) \in \mathbb{Z}_1$ because the point (x_s, u_s) satisfies the conditions in Equation (20) for $i = 0$. Thus we have $\mathbb{Z}_0 \subseteq \mathbb{Z}_1$. Then we can prove by induction. Suppose that $\mathbb{Z}_{i-1} \subseteq \mathbb{Z}_i$ for all $i \le M$. Consider $i = M - 1$ in Equation (20) and $(x, u) \in \mathbb{Z}_M$, we have $f(x, u) \in proj_{\mathbb{X}}(\mathbb{Z}_{M-1}) \subseteq proj_{\mathbb{X}}(\mathbb{Z}_M)$. Now consider $i = M$ in Equation (20) and $(x, u) \in \mathbb{Z}_M$. Since $f(x, u) \in proj_{\mathbb{X}}(\mathbb{Z}_M)$ for any $(x, u) \in \mathbb{Z}_M$, $\mathbb{Z}_M$ is a subset of $\mathbb{Z}_{M+1}$. This proves that the sets $\mathbb{Z}_i \subseteq \mathbb{Z}_{i+1}$ are nested for all $i \in \mathbb{I}_0^{M-1}$. The nestedness of the set $\mathbb{Z}_i$ implies that whenever $x(n) \in \mathbb{Z}_t' = \mathbb{Z}_t^M$, there is a feasible sequence bounded in $\mathbb{Z}_t'$ that reaches the optimal steady state in M steps. This further implies that the exact penalty for tracking the modified target zone $\mathbb{Z}_t'$ will act as hard constraint and prevent the system from leaving the modified target zone $\mathbb{Z}_t'$.

(ii) Let the optimal solution at time instant n be denoted by $u^*(i|n)$ and $x^*(i|n)$, $i \in \mathbb{I}_0^{N-1}$. The corresponding terminal state is $x^*(N|n) = x_s$ because of the terminal constraint Equation (8). A feasible solution at time instant $n + 1$ can be constructed as follows: $[u^*(1|n), u^*(2|n), \cdots, u^*(N-1|n), u_s]$, with the corresponding state trajectory: $[x^*(1|n), x^*(2|n), \cdots, x^*(N-1|n), x_s]$. Let $V(n)$ denote the optimal value function at time n, the value function $V(n+1)$ is upper bounded by the above feasible solution, which yields:

$$\begin{aligned} V(n+1) &\le \ell_e(x^*(1|n), u^*(1|n)) + \ell_e(x^*(2|n), u^*(2|n)) + \ldots + \ell_e(x_s, u_s) \\ &= V(n) - \ell_e(x^*(0|n), u^*(0|n)) + \ell_e(x_s, u_s) \end{aligned}$$

Rearranging the above and replacing $x^*(0|n), u^*(0|n)$ with $x(n), u(n)$, we have:

$$\ell_e(x(n), u(n)) - l_e(x_s, u_s) \le V(n) - V(n+1)$$

Summing both sides from $n = K$ to ∞:

$$\sum_{n=K}^{\infty} \left(\ell_e(x(n), u(n)) - \ell_e(x_s, u_s)\right) \le V(K) - V(\infty) \tag{21}$$

If $(x(K), u(K)) \in \mathbb{Z}_t'$, from Algorithm 1 and Equation (20), there exist state and input trajectories which satisfy:

$$\begin{aligned} &\ell_e(x(i|K), u(i|K)) \le \ell_e(x_s, u_s) + \alpha, && \left(x(i|K), u(i|K)\right) \in \mathbb{Z}_t' && i \in \mathbb{I}_0^{M-1} \\ &\ell_e(x(i|K), u(i|K)) = \ell_e(x_s, u_s), && \left(x(i|K), u(i|K)\right) = (x_s, u_s) && i \in \mathbb{I}_M^{N-1} \end{aligned}$$

The above implies that

$$V(K) \le N \cdot \ell_e(x_s, u_s) + M \cdot \alpha \tag{22}$$

Moreover, under Assumptions 1 and 2, we know from Theorem 1 that $\lim_{n \to \infty} x(n) = x_s$. This means that the optimal solution $\left(x^*(i|n), u^*(i|n)\right) = (x_s, u_s)$ as $n \to \infty$. Thus we have

$$V(\infty) = N \cdot \ell_e(x_s, u_s) \tag{23}$$

Theorem 4 (ii) is proved by substituting Equations (22) and (23) into Equation (21). □

Remark 6. *The tuning parameters M and α in Algorithm 1 may have significant impacts shaping the modified target zone. The parameter α can be thought of as the instantaneous acceptable performance loss and $M \cdot \alpha$ the total transient economic performance loss in the modified target zone. Under the same M, larger α may, but not necessarily, result in larger modified target zone. The parameter α is useful only if $\alpha \leq \max\{\ell_e(x,u) \mid (x,u) \in \mathbb{Z}_t'\}$. Similarly, under the same α, larger M may result in larger modified target zone. It is possible that the modified target zone is finitely determined. That is, for some $K \in \mathbb{I}_0^N$, $\mathbb{Z}_M = \mathbb{Z}_K$ for all $M \geq K$. In this case Algorithm 1 may be stopped after K steps. In the extreme case where $M = 1$, the modified target zone is a singleton of the optimal steady state (x_s, u_s). The proposed approach becomes the conventional set-point tracking MPC. On the other hand if both M and α are sufficiently large then the modified target zone equals or approaches the maximal control invariant set in the target zone. We note that the modified target zone based on the tuning parameters M and α essentially provides a means to make parameter tuning of the original problem more transparent.*

Remark 7. *Note that step (3) of Algorithm 1 involves set projection and set intersection operations. These operations could be computationally difficult for generic nonlinear systems. For linear systems with polyhedral target set $\mathbb{Z}_t$ and polyhedral performance level set $\{(x,u) : \ell_e(x,u) \leq \alpha\}$, Equation (20) can be computed using Fourier–Motzkin elimination as well as the redundancy removal method in ([21]). Note also that large M or a large number of system states may result in complex representations of the modified target zone (a set characterized by many inequalities) in Algorithm 1. Therefore computation complexity of the modified EMPC also needs to be taken into consideration while choosing parameters M and α.*

5. Simulation

5.1. Example 1

The first example is a linear scalar system:

$$x(n+1) = 1.25x(n) + u(n)$$

With state and input constraints $\mathbb{X} = [-5, 5]$, $\mathbb{U} = [-5, 5]$ respectively. The target set is $\mathbb{Z}_t = \{(x,u) \mid x \in [-5,5], u \in [-1,1]\}$. The economic cost is $\ell_e = (u - 0.9)^2$ which corresponds to an optimal steady state $(x_s, u_s) = (-3.6, 0.9)$. Two different initial states: $x(0) = -5$ and $x(0) = 5$ are considered to indicate the asymmetric closed-loop performance. The control horizon is $N = 20$ when not specified. The simulations and discussions are carried out as follows: first, we will simulate the EMPC in Section 3 which tracks the original target zone. Exact zone tracking penalty is used and special attention is paid to the difficulties in parameter tunning. Then we will simulate EMPC with modified target zone in Section 4 to demonstrate the advantage of the proposed approach.

5.1.1. EMPC Tracking the Original Target Zone

Figure 1 shows the closed-loop input trajectories of the EMPC of Equation (10) with $c_1 = 10^4$ and $c_2 = 10^2, 10^3, 10^4, 10^5$ respectively. The corresponding economic performances are shown in Figure 2. It is seen that all closed-loop trajectories reach the target zone in finite steps and asymptotically converge to the optimal value $u_s = 0.9$. As the quadratic zone tracking penalty c_2 increases, the control action gets milder, the magnitude of zone tracking violation becomes smaller but it takes longer to reach the target zone.

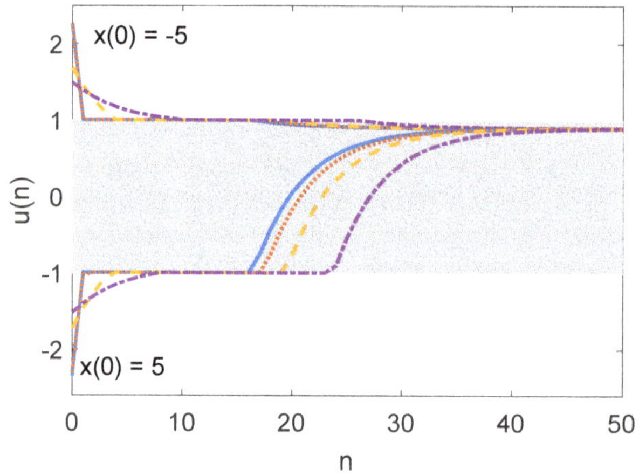

Figure 1. Closed-loop input trajectories of EMPC of Equation (10) with $c_1 = 10^4$ and $c_2 = 10^2$ (solid), $c_2 = 10^3$ (dotted), $c_2 = 10^4$ (dashed), $c_2 = 10^5$ (dash dotted), respectively. Shaded area depicts the input target zone. The upper and lower part correspond to initial state $x(0) = -5$ and $x(0) = 5$, respectively.

Another interesting observation is the target boundary riding. That is, the closed-loop input trajectories reach the target zone and stay on the zone boundary (at $u = -1$ or $u = 1$) for some time before approaching the optimal steady state value. Similar phenomena of zone boundary riding due to l_1 norm penalty was observed in [7]. The occurrence of boundary riding is due to the exact zone tracking penalty and inconsistency between the zone tracking and economic objective (which also includes setpoint tracking).

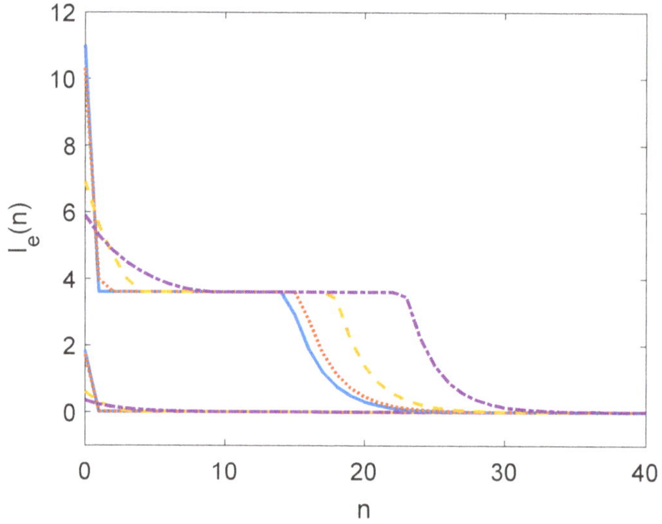

Figure 2. Closed-loop economic performance of EMPC of Equation (10) with $c_1 = 10^4$ and $c_2 = 10^2$ (solid), $c_2 = 10^3$ (dotted), $c_2 = 10^4$ (dashed), $c_2 = 10^5$ (dash dotted), respectively. The upper and lower part correspond to initial state $x(0) = 5$ and $x(0) = -5$, respectively.

Comparing the trajectories emitting from $x(0) = -5$ and $x(0) = 5$ with the same zone tracking penalty, we can see that they are (almost) symmetric outside or on the zone boundary, and asymmetric inside the target zone. Again, this is due to the large zone tracking penalty. The transient economic performance of different trajectories are summarized in Table 1. From Figure 2 and Table 1, we can see that economic performance is less of concern if the initial state is $x(0) = -5$. In this case we may pick the zone tracking penalty regardless of the transient economic performance. If smaller zone tracking violation is desirable, a large quadratic penalty, $c_2 = 10^5$, may be picked. However, if the initial state is $x(0) = 5$, transient performance varies significantly under different quadratic penalties. In this case it might be desirable to pick smaller quadratic zone tracking penalty to achieve better transient economic performance. This shows that the difficulty of parameter tuning is intrinsic to the system.

Table 1. Transient economic performance $\sum_{n=0}^{50} \ell_e(x(n), u(n))$ of EMPC of Equation (10).

	$c_2 = 10^2$	$c_2 = 10^3$	$c_2 = 10^4$	$c_2 = 10^5$
$x(0) = -5$	2.0195	2.0225	1.2560	1.2465
$x(0) = 5$	76.1218	79.5542	86.5742	103.0781

Figure 3 shows the input trajectories under different control horizons $N = 10, 20, 30, 40$. It is seen that with poor choices of the penalties c_1 and c_2, increasing the control horizon N may result in longer duration of target boundary riding and deteriorated economic performance. This also indicates difficulty in parameter tuning and controller design since one would expect larger control horizon to result in better performance.

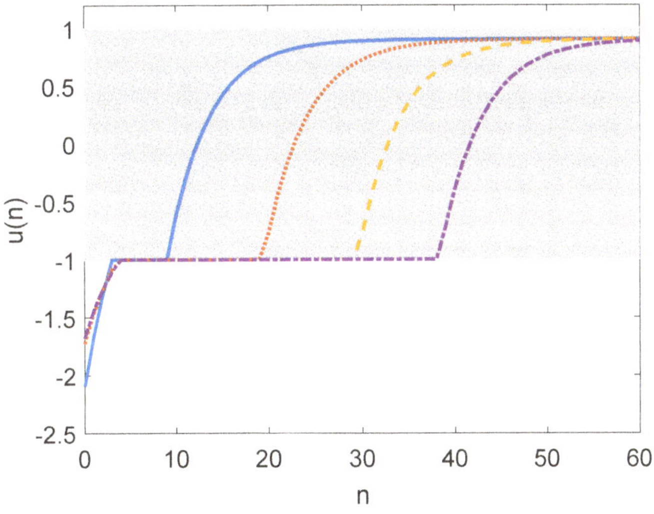

Figure 3. Closed-loop input trajectories of EMPC in Equation (10) with $x(0) = 5$, $c_1 = 10^4$ and $c_2 = 10^4$, $N = 10$ (solid), $N = 20$ (dotted), $N = 30$ (dashed), $N = 40$ (dash dotted), respectively.

5.1.2. EMPC Tracking the Modified Target Zone

Apply Algorithm 1 to the example with $M = 10$ and $\alpha = 1$, the modified target zone is obtained as follows:

$$\mathbb{Z}'_t = \{(x, u) \mid Ex + Fu \leq G\} \tag{24}$$

where $E = [1.25, -1.25, 0, 0]^T$, $F = [1, -1, 1, -1]^T$, $G = [-0.7436, 3.9571, 1.0000, 0.1000]$. The constraint set $\mathbb{Z}$, target zone $\mathbb{Z}_t$ and modified target zone $\mathbb{Z}_t'$ are illustrated in Figure 4.

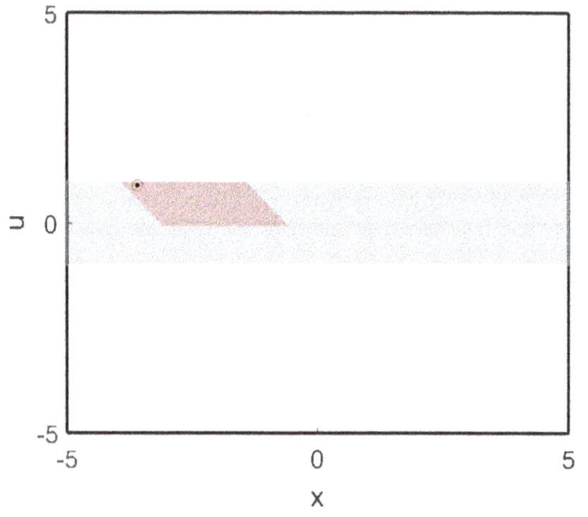

Figure 4. The constraint set $\mathbb{Z}$ (box), target zone $\mathbb{Z}_t$ (shaded rectangle) and modified target zone $\mathbb{Z}_t'$ (parallelogram). The circle indicates the optimal steady state (x_s, u_s).

Figures 5 and 6 show the closed-loop input trajectories and economic performances of the EMPC tracking the modified target zone $\mathbb{Z}_t'$ in Equation (24). Comparing Figures 5 and 6 with Figures 1 and 2, we can see that EMPC tracking the modified target zone $\mathbb{Z}_t'$ leads to more balanced economic performance in the target zone while accomplishing fast zone tracking for the target zone $\mathbb{Z}_t$. Boundary riding takes place at the boundary of the modified target zone $\mathbb{Z}_t'$ which corresponds to the economic performance bound specified by $\ell_e(x, u) \leq \alpha = 1$. This makes α a good tuning parameter.

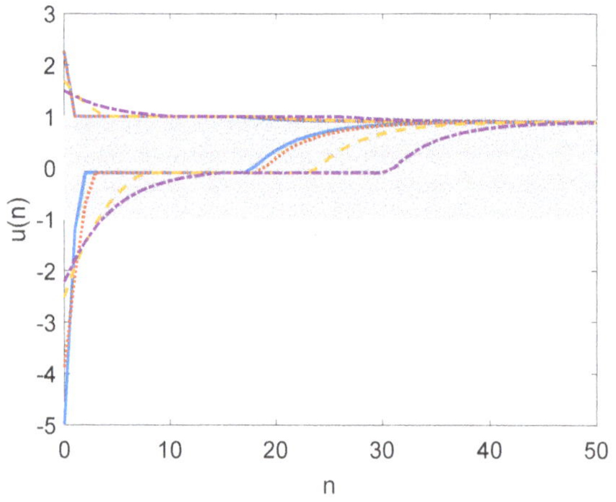

Figure 5. Closed-loop input trajectories of EMPC of Equation (10) with modified zone $\mathbb{Z}_t'$ in Equation (24), with $c_1 = 10^4$ and $c_2 = 10^2$ (solid), $c_2 = 10^3$ (dotted), $c_2 = 10^4$ (dashed), $c_2 = 10^5$ (dash dotted), respectively. Shaded area depicts the input target zone. The upper and lower part correspond to initial state $x(0) = -5$ and $x(0) = 5$, respectively.

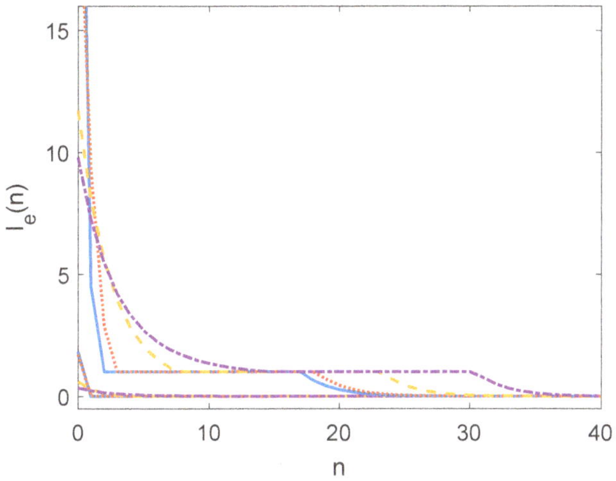

Figure 6. Closed-loop economic performance of EMPC of Equation (10) with modified zone $\mathbb{Z}_t'$ in Equation (24), with $c_1 = 10^4$ and $c_2 = 10^2$ (solid), $c_2 = 10^3$ (dotted), $c_2 = 10^4$ (dashed), $c_2 = 10^5$ (dash dotted), respectively. The upper and lower part correspond to initial state $x(0) = 5$ and $x(0) = -5$, respectively.

Table 2 compares the transient economic performance of the EMPC tracking the target zone $\mathbb{Z}_t$ and EMPC tracking the modified target zone $\mathbb{Z}_t'$. It is seen that both EMPCs achieve almost the same economic performance for the initial state $x(0) = -5$. When $x(0) = 5$, EMPC tracking the modified target zone $\mathbb{Z}_t'$ results in improved transient economic performance with less variations under different tuning parameter c_2. This implies that we can now use c_2 to tune for zone tracking (magnitude versus duration) and use α and M associated with the modified target zone to tune for economic performance. Tracking the modified target zone effectively decomposes the zone tracking and economic objectives.

Table 2. Comparison of the transient economic performance $\sum_{n=0}^{50} \ell_e(x(n), u(n))$ of EMPC tracking the target zone $\mathbb{Z}_t$ and EMPC tracking the modified zone $\mathbb{Z}_t'$.

Tracking $\mathbb{Z}_t$	$c_2 = \mathbf{10^2}$	$c_2 = \mathbf{10^3}$	$c_2 = \mathbf{10^4}$	$c_2 = \mathbf{10^5}$
$x(0) = -5$	2.0195	2.0225	1.2560	1.2465
$x(0) = 5$	76.1218	79.5542	86.5742	103.0781
Tracking $\mathbb{Z}_t'$	$c_2 = \mathbf{10^2}$	$c_2 = \mathbf{10^3}$	$c_2 = \mathbf{10^4}$	$c_2 = \mathbf{10^5}$
$x(0) = -5$	2.0195	2.0225	1.2560	1.2465
$x(0) = 5$	57.4483	52.7305	54.9608	64.2366

5.2. Example 2

The second example is a building heating control system taken from [22]. The heating system is modeled as the following discrete-time linear system:

$$\begin{aligned} x_{k+1} &= Ax_k + Bu_k + Ed_k \\ y_k &= Cx_k \end{aligned} \tag{25}$$

where $x = [T_r\ T_f\ T_w]^T$, $u = W_c$ and $d = [T_a\ \phi_s]^T$ are the state, input and disturbance vectors, respectively. The process variables are described in Table 3. The sampling time of the system is 0.5 h. The system matrices are

$$A = \begin{bmatrix} 0.6822 & 0.3028 & 0.0007 \\ 0.0740 & 0.9213 & 0.0040 \\ 0.0007 & 0.0159 & 0.9834 \end{bmatrix} \qquad E = \begin{bmatrix} 143.1277 & 4.6253 \\ 6.9283 & 0.3671 \\ 0.0401 & 0.0025 \end{bmatrix} \times 10^{-4}$$

$$B = \begin{bmatrix} 0.6822 & 0.3028 & 0.0007 \end{bmatrix}^T \times 10^{-4} \quad C = \begin{bmatrix} 1 & 0 & 0 \end{bmatrix}$$

Table 3. Building heating control system variable description.

Variable	Unit	Description
T_r	°C	Room air temperature
T_f	°C	Floor temperature
T_w	°C	Water temperature in floor heating pipes
W_c	W	Heat pump compressor input power
T_a	°C	Ambient temperature
ϕ_s	W	Solar radiation power

The control objective is to keep the room temperature between 20 °C and 21 °C while reducing energy consumption from the heat pump compressor. The target zone is thus $\mathbb{Z}_t = \{x_1 \mid 20 \le x_1 \le 21\}$, the economic cost is $\ell_e = p \cdot u$ where p is the electricity price. Without loss of generality in the simulation results, we assume that the electricity price $p = 0.1$ USD/kW·h. We assume constant ambient temperature and solar radiation $d = [5\ 0]^T$. The control input is subject to the constraint $0 \le u \le 2000$ W. The optimal steady state which solves Equation (3) is: $x_s = [20.0\ 20.7\ 35.7]^T$, $u_s = 140.0$.

In the simulation, we compare three different control schemes: (i) a set-point tracking MPC which tracks the optimal steady state with a stage cost $\ell_t = \|x - x_s\|_Q^2 + \|u - u_s\|_R^2$ where $Q = diag(10^6, 1, 1)$ and $R = 1$; (ii) the EMPC of Equation (10) with $c_1 = 10^4$ and $c_2 = 0$ which is equivalent to the EMPC in [22] with a large zone tracking penalty; and (iii) EMPC with modified target zone by Algorithm 1 with $\alpha = 200$ and $M = 12$. The initial state is $x(0) = [19.00,\ 19.64,\ 33.89]^T$. The control horizons of all controllers are $N = 96$ (48 h). The room temperature and heat input profiles are shown in Figure 7, the phase space plot of the system is shown in Figure 8, transient economic performances or the additional electricity cost $\sum_{n=0}^{72} \ell_e(x(n), u(n)) - \ell_e(x_s, u_s)$ of the three control schemes are summarized in Table 4. It is seen that EMPC with prioritized zone tracking is able to reach the desired room temperature zone in the shortest period of time of 10 h but at a higher electricity cost of \$411.1 with a significant overshoot. The conventional setpoint tracking MPC has the slowest response and reaches the desired temperature in 24 h with an additional electricity cost of \$363.3. EMPC with the modified target zone yields a balanced solution which reaches the desired room temperature in 16 h with an additional electricity cost of \$369.8.

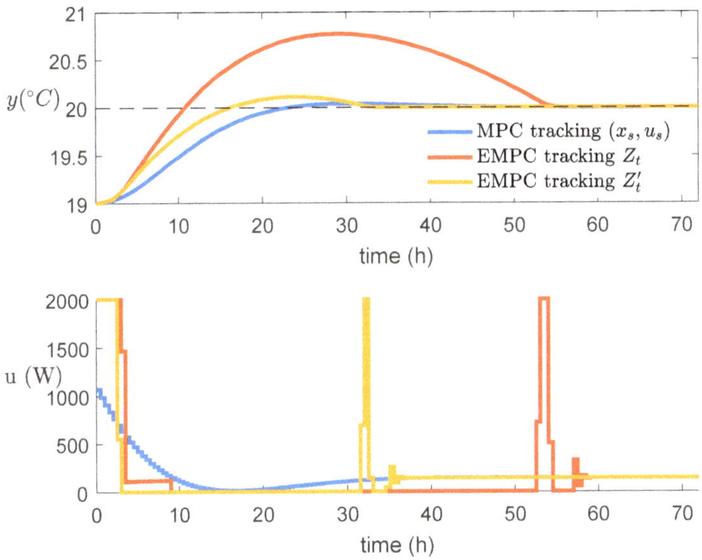

Figure 7. Room temperature and heat input profiles under different control schemes.

Table 4. Additional electricity cost of the three control schemes.

	MPC Tracking (x_s, u_s)	EMPC Tracking $\mathbb{Z}_t$	EMPC Tracking $\mathbb{Z}_t'$
Additional electricity cost (USD)	363.3	411.1	369.8

Moreover, we investigated the closed-loop performance of the EMPC with modified target zones under varying parameters α and M in Algorithm 1. It was found that in this example, the closed-loop behavior is more sensitive to M than to α. Figure 9 shows the room temperature and heat input profiles of EMPC with modified target zone $\mathbb{Z}_t'$ by Algorithm 1 with $\alpha = 2000$ and $M = 1, 6, 12, 18$. It is seen that as M increases, the time to reach the desired room temperature deceases from 24 to 15 h, which suggests that M is a good tuning parameter.

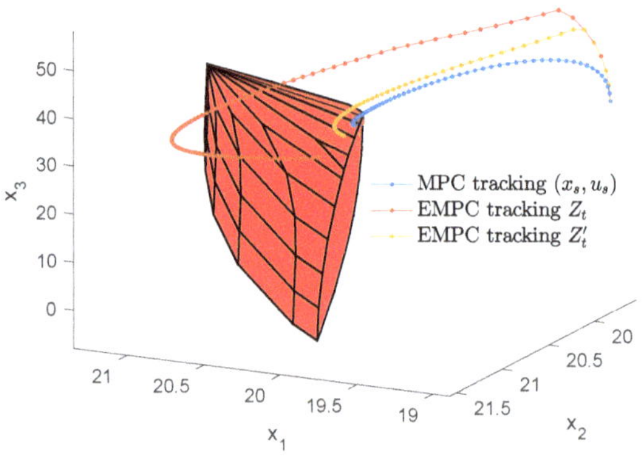

Figure 8. Phase space plot of different control schemes. The polyhedron depicts the modified target zone $\mathbb{Z}_t'$.

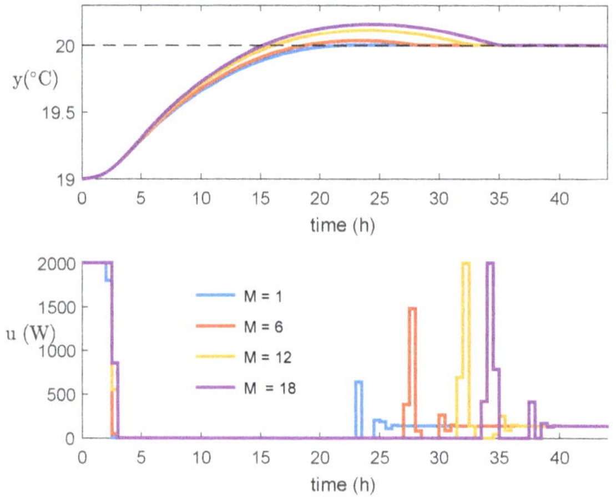

Figure 9. Room temperature and heat input profiles of EMPC with modified target zone $\mathbb{Z}_t'$ by Algorithm 1 with $\alpha = 2000$ and $M = 1, 6, 12, 18$.

6. Conclusions

In this work, we proposed an EMPC framework with integrated zone control. The proposed EMPC design is essentially a multi-objective optimization problem in which the magnitude and duration of zone objective violation as well as the economic performance are traded off. Asymptotic stability of the optimal steady state and prioritized zone tracking with exact penalty function were discussed. An algorithm to modify the target zone is proposed which simplifies the parameter tuning procedure and decouples the zone tracking and economic objective. Future research will consider the proposed approach for time-varying and uncertain systems.

Author Contributions: Su Liu and Jinfeng Liu conceived the idea of the proposed controller design. Su developed the theory, carried out the simulations and wrote the paper. Jinfeng Liu supervised the research findings and reviewed the manuscript.

Conflicts of Interest: The authors declare no conflict of interest.

References

1. Lu, J.Z. Challenging control problems and emerging technologies in enterprise optimization. *Control Eng. Pract.* **2003**, *11*, 847–858. [CrossRef]
2. Mayne, D.Q.; Rawlings, J.B.; Rao, C.V.; Scokaert, P.O.M. Constrained model predictive control: Stability and optimality. *Automatica* **2000**, *36*, 789–814. [CrossRef]
3. Rawlings, J.B.; Angeli, D.; Bates, C.N. Fundamentals of economic model predictive control. In Proceedings of the 51th IEEE Conference on Decision and Control, Grand Wailea, HI, USA, 10–13 December 2012; pp. 3851–3861.
4. Grüne, L. Economic receding horizon control without terminal constraints. *Automatica* **2013**, *49*, 725–734. [CrossRef]
5. Ellis, M.; Durand, H.; Christofides, P.D. A tutorial review of economic model predictive control methods. *J. Process Control* **2014**, *24*, 1156–1178. [CrossRef]
6. Liu, S.; Liu, J. Economic model predictive control with extended horizon. *Automatica* **2016**, *73*, 180–192. [CrossRef]
7. Scokaert, P.; Rawlings, J.B. Feasibility issues in linear model predictive control. *AIChE J.* **1999**, *45*, 1649–1659. [CrossRef]
8. Kerrigan, E.C.; Maciejowski, J.M. Soft constraints and exact penalty functions in model predictive control. In Proceedings of the UKACC International Conference Control, Cambridge, UK, 4–7 October 2000.
9. De Oliveira, N.M.C.; Biegler, L.T. Constraint handing and stability properties of model-predictive control. *AIChE J.* **1994**, *40*, 1138–1155. [CrossRef]
10. Zeilinger, M.N.; Morari, M.; Jones, C.N. Soft Constrained Model Predictive Control With Robust Stability Guarantees. *IEEE Trans. Autom. Control* **2014**, *59*, 1190–1202. [CrossRef]
11. Askari, M.; Moghavvemi, M.; Almurib, H.A.F.; Haidar, A.M.A. Stability of Soft-Constrained Finite Horizon Model Predictive Control. *IEEE Trans. Ind. Appl.* **2017**, *53*, 5883–5892. [CrossRef]
12. Ferramosca, A.; Limon, D.; González, A.H.; Odloak, D.; Camacho, E. MPC for tracking zone regions. *J. Process Control* **2010**, *20*, 506–516. [CrossRef]
13. Lez, A.H.G.; Marchetti, J.L.; Odloak, D. Robust model predictive control with zone control. *IET Control Theory Appl.* **2009**, *3*, 121–135.
14. González, A.H.; Odloak, D. A stable MPC with zone control. *J. Process Control* **2009**, *19*, 110–122. [CrossRef]
15. Blanchini, F. Set invariance in control. *Automatica* **1999**, *35*, 1747–1767. [CrossRef]
16. Amrit, R.; Rawlings, J.B.; Angeli, D. Economic optimization using model predictive control with a terminal cost. *Annu. Rev. Control* **2011**, *35*, 178–186. [CrossRef]
17. Diehl, M.; Amrit, R.; Rawlings, J.B. A Lyapunov function for economic optimizing model predictive control. *IEEE Trans. Autom. Control* **2011**, *56*, 703–707. [CrossRef]
18. Rao, C.V.; Rawlings, J.B. Linear programming and model predictive control. *J. Process Control* **2000**, *10*, 283–289. [CrossRef]
19. Fletcher, R. *Practical Methods of Optimization*; John Wiley & Sons: New York, NY, USA, 2013.
20. Kerrigan, E.C.; Maciejowski, J.M. Invariant sets for constrained nonlinear discrete-time systems with application to feasibility in model predictive control. In Proceedings of the 39th IEEE Conference on Decision and Control, Sydney, Australia, 12–15 December 2000; Volume 5, pp. 4951–4956.
21. Keerthi, S.; Gilbert, E. Computation of minimum-time feedback control laws for discrete-time systems with state-control constraints. *IEEE Trans. Autom. Control* **1987**, *32*, 432–435. [CrossRef]
22. Halvgaard, R.; Poulsen, N.K.; Madsen, H.; Jørgensen, J.B. Economic model predictive control for building climate control in a smart grid. In Proceedings of the IEEE Innovative Smart Grid Technologies, Washington, DC, USA, 16–20 January 2012; pp. 1–6.

mathematics

MDPI

Article

Safeness Index-Based Economic Model Predictive Control of Stochastic Nonlinear Systems

Zhe Wu [1], Helen Durand [2] and Panagiotis D. Christofides [1,3,*]

1 Department of Chemical and Biomolecular Engineering, University of California, Los Angeles, CA 90095-1592, USA; wuzhe@g.ucla.edu
2 Department of Chemical Engineering and Materials Science, Wayne State University, Detroit, MI 48202, USA; helen.durand@wayne.edu
3 Department of Electrical and Computer Engineering, University of California, Los Angeles, CA 90095-1592, USA
* Correspondence: pdc@seas.ucla.edu

Received: 28 March 2018; Accepted: 27 April 2018; Published: 3 May 2018

Abstract: Process operational safety plays an important role in designing control systems for chemical processes. Motivated by this, in this work, we develop a process Safeness Index-based economic model predictive control system for a broad class of stochastic nonlinear systems with input constraints. A stochastic Lyapunov-based controller is first utilized to characterize a region of the state-space surrounding the origin, starting from which the origin is rendered asymptotically stable in probability. Using this stability region characterization and a process Safeness Index function that characterizes the region in state-space in which it is safe to operate the process, an economic model predictive control system is then developed using Lyapunov-based constraints to ensure economic optimality, as well as process operational safety and closed-loop stability in probability. A chemical process example is used to demonstrate the applicability and effectiveness of the proposed approach.

Keywords: process operational safety; economic model predictive control; Safeness Index; nonlinear systems; chemical processes; probabilistic uncertainty

1. Introduction

Process operational safety has become crucially important in the chemical industry since the failure of process safety devices/human error often leads to disastrous incidents causing human and capital loss [1]. Motivated by this, recently, a new class of economic model predictive control systems (EMPC), in which the cost function penalizes process economics instead of the distances from the steady-state in a general quadratic form, was utilized to account for process operational safety and economic optimality based on a function called the Safeness Index [2,3]. These new EMPC methods complement previous efforts on economic model predictive control (e.g., [4–7]), which were not concerned explicitly with process operational safety. Specifically, in [2], a Safeness Index function that indicates the level of safety of a given state was utilized to characterize a safe operating region and used as a constraint in the EMPC design such that the closed-loop state of a nonlinear process is guaranteed to be driven into the safe operating region in finite time in the presence of sufficiently small bounded disturbances and, if the Safeness Index takes a special form related to a Lyapunov function used in the EMPC design, to never again exit that safe operating region while maximizing the economics of the process. However, in general, the Safeness Index does not have to take this special form and may therefore leave the safe operating region for finite periods of time (this may be acceptable depending on how the notion of a "safe" region of operation is selected; e.g., perhaps a "safe" region of operation means it is safe to operate in for all times, but that if the state is not in that region for short periods of time, there is not an immediate concern). Therefore, with a general form of the Safeness Index, the hard constraint on this function in the EMPC design of [2] with a Safeness

Index-based constraint may not be feasible. Due to the potential infeasibility issue caused by the hard constraint, the potential for the state to leave the safe operating region unless the Safeness Index has a specific form and the fact that disturbances may not be sufficiently small to guarantee that the closed-loop state re-enters this safe operating region, the EMPC design with a Safeness Index-based constraint may be limited in terms of its applicability to stochastic nonlinear systems.

On the other hand, MPC and EMPC of stochastic nonlinear systems have received a lot of attention recently (e.g., [8,9]). Uncertainty in the process model may be considered to have a worst-case upper and lower bound, or it may be considered to have unbounded variation and therefore be treated in a probabilistic manner. Since the variation of disturbances is not bounded in a stochastic nonlinear system, the Lyapunov-based economic model predictive control (LEMPC) framework [4] developed for nonlinear systems with small bounded disturbances is unable to guarantee closed-loop stability (i.e., the state of the closed-loop system stays within a well-characterized region of the state-space); instead, probabilistic closed-loop stability results are expected in this case. To that end, in [10], the Markov-chain Monte Carlo technique was used to derive the probabilistic convergence to a near-optimal solution for a constrained stochastic optimization problem. In [9], a Lyapunov-based model predictive control (LMPC) method was proposed for stochastic nonlinear systems to drive the state to a steady-state within an explicitly characterized region of attraction in probability. Recently, the work [11] developed a Lyapunov-based EMPC method for stochastic nonlinear systems by utilizing the probability distribution of the disturbance term to derive closed-loop stability and recursive feasibility results in probability.

In the same direction, this work focuses on the design of Safeness Index-based economic model predictive control systems for a broad class of stochastic nonlinear systems with input constraints. Specifically, under the assumption of the stabilizability of the origin of the stochastic nonlinear system via a stochastic Lyapunov-based control law, a process Safeness Index function and the level sets of multiple Lyapunov functions are first utilized to characterize a safe operating region in state-space, starting from which recursive feasibility and process operational safety are derived in probability for the stochastic nonlinear system under an economic model predictive controller. This economic model predictive control method is then designed that utilizes stochastic Lyapunov-based constraints to achieve economic optimality, as well as feasibility and process operational safety in probability in the well-characterized safe operating region.

The rest of the manuscript is organized as follows: in the Preliminaries, the notation, the class of systems and the stabilizability assumptions are given. In the Main Results, the process Safeness Index and the Safeness Index-based LEMPC are introduced. Subsequently, the Safeness Index-based LEMPC using multiple level sets of Lyapunov functions (to broaden the state-space set for which it is recursively feasible) is developed for the nominal system. Based on this, the corresponding stochastic Safeness Index-based LEMPC and its probabilistic process operational safety and feasibility properties are developed for the nonlinear stochastic system. Finally, a nonlinear chemical process example is used to demonstrate the application of the proposed stochastic Safeness Index-based LEMPC.

2. Preliminaries

2.1. Notations

Throughout the paper, we use the notation $(\Omega, \mathcal{F}, \mathbf{P})$ to denote a probability space. The notation $|\cdot|$ is used to denote the Euclidean norm of a vector, and the notation $|\cdot|_Q$ denotes the weighted Euclidean norm of a vector (i.e., $|x|_Q = x^T Q x$ where Q is a positive definite matrix). x^T denotes the transpose of x. $\mathbf{R}_+$ denotes the set $[0, \infty)$. The notation $L_f V(x)$ denotes the standard Lie derivative $L_f V(x) := \frac{\partial V(x)}{\partial x} f(x)$. Given a set $\mathcal{D}$, we denote the boundary of $\mathcal{D}$ by $\partial\mathcal{D}$, the closure of $\mathcal{D}$ by $\overline{\mathcal{D}}$ and the interior of $\mathcal{D}$ by $\mathcal{D}^o$. Set subtraction is denoted by "$\backslash$", i.e., $A \backslash B := \{x \in \mathbf{R}^n : x \in A, x \notin B\}$. A continuous function $\alpha : [0, a) \to [0, \infty)$ is said to be a class $\mathcal{K}$ function if $\alpha(0) = 0$ and it is strictly increasing. The function $f(x)$ is said to be a class C^k function if the i-th derivative of f exists and is continuous for all $i = 1, 2, ..., k$. Consider a stochastic process $x(t, w) : [0, \infty) \times \Omega \to \mathbf{R}^n$ on $(\Omega, \mathcal{F}, \mathbf{P})$. For each $w \in \Omega$, $x(\cdot, w)$ is a realization or trajectory of the stochastic process, and we abbreviate $x(t, w)$ as $x_w(t)$. $\mathbf{E}(A)$, $\mathbf{P}(A)$, $\mathbf{E}(A \mid \cdot)$ and $\mathbf{P}(A \mid \cdot)$ are the expectation, the probability, the conditional expectation and the conditional probability of the occurrence of the event A, respectively. The hitting time τ_X of a set X is the first time that the state trajectory hits the boundary of X. Additionally, we define $\tau_{X,T}(t) = \min\{\tau_X, T, t\}$, where T is the operation time.

2.2. Class of Systems

Consider a class of continuous-time stochastic nonlinear systems described by the following system of stochastic differential equations:

$$dx(t) = f(x(t))dt + g(x(t))u(t)dt + h(x(t))dw(t) \tag{1}$$

where $x \in \mathbf{R}^n$ is the stochastic state vector and $u \in \mathbf{R}^m$ is the input vector. The available control action is defined by $U := \{u \in \mathbf{R}^m \mid u_i^{\min} \le u \le u_i^{\max}, i = 1, 2, ..., m\}$. The disturbance $w(t)$ is a standard q-dimensional independent Wiener process defined on the probability space $(\Omega, \mathcal{F}, \mathbf{P})$. $f(\cdot)$, $g(\cdot)$, and $h(\cdot)$ are sufficiently smooth vector and matrix functions of dimensions $n \times 1$, $n \times m$ and $n \times q$, respectively. It is assumed that the steady-state of the system with $w(t) \equiv 0$ is $(x_s^*, u_s^*) = (0, 0)$. The initial time t_0 is defined as zero ($t_0 = 0$). We also assume that $h(0) = 0$ such that the disturbance term $h(x(t))dw(t)$ of Equation (1) vanishes at the origin.

Definition 1. *Given a C^2 Lyapunov function $V : \mathbf{R}^n \to \mathbf{R}_+$, the infinitesimal generator ($\mathcal{L}V$) of the system of Equation (1) is defined as follows:*

$$\mathcal{L}V(x) = L_f V(x) + L_g V(x) u + \frac{1}{2} Tr\{h(x)^T \frac{\partial^2 V}{\partial x^2} h(x)\} \tag{2}$$

We assume that $L_f V(x)$, $L_g V(x)$ and $h(x)^T \frac{\partial^2 V}{\partial x^2} h(x)$ are locally Lipschitz throughout the work.

Definition 2. *Assuming that the equilibrium of the uncontrolled system $dx(t) = f(x(t))dt + h(x(t))dw(t)$ is at the origin, then the origin is said to be asymptotically stable in probability, if for any $\epsilon > 0$, the following conditions hold ([12]):*

$$\lim_{x(0) \to 0} \mathbf{P}(\lim_{t \to \infty} x(t) = 0) = 1 \tag{3a}$$

$$\lim_{x(0) \to 0} \mathbf{P}(\sup_{t \ge 0} |x(t)| > \epsilon) = 0 \tag{3b}$$

Proposition 1. *Given the uncontrolled system $dx(t) = f(x(t))dt + h(x(t))dw(t)$, if for all $x \in D_0 \subset \mathbf{R}^n$, where D_0 is an open neighborhood of the origin, $\mathcal{L}V < 0$ holds $\forall t \in (0, \infty)$, then $\mathbf{E}(V(x(t))) < V(x(0))$, $\forall t \in (0, \infty)$, and the origin of the uncontrolled system is asymptotically stable in probability ([12]).*

2.3. Stabilizability Assumptions

We assume there exists a stochastic stabilizing feedback control law $u = \Phi_s(x) \in U$ (e.g., [13,14]) such that the origin of the system of Equation (1) can be rendered asymptotically stable in probability for all $x \in D \subset \mathbf{R}^n$, where D is an open neighborhood of the origin, in the sense that there exists a positive definite C^2 stochastic control Lyapunov function V that satisfies the following inequality:

$$\begin{aligned} \mathcal{L}V &= L_f V(x) + L_g V(x) \Phi_s(x) + \frac{1}{2} Tr\{h^T \frac{\partial^2 V}{\partial x^2} h\} \\ &\le -\alpha_1(|x|) \end{aligned} \tag{4}$$

where $\alpha_1(\cdot)$ is a class $\mathcal{K}$ function.

Based on the controller $\Phi_s(x)$, we characterize the set $\phi_d := \{x \in \mathbf{R}^n \mid \mathcal{L}V + \kappa V(x) \leq 0, u = \Phi_s(x) \in U, \kappa > 0\}$. We also choose a level set $\Omega_\rho := \{x \in \phi_d \mid V(x) \leq \rho\}$ of $V(x)$ inside ϕ_d as the stability region for the system of Equation (1). Therefore, the origin of the system of Equation (1) is rendered asymptotically stable via the controller $\Phi_s(x)$ in probability if $x(0) = x_0 \in \Omega_\rho$.

In this work, we develop an economic MPC design that takes advantage of the Safeness Index function [2] in its design to achieve probabilistic process operational safety in the following sense:

Definition 3. *Consider the system of Equation (1) with input constraints $u \in U$. If there exists a control law $u = \Phi \in U$ such that the state trajectories of the system for any initial state $x(0) = x_0 \in \mathcal{S}$ satisfy $x(t) \in \mathcal{S}$, $\forall\, t \geq 0$ with the probability p, where $\mathcal{S}$ is a safe operating region in state-space that excludes the unsafe region $\mathcal{D}$, we say that the control law Φ maintains the process state within a safe operating region $\mathcal{S}$ with probability p.*

Remark 1. *In general, the safe operating region $\mathcal{S}$ is characterized as a subset of the stability region (because process operation is safe provided that the system is operated within a closed-loop stability region) for the closed-loop system of Equation (1) to account for the additional safety constraints. Therefore, if there exists a control law $u = \Phi(x) \in U$ that maintains the process state within $\mathcal{S}$ with the probability p, it also maintains the process state within the stability region at least with probability p. This implies that the probability of process operational safety of the system of Equation (1), which we will discuss in the following sections, also gives a lower bound on probabilistic closed-loop stability.*

3. Main Results

In this section, the process Safeness Index and the optimization problem of Safeness Index-based LEMPC designed for the nominal system of Equation (1) with $w(t) \equiv 0$ are first presented. Based on that, the Safeness Index-based LEMPC using multiple level sets of Lyapunov functions is developed for the nominal system of Equation (1) to guarantee recursive feasibility and to guarantee that the closed-loop state does not enter an unsafe operating region $\mathcal{D}$. Subsequently, the stochastic Safeness Index-based LEMPC is developed for the system of Equation (1) to account for the disturbances $w(t)$ with unbounded variation. The stochastic safety and feasibility in probability of the closed-loop system of Equation (1) are finally investigated under the sample-and-hold implementation of the proposed stochastic Safeness Index-based LEMPC.

3.1. Process Safeness Index

In [2], the Safeness Index function $S(x)$ was developed to indicate the level of safety of a given state, through which process operational safety was integrated with process control system design to account for the process operational safety considerations resulting from multivariable interactions or interactions between units. There are various methods of determining the functional form of $S(x)$, for example by utilizing first-principles process models or using systematic safety analysis tools such as HAZOP and fault tree analysis.

Based on the functional form of $S(x)$, the closed-loop state predictions are required to be maintained within a safe region $\mathcal{S}$ (where $S(x)$ is below the threshold on the Safeness Index S_{TH}) by using the Safeness Index-based constraint within the process control design. Additionally, the safety systems (e.g., the alarm, emergency shut-down and relief systems) can be triggered if the threshold S_{TH} is sufficiently exceeded, which implies that the process operation becomes unsafe and further actions are required.

3.2. Safeness Index-Based LEMPC

Safeness Index-based LEMPC optimizes an economic cost function $L_e(\cdot,\cdot)$ and maintains the closed-loop state of the nominal system of Equation (1) with $w(t) \equiv 0$ in a safe operating region by utilizing the Safeness Index function as a hard constraint within the LEMPC design. Specifically, the formulation of the Safeness Index-based LEMPC is as follows:

$$\max_{u(t)\in ST(\Delta)} \int_{t_k}^{t_k+\tau_P\Delta} L_e(\tilde{x}(\tau),u(\tau))\,d\tau \tag{5a}$$

$$\text{s.t. } \dot{\tilde{x}}(t) = f(\tilde{x}(t)) + g(\tilde{x}(t))u(t) \tag{5b}$$

$$u(t) \in U,\ \forall\, t \in [t_k, t_k+\tau_P\Delta) \tag{5c}$$

$$\tilde{x}(t_k) = x(t_k) \tag{5d}$$

$$V(\tilde{x}(t)) < \rho_e',\ \forall\, t \in [t_k, t_k+\tau_P\Delta) \quad \text{if } x(t_k) \in \Omega^{o}_{\rho_e'} \tag{5e}$$

$$S(\tilde{x}(t)) \leq S_{TH},\ \forall\, t \in [t_k, t_k+\tau_P\Delta) \quad \text{if } S(x(t_k)) \leq S_{TH} \tag{5f}$$

$$\dot{V}(x(t_k),u(t_k)) \leq \dot{V}(x(t_k),\Phi_n(x(t_k))), \quad \text{if } x(t_k) \in \Omega_{\rho'}\backslash\Omega^{o}_{\rho_e'} \text{ or } S(x(t_k)) > S_{TH} \tag{5g}$$

where $\tilde{x}$ is the predicted state trajectory, $ST(\Delta)$ is the set of piecewise constant functions with sampling period Δ, τ_P is the number of sampling periods of the prediction horizon and $\dot{V} = L_f V(x) + L_g V(x)u$. $\Phi_n(x)$ is the stabilizing feedback control law designed for the nominal system of Equation (1) with $w(t) \equiv 0$ such that the origin of the system of Equation (1) can be rendered asymptotically stable. Under the controller $\Phi_n(x)$, we first characterize the set $\phi_n := \{x \in \mathbf{R}^n \mid \dot{V} + \kappa V(x) \leq 0, u = \Phi_n(x) \in U, \kappa > 0\}$ and choose the level set $\Omega_{\rho'} := \{x \in \phi_n \mid V(x) \leq \rho'\}$ inside ϕ_n as the stability region. $\Omega_{\rho_e'} := \{x \in \mathbf{R}^n \mid V(x) \leq \rho_e'\}$ where $0 < \rho_e' < \rho'$ is further designed to make the region $\Omega_{\rho'}$ a forward invariant set in the presence of sufficiently small bounded disturbances.

The constraint of Equation (5e) allows the cost function of Equation (5a) to be maximized while keeping the predicted closed-loop state within $\Omega^{o}_{\rho_e'}$ if $x(t_k) \in \Omega^{o}_{\rho_e'}$. The safety constraint of Equation (5f) is applied to maintain the predictions of the closed-loop state within the safe operating region $\mathcal{S} := \{x \in \mathbf{R}^n \mid S(x) \leq S_{TH}\}$ if $x(t_k) \in \mathcal{S}$. On the other hand, if $x(t_k) \in \Omega_{\rho'}\backslash\Omega^{o}_{\rho_e'}$ or $x(t_k)$ is outside of $\mathcal{S}$, the constraint of Equation (5g) is activated to decrease $V(x)$ such that $x(t)$ will move towards the origin within the current sampling period.

Remark 2. *Since the safe operating region $\mathcal{S}$ is not necessarily a forward invariant set based on the formulation of the Safeness Index function, the threshold S_{TH} set on the Safeness Index may define a region that is irregularly shaped, for example the grey region in Figure 1 [2] corresponding to a chemical reactor example similar to the one in the section "Application to a Chemical Process Example" of this manuscript. Therefore, the existence of feasible solutions (i.e., the satisfaction of the constraints of Equation (5)) of the Safeness Index-based LEMPC is not guaranteed in $\mathcal{S}$ due to the constraint of Equation (5f). Additionally, $S(x(t))$ may not even decrease under the constraint of Equation (5g) due to the same reason (that $\mathcal{S}$ is not an invariant set). Considering the above feasibility issue in the formulation of the Safeness Index-based LEMPC, a new Safeness Index-based LEMPC is developed in the following subsection by using multiple Lyapunov functions to characterize the safe operating region $\mathcal{S}$.*

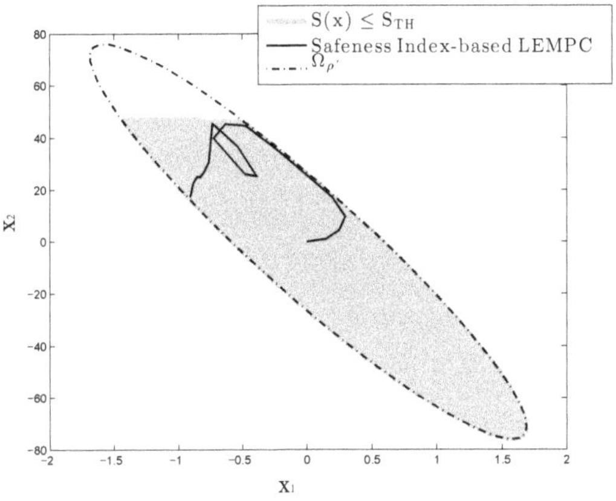

Figure 1. A schematic representing the safe operating region $\mathcal{S}$ (the gray region) with an example closed-loop trajectory under the Safeness Index-based Lyapunov-based economic model predictive control (LEMPC) design of Equation (5) for the initial condition $(0, 0)$.

3.3. *Safeness Index-Based LEMPC Using Multiple Level Sets*

The improved Safeness Index-based LEMPC for the nominal system of Equation (1) with $w(t) \equiv 0$ is developed utilizing the level sets of two Lyapunov functions V_1 and V_2 to characterize the safe and unsafe operating regions. Throughout this work, we assume that the shape of the stability regions, $\mathcal{D}$, and their intersection are amenable to the treatment in this work, such as the use of only two Lyapunov functions in the LEMPC design and also the types of overlap of the stability regions described. Specifically, as shown in Figure 2, we define two level sets: $\Omega_{\rho'} := \{x \in \phi'_n \mid V_1(x) \leq \rho'\}$ and $\mathcal{U}_{s'} := \{x \in \phi'_n \mid V_2(x) \leq s'\}$ inside $\phi'_n := \{x \in \mathbf{R}^n \mid \dot{V}_i + \kappa V_i(x) \leq 0, i = 1, 2, u = \Phi_n(x) \in U, \kappa > 0\}$, from which the origin of the nominal system of Equation (1) is rendered asymptotically stable.

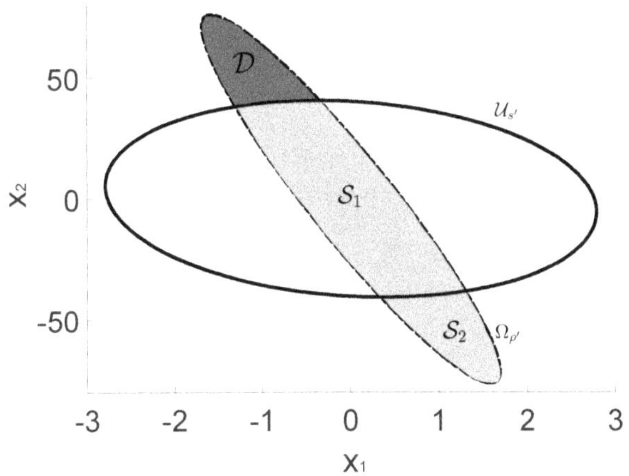

Figure 2. A schematic representing the unsafe region $\mathcal{D}$ (dark gray) and the safe operating region $\mathcal{S} := \mathcal{S}_1 \cup \mathcal{S}_2$ (light gray).

$\Omega_{\rho'}$ represents the stability region as it is in the Safeness Index-based LEMPC of Equation (5), and $\mathcal{U}_{s'}$ is designed to exclude the unsafe region $\mathcal{D}$ where $S(x) > S_{TH}$. Therefore, the safe operating region $\mathcal{S}$ becomes the union of $S_1 := \Omega_{\rho'} \cap \mathcal{U}_{s'}$ and $S_2 := \Omega_{\rho'} \backslash (S_1 \cup \mathcal{D})$ in Figure 2. This new Safeness Index-based LEMPC design is formulated by the following optimization problem:

$$\max_{u \in S T(\Delta)} \int_{t_k}^{t_k+\tau_P\Delta} L_e(\tilde{x}(t), u(t))\, dt \tag{6a}$$

$$\text{s.t. } \dot{\tilde{x}}(t) = f(\tilde{x}(t)) + g(\tilde{x}(t))u(t) \tag{6b}$$

$$\tilde{x}(t_k) = x(t_k) \tag{6c}$$

$$u(t) \in U, \quad \forall t \in [t_k, t_k + \tau_P\Delta) \tag{6d}$$

$$V_1(\tilde{x}(t)) < \rho'_e, \quad \text{if } x(t_k) \in \Omega^{\circ}_{\rho'_e}, \quad \forall t \in [t_k, t_k + \tau_P\Delta) \tag{6e}$$

$$V_2(\tilde{x}(t)) < s'_e, \quad \text{if } x(t_k) \in \mathcal{U}^{\circ}_{s'_e}, \quad \forall t \in [t_k, t_k + \tau_P\Delta) \tag{6f}$$

$$\dot{V}_i(x(t_k), u(t_k)) \leq \dot{V}_i(x(t_k), \Phi_n(x(t_k))), i = 1, 2 \quad \text{if } x(t_k) \in \Omega_{\rho'} \backslash \Omega^{\circ}_{\rho'_e}, \text{ or } x(t_k) \in \mathcal{U}_{s'} \backslash \mathcal{U}^{\circ}_{s'_e} \tag{6g}$$

where the notation follows that in Equation (5). $\Omega_{\rho'_e}$ and $\mathcal{U}_{s'_e}$ are again chosen as the level sets inside ϕ'_n to make $\Omega_{\rho'}$ and $\mathcal{U}_{s'}$ forward invariant sets, respectively. In the optimization problem of Equation (6), the objective function of Equation (6a) is the integral of $L_e(\tilde{x}(t), u(t))$ over the prediction horizon. The constraint of Equation (6b) is the nominal system of Equation (1) with $w(t) \equiv 0$ that is used to predict the states of the closed-loop system. Equation (6c) defines the initial condition $\tilde{x}(t_k)$ of the optimization problem determined from a state measurement $x(t_k)$ at $t = t_k$. Equation (6d) represents the input constraints applied over the entire prediction horizon. The constraint of Equation (6e) maintains the predicted states in $\Omega^{\circ}_{\rho'_e}$ when the current state $x(t_k) \in \Omega^{\circ}_{\rho'_e}$. Similarly, the constraint of Equation (6f) maintains the predicted states in $\mathcal{U}^{\circ}_{s'_e}$ when $x(t_k) \in \mathcal{U}^{\circ}_{s'_e}$. The contractive constraint of Equation (6g) is activated to decrease both V_1 and V_2 such that the closed-loop state enters the smaller level sets of V_1 and V_2 (i.e., towards the interior of $\mathcal{S}_1$). Therefore, under the Safeness Index-based LEMPC of Equation (6), if $x(t_k) \in \mathcal{S}_1$, the constraints of Equations (6e)–(6g) maintain the closed-loop state in $\mathcal{S}_1$. If $x(t_k) \in \mathcal{S}_2$, the constraints of Equations (6e) and (6g) are applied to maintain the closed-loop state in $\Omega_{\rho'}$, under which $x(t)$ will stay in $\mathcal{S}_2$ or enter $\mathcal{S}_1$ in some time.

Remark 3. *Based on the Safeness Index function $S(x)$ and its threshold S_{TH}, the level set $\mathcal{U}_{s'}$ of the Lyapunov function V_2 is chosen to exclude the unsafe region $\mathcal{D}$ that is originally in the level set $\Omega_{\rho'}$ of the Lyapunov function V_1 as shown in Figure 2. Since $\mathcal{U}_{s'}$ and $\Omega_{\rho'}$ are both forward invariant sets for the nominal system (or the system with sufficiently small bounded disturbances) of Equation (1) under the controller $\Phi_n(x) \in U$ that satisfies $\dot{V}_i + \kappa V_i(x) \leq 0, i = 1, 2, \kappa > 0$, it follows that under the corresponding constraint of Equation (6g), the overlapping region $\mathcal{S}_1$ is also an invariant set. Therefore, the infeasibility problem caused by the Safeness Index constraint of Equation (5f) is solved by introducing the second level set $\mathcal{U}_{s'}$ into the LEMPC design. For the remaining part of the safe operating region $\mathcal{S}_2$, the constraints of Equations (6e) and (6g) are utilized to ensure that the closed-loop state stays in $\Omega_{\rho'}$ all the time, which is similar to closed-loop stability under the traditional LEMPC [4]. Since the sampling period Δ has to be sufficiently small in the sample-and-hold implementation of the Safeness Index-based LEMPC of Equation (6), we can utilize a sufficiently small Δ such that $x(t_{k+1})$ is unable to jump into $\mathcal{D}$ within one sampling period if $x(t_k) \in \mathcal{S}_2$. This implies that at the next sampling time, the state $x(t_{k+1})$ either stays in $\mathcal{S}_2$ or enters $\mathcal{S}_1$ via the boundary between $\mathcal{S}_1$ and $\mathcal{S}_2$. In both cases, it is considered that the state is maintained in the safe operating region according to Definition 3.*

Remark 4. *Besides the above development of Safeness Index-based LEMPC using multiple Lyapunov functions, there are also other methods that can guarantee the feasibility of the Safeness Index-based constraint in the LEMPC design. For example, in the optimization problem of Equation (5), we can choose a more conservative level set of $V(x)$ (i.e., a small level set inside $\Omega_{\rho'}$ that excludes $\mathcal{D}$) as the safe operating region. However, if the unsafe region characterized by the Safeness Index function is a set of points inside the stability region and is difficult to exclude by a single level set like $\mathcal{U}_{s'}$, we may want to use control Lyapunov barrier functions to design the constraints that account for the unsafe region in state-space [15] and overcome the infeasibility problem.*

3.4. Stochastic Safeness Index-Based LEMPC

Inspired by the Safeness Index-based LEMPC design of Equation (6), the stochastic Safeness Index-based LEMPC design is given by the following optimization problem:

$$\max_{u \in S T(\Delta)} \int_{t_k}^{t_k+\tau_P \Delta} L_e(\tilde{x}(t), u(t))\, dt \tag{7a}$$

$$\text{s.t. } \dot{\tilde{x}}(t) = f(\tilde{x}(t)) + g(\tilde{x}(t))u(t) \tag{7b}$$

$$\tilde{x}(t_k) = x(t_k) \tag{7c}$$

$$u(t) \in U, \quad \forall\, t \in [t_k, t_k + \tau_P \Delta) \tag{7d}$$

$$V_1(\tilde{x}(t)) < \rho_e, \quad \text{if } x(t_k) \in \Omega^{\mathrm{o}}_{\rho_e}, \quad \forall t \in [t_k, t_k + \tau_P \Delta) \tag{7e}$$

$$V_2(\tilde{x}(t)) < s_e, \quad \text{if } x(t_k) \in \mathcal{U}^{\mathrm{o}}_{s_e}, \quad \forall t \in [t_k, t_k + \tau_P \Delta) \tag{7f}$$

$$\mathcal{L}V_i(x(t_k), u(t_k)) \leq \mathcal{L}V_i(x(t_k), \Phi_s(x(t_k))), i = 1, 2 \quad \text{if } x(t_k) \in \Omega_\rho \backslash \Omega^{\mathrm{o}}_{\rho_e}, \text{ or } x(t_k) \in \mathcal{U}_s \backslash \mathcal{U}^{\mathrm{o}}_{s_e} \tag{7g}$$

where the notation follows that in Equation (6) except using ρ, ρ_e, s, s_e, $\Phi_s(x)$ and $\mathcal{L}V$ to replace ρ', ρ'_e, s', s'_e, $\Phi_n(x)$ and $\dot{V}$, respectively. For the system of Equation (1) with multiple Lyapunov functions, ϕ_d is characterized as: $\phi_d = \{x \in \mathbf{R}^n \mid \mathcal{L}V_i + \kappa V_i(x) \leq 0, i = 1, 2, u = \Phi_s(x) \in U, \kappa > 0\}$. Ω_ρ, Ω_{ρ_e}, $\mathcal{U}_s$ and $\mathcal{U}_{s_e}$ are level sets of V_1 and V_2 inside ϕ_d, where $0 < \rho_e < \rho$ and $0 < s_e < s$. Similar to the LEMPC designs of Equations (5) and (6), the optimal input trajectory determined by the optimization problem of the stochastic Safeness Index-based LEMPC is denoted by $u^*(t)$, which is calculated over the entire prediction horizon $t \in [t_k, t_k + \tau_P \Delta)$. The control action computed for the first sampling period of the prediction horizon $u^*(t_k)$ is sent to the actuators to be applied over the sampling period, and the optimization problem of Equation (7) is re-solved at the next sampling time.

The constraint of Equation (7e) maintains the predicted state in $\Omega^{\mathrm{o}}_{\rho_e}$ when the current state $x(t_k) \in \Omega^{\mathrm{o}}_{\rho_e}$ and the constraint of Equation (7f) maintains the predicted state in $\mathcal{U}^{\mathrm{o}}_{s_e}$ when the current state $x(t_k) \in \mathcal{U}^{\mathrm{o}}_{s_e}$. However, if $x(t_k) \in \Omega_\rho \backslash \Omega^{\mathrm{o}}_{\rho_e}$ or $x(t_k) \in \mathcal{U}_s \backslash \mathcal{U}^{\mathrm{o}}_{s_e}$, the constraint of Equation (7g) is activated to decrease $V_1(x)$ and $V_2(x)$ such that it is possible that $x(t)$ moves back to $\Omega^{\mathrm{o}}_{\rho_e} \cap \mathcal{U}^{\mathrm{o}}_{s_e}$.

Since there exists a disturbance $w(t)$ with unbounded variation $dw(t)$ in the system of Equation (1), process operational safety (i.e., the closed-loop state is bounded in the safe operating region $\mathcal{S}$) can only be ensured in probability. Therefore, in the following sections, we will establish the probabilities of process operational safety of the system of Equation (1) under the stochastic Safeness Index-based LEMPC of Equation (7).

3.5. Sample-And-Hold Implementation

We first investigate the impact of the sample-and-hold implementation of Equation (7) on the stability of the closed-loop system of Equation (1) following similar arguments to those in [9,11]. Specifically, the probabilities of the sets Ω_ρ and $\mathcal{U}_s$ remaining invariant under the sample-and-hold implementation of the Safeness Index-based LEMPC of Equation (7) with a sampling period Δ are given as follows.

Theorem 1. *Consider the system of Equation (1) with Ω_ρ and $\mathcal{U}_s$ inside ϕ_d under the control actions u computed by the LEMPC of Equation (7). Let $u(t) = u(t_k), \forall t \in [t_k, t_k + \Delta)$. Then, given any probability $\lambda \in (0,1]$, there exist positive real numbers $\rho_s < \rho_e < \rho$ and $\rho_s < s_e < s$ where Ω_{ρ_s} and $\mathcal{U}_{\rho_s}$ are level sets of V_1 and V_2, respectively, around the origin where $\mathcal{L}V_i$, $i = 1,2$ are not required to remain negative for the nominal system of Equation (1) under the sample-and-hold implementation of $u(t)$, and there also exists a sampling period $\Delta^* := \Delta^*(\lambda)$, such that if $\Delta \in (0, \Delta^*]$, then:*

$$\mathbf{P}(\sup_{t\in[0,\Delta]} V_1(x(t)) < \rho) \geq 1 - \lambda,\ \forall x(0) \in \Omega^o_{\rho_e} \tag{8}$$

$$\mathbf{P}(\sup_{t\in[0,\Delta]} V_2(x(t)) < s) \geq 1 - \lambda,\ \forall x(0) \in \mathcal{U}^o_{s_e} \tag{9}$$

$$\begin{aligned} &\mathbf{P}(\sup_{t\in[0,\Delta]} \mathcal{L}V_i(x(t)) < -\epsilon < 0) \geq 1 - \lambda,\ i = 1,2, \\ &\forall x(0) \in (\Omega_\rho \cup \mathcal{U}_s)\backslash(\Omega^o_{\rho_s} \cap \mathcal{U}^o_{\rho_s}) \end{aligned} \tag{10}$$

Proof. Let $A_B := \{w : \sup_{t\in[0,\Delta^*]} |w(t)| \leq B\}$. Using the results for standard Brownian motion [16], given any probability $\lambda \in (0,1]$, there exists a sufficiently small B, s.t. $P(A_B) = 1 - \lambda$. For each realization $x_w(t)$ with $x(0) \in \Omega_\rho \cup \mathcal{U}_s$ and $w \in A_B$, there almost surely exists a positive real number k_1, s.t. $\sup_{t\in[0,\Delta^*]} |x_w(t) - x(0)| \leq k_1(\Delta^*)^r$, where $r < 1/2$, according to the local Hölder continuity. Therefore, the probability of the event $A_W := \{w : \sup_{t\in[0,\Delta^*]} |x(t) - x(0)| \leq k_1(\Delta^*)^r\}$ is:

$$\mathbf{P}(A_W) \geq 1 - \lambda \tag{11}$$

We first prove that the probabilities of Equations (8) and (9) hold for the first sampling period. It should be noted that the probabilities of Equations (8)–(10) can be generalized to any sampling period $t \in [t_k, t_k + \Delta]$ with the measurement of $x(t_k)$ playing the role of $x(0)$ in Equations (8)–(10).

Since $V_i(x)$, $i = 1,2$ satisfies the local Lipschitz condition, there exist positive real numbers k_{2i}, $i = 1,2$, such that $|V_i(x(t)) - V_i(x(0))| \leq k_{2i}|x(t) - x(0)|$, $i = 1,2$. Therefore, for all $w \in A_W$, if $\Delta^* < \Delta_1 = (\frac{\rho-\rho_e}{k_{21}k_1})^{(\frac{1}{r})}$, it follows that $|V_1(x_w(t)) - V_1(x(0))| < \rho - \rho_e, \forall t \leq \Delta^*$. Furthermore, $\forall x(0) \in \Omega^o_{\rho_e}$, it is obtained that $V_1(x_w(t)) < \rho$, $\forall t \leq \Delta^*$ since $-(\rho - \rho_e) < V_1(x(t)) - V_1(x(0)) < \rho - \rho_e$ and $\sup_{x(0)\in\Omega^o_{\rho_e}} V_1(x(0)) = \rho_e$. Therefore, if $x(0) \in \Omega^o_{\rho_e}$, the probability of $x(t)$ staying inside Ω_ρ is $\mathbf{P}(\sup_{t\in[0,\Delta^*]} V_1(x(t)) < \rho) \geq 1 - \lambda$. Similarly, if $\Delta^* < \Delta_2 = (\frac{s-s_e}{k_{22}k_1})^{(\frac{1}{r})}$, for any $x(0) \in \mathcal{U}^o_{s_e}$, the probability of $x(t)$ staying inside $\mathcal{U}_s$ is $\mathbf{P}(\sup_{t\in[0,\Delta^*]} V_2(x(t)) < s) \geq 1 - \lambda$.

We now prove the probability of Equation (10) by using the equation $\mathcal{L}V_i(x(t)) = \mathcal{L}V_i(x(0)) + (\mathcal{L}V_i(x(t)) - \mathcal{L}V_i(x(0)))$, $\forall t \in [0, \Delta^*]$, $i = 1,2$. It is shown that there exists a positive real number ϵ such that $\mathcal{L}V_i(x(t)) < -\epsilon$ holds $\forall x(0) \in (\Omega_\rho \cup \mathcal{U}_s)\backslash(\Omega^o_{\rho_s} \cap \mathcal{U}^o_{\rho_s})$ for the nominal system of Equation (1) based on the definition of the value of $\mathcal{L}V_i$ in ϕ_d. However, $\mathcal{L}V_i(x(t)) < -\epsilon$ only holds in probability for the system in the presence of the disturbances $w(t)$. Based on the local Lipschitz conditions of $L_f V_i(x)$, $L_g V_i(x)$ and $h(x(t))^T \frac{\partial^2 V_i(x(t))}{\partial x^2} h(x(t))$, there exist positive real numbers k_3, k_4, k_5, such that $|L_f V_i(x(t)) - L_f V_i(x(0))| \leq k_3|x(t) - x(0)|$, $|L_g V_i(x(t)) - L_g V_i(x(0))| \leq k_4|x(t) - x(0)|$, $|\frac{1}{2}Tr\{h(x(t))^T \frac{\partial^2 V_i(x(t))}{\partial x^2} h(x(t))\} - \frac{1}{2}Tr\ \{h(x(0))^T \frac{\partial^2 V_i(x(0))}{\partial x^2} h(x(0))\}| \leq k_5|x(t) - x(0)|, i = 1,2$.

Let $0 < \epsilon < \kappa\rho_s$ and $\Delta^* < \Delta_3 = (\frac{\kappa\rho_s - \epsilon}{k_1(k_3+k_4+k_5)})^{(\frac{1}{r})}$. It follows from $\mathcal{L}V_i(x(t)) \leq \mathcal{L}V_i(x(0)) + |\mathcal{L}V_i(x(t)) - \mathcal{L}V_i(x(0))| < \mathcal{L}V_i(x(0)) + \kappa\rho_s - \epsilon$ (which follows from the application of the Lipschitz properties of the components of $\mathcal{L}V_i$ with $\Delta^* < \Delta_3$) and the fact that $x(0) \in (\Omega_\rho \cup \mathcal{U}_s)\backslash(\Omega^o_{\rho_s} \cap \mathcal{U}^o_{\rho_s})$ and $\mathcal{L}V_i(x_0) < -\kappa V_i(x(0))$, that $\forall w \in A_W$, $\mathcal{L}V_i(x_w(t)) < -\epsilon < 0$, $\forall t \leq \Delta^*$, $i = 1,2$ holds. Therefore, by choosing the sampling period $\Delta \in (0, \Delta^*]$, given any initial condition $x(0) \in (\Omega_\rho \cup \mathcal{U}_s)\backslash(\Omega^o_{\rho_s} \cap \mathcal{U}^o_{\rho_s})$, the probability that $\mathcal{L}V_i(x(t)) < -\epsilon$ is as follows: $\mathbf{P}(\sup_{t\in[0,\Delta^*]} \mathcal{L}V_i(x(t)) < -\epsilon,\ i = 1,2) \geq 1-\lambda$. Finally, let $\Delta^* \leq \min\{\Delta_1, \Delta_2, \Delta_3\}$, and the probabilities of Equations (8)–(10) are all satisfied for $\Delta \in (0, \Delta^*]$. □

3.6. Stability in Probability

Based on the results from the above section, the probabilistic process operational safety of the closed-loop system of Equation (1) under the Safeness Index-based LEMPC of Equation (7) applied in a sample-and-hold fashion is established by the following theorem.

Theorem 2. *Consider the system of Equation (1) under the stochastic Safeness Index-based LEMPC of Equation (7) applied in a sample-and-hold implementation (i.e., $u(t) = u(i\Delta),\ \forall\ i\Delta \leq t < (i+1)\Delta$, $i = 0,1,2,...$). Then, given $\rho_e \in (0,\rho)$, $s_e \in (0,s)$ and probability $\lambda \in (0,1]$, there exist a sampling time $\Delta \in (0, \Delta^*(\lambda)]$ and probabilities $\beta, \beta', \gamma, \gamma' \in [0,1]$:*

$$\frac{\sup_{x\in\partial\Omega_{\rho_e}} V_1(x)}{\inf_{x\in\mathbf{R}^n\backslash\Omega_\rho} V_1(x)} \leq \beta \tag{12a}$$

$$\frac{\sup_{x\in\partial\mathcal{U}_{s_e}} V_2(x)}{\inf_{x\in\mathbf{R}^n\backslash\mathcal{U}_s} V_2(x)} \leq \beta' \tag{12b}$$

$$\max\{\frac{V_1(x(0))}{\rho}, \beta\} \leq \gamma \tag{12c}$$

$$\max\{\frac{V_2(x(0))}{s}, \beta'\} \leq \gamma' \tag{12d}$$

such that the following probabilities hold:

$$\begin{aligned} &\mathbf{P}(\sup_{t\in[0,\Delta)} V_1(x(t)) < \rho,\ \sup_{t\in[0,\Delta)} V_2(x(t)) < s) \\ &\geq (1-\beta)(1-\beta')(1-\lambda), \quad \forall x(0) \in \mathcal{S}_{1e} \end{aligned} \tag{13}$$

$$\begin{aligned} &\mathbf{P}(\sup_{t\in[0,\Delta)} V_1(x(t)) < \rho) \\ &\geq (1-\beta)(1-\lambda), \quad \forall x(0) \in \mathcal{S}_{2e} \end{aligned} \tag{14}$$

$$\begin{aligned} &\mathbf{P}(\tau_{\mathbf{R}^n\backslash\mathcal{S}_{1e}}(\Delta) \leq \tau_{\mathcal{S}_1}(\Delta)) \\ &\geq (1-\gamma)(1-\gamma')(1-\lambda), \quad \forall x(0) \in \mathcal{S}_1\backslash\mathcal{S}_{1e}^o \end{aligned} \tag{15}$$

where $\mathcal{S}_e := \mathcal{S}_{1e} \cup \mathcal{S}_{2e}$ is a subset of $\mathcal{S}$ that subtracts the risk margins $\rho - \rho_e$ and $s - s_e$. The relationship among the sets $\mathcal{S}_{1e} := \Omega_{\rho_e} \cap \mathcal{U}_{s_e}$ and $\mathcal{S}_{2e} := \mathcal{S}_e\backslash\mathcal{S}_{1e}$ and the unsafe region $\mathcal{D}$ are shown in Figure 3.

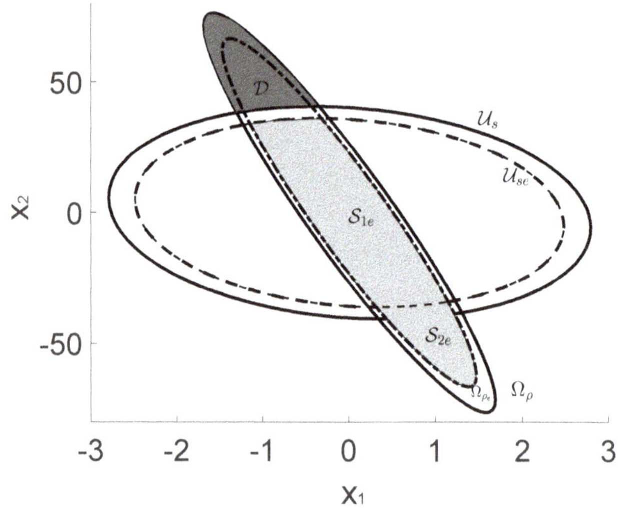

Figure 3. A schematic representing the unsafe region $\mathcal{D}$ (dark gray) and the region $\mathcal{S}_e := \mathcal{S}_{1e} \cup \mathcal{S}_{2e}$ (light gray), which is the safe operating region $\mathcal{S}$ subtracting the risk margins $\rho - \rho_e$ and $s - s_e$.

Proof. The proof consists of three parts. We first show that under the Safeness Index-based LEMPC of Equation (7), any state trajectory initiated from $x(0) \in \mathcal{S}_{1e}$ has the probability defined by Equation (13) of staying in $\mathcal{S}_1 := \Omega_\rho \cap \mathcal{U}_s$. However, if $x(0) \in \mathcal{S}_{2e}$, we prove that under the Safeness Index-based LEMPC of Equation (7), there exists the probability of Equation (14) for the state of the closed-loop system to stay in Ω_ρ and with a sufficiently small Δ to stay in the part of Ω_ρ that excludes $\mathcal{D}$. Finally, if $x(0)$ is inside $\mathcal{S}_1 \backslash \mathcal{S}^o_{1e}$, we can show that the closed-loop state trajectory reaches the boundary of $\mathcal{S}_{1e}$ first before it leaves $\mathcal{S}_1$ (implying it does not enter $\mathcal{D}$) with the probability of Equation (15). However, if $x(0) \in \Omega_\rho \backslash \Omega^o_{\rho_e}$ and $x(0) \notin (\mathcal{U}_s \cup \mathcal{D})$ (i.e., the white risk margin around $\mathcal{S}_{2e}$ in Figure 3), we show that it does not enter $\mathcal{D}$, $\forall t \in [0, \Delta)$ in probability, as well. Additionally, for the sake of simplicity, we denote the probabilities and expectations conditional on the event of A_W given in the section "Sample-And-Hold Implementation" as $\mathbf{P}^*(\cdot)$ and $\mathbf{E}^*(\cdot)$.

Part 1: To show that Equation (13) holds for all $x(0) \in \mathcal{S}_{1e}$, we consider both the case that $x(0) \in S^o_{1e}$ and that $x(0) \in \partial S_{1e}$. The former case is handled by Equations (8) and (9). Specifically, if $x(0) \in S^o_{1e}$, then both $x(0) \in \Omega^o_{\rho_e}$ and also $x(0) \in \mathcal{U}^o_{s_e}$. Then, $\mathbf{P}(\sup_{t \in [0,\Delta)} V_1(x(t)) < \rho, \sup_{t \in [0,\Delta)} V_2(x(t)) < s) \geq 1 - \lambda$ (Equations (8) and (9) for $\Delta^* \leq \min\{\Delta_1, \Delta_2\}$). Since $(1-\lambda) \geq (1-\beta)(1-\beta')(1-\lambda)$ for $\beta \in [0,1]$, $\beta' \in [0,1]$, Equation (13) holds when $x(0) \in S^o_{1e}$. When $x(0) \in \partial\mathcal{S}_{1e}$, Equation (13) is also satisfied. To show this, we first assume $x(0) \in \partial\Omega_{\rho_e}$ and prove that the probability of $x(t)$ staying in Ω_ρ within one sampling period conditioned on the event of A_W is $(1-\beta)$. When $x(0) \in \partial\Omega_{\rho_e}$, Equation (7g) will be utilized in the LEMPC of Equation (7). Under the constraint of Equation (7g), the optimization problem of Equation (7) is solved such that $\mathcal{L}V_1$ is forced to be negative for any $x(t_k) \in \Omega_\rho \backslash \Omega^o_{\rho_e}$, which implies that Equation (10) holds (i.e., $\mathcal{L}V_1 < -\epsilon$ for $t \in [0, \Delta]$ with the probability of the event A_W). Using Dynkin's formula [17], the following equation can be derived:

$$\begin{aligned} &\mathbf{E}^*(V_1(x(\tau_{\Omega_\rho \backslash \Omega^o_{\rho_e}}(t)))) \\ &\quad = V_1(x(0)) + \mathbf{E}^*\left(\int_0^{\tau_{\Omega_\rho \backslash \Omega^o_{\rho_e}}(t)} \mathcal{L}V_1(x(s))ds\right) \end{aligned} \tag{16}$$

The following probability is derived using similar arguments as in [11], for all $x(0) \in \partial\Omega_{\rho_e}$:

$$\mathbf{P}^*(V_1(x(t)) \geq \rho, \text{ for some } t \in [0, \Delta)) \leq \frac{V_1(x(0))}{\inf_{x \in \mathbf{R}^n \setminus \Omega_\rho} V_1(x)} \tag{17}$$

Bounding Equation (17) with Equation (12a) and taking the complementary events, the following probability is obtained:

$$\inf_{x(0) \in \partial\Omega_{\rho_e}} \mathbf{P}^*(V_1(x(t)) < \rho, \forall t \in [0, \Delta)) \geq (1 - \beta) \tag{18}$$

Using the same steps as performed above, we can prove that $\forall x(0) \in \partial\mathcal{U}_{s_e}$, the probability of $x(t)$ staying in $\mathcal{U}_s$ within one sampling period conditioned on the event of A_W is as follows:

$$\inf_{x(0) \in \partial\mathcal{U}_{s_e}} \mathbf{P}^*(V_2(x(t)) < s, \forall t \in [0, \Delta)) \geq (1 - \beta') \tag{19}$$

Since the set of initial conditions $x(0) \in \mathcal{S}_{1e}$ is the intersection of Ω_{ρ_e} and $\mathcal{U}_{s_e}$, by combining the probabilities of Equations (18) and (19) together and using Equation (10), the probability of Equation (13) is obtained via the definition of conditional probability.

Part 2: If $x(0) \in \mathcal{S}_{2e} \subset \Omega_{\rho_e}$, then either $x(0) \in \Omega^o_{\rho_e}$ or $x(0) \in \partial\Omega_{\rho_e}$. If $x(0) \in \mathcal{S}_{2e}$ and $\Omega^o_{\rho_e}$, then Equation (8) holds and $\mathbf{P}(\sup_{t \in [0,\Delta)} V_1(x(t)) < \rho) \geq 1 - \lambda \geq (1 - \beta)(1 - \lambda)$ for $\beta \in [0, 1]$, and Equation (14) therefore holds. If instead $x(0) \in \mathcal{S}_{2e}$ and $\partial\Omega_{\rho_e}$, then the results of Part 1 indicate that Equation (18) holds. Applying the definition of conditional probability, this also gives that Equation (14) holds. Moreover, we show that $x(t)$ is maintained inside the safe operating region $\mathcal{S}$ within one sampling period with the probability of Equation (14) (i.e., $\forall t \in [0, \Delta)$, $x(t)$ will not jump into $\mathcal{D}$ in probability). It is shown in the section "Sample-And-Hold Implementation" that $\forall t \in [0, \Delta)$, the change of $V_i(x)$ is limited (i.e., $|V_1(x(t)) - V_1(x(0))| < k_1 k_{21} \Delta^*, \forall t \leq \Delta^*$ and $|V_2(x(t)) - V_2(x(0))| < k_1 k_{22} \Delta^*, \forall t \leq \Delta^*$) with a sufficiently small sampling period Δ^* (maybe smaller than the one derived by $\Delta^* \leq \min\{\Delta_1, \Delta_2, \Delta_3\}$). Therefore, if $x(0) \in \mathcal{S}_{2e} \subset \Omega_{\rho_e}$, $x(t)$ cannot move across the entire level set $\mathcal{U}_s$ and jump into $\mathcal{D}$ within a sufficiently small Δ with the probability $(1 - \lambda)$. Instead, the closed-loop state at the next sampling time either stays in $\mathcal{S}_{2e}$ or moves into $\mathcal{S}_{1e}$ in probability. If $x(t)$ enters $\mathcal{S}_{1e}$, the probability of Equation (13) will be used to estimate the probability of closed-loop process operational safety thereafter. Because $(1 - \lambda) \geq (1 - \beta)(1 - \lambda)$, for $\beta, \lambda \in (0, 1]$, Equation (14) establishes the probability of $x(t)$ staying in the safe operating region $\mathcal{S}$ within one sampling period $\forall x(0) \in \mathcal{S}_{2e}$.

Part 3: If $x(0) \in \mathcal{S}_1 \setminus \mathcal{S}^o_{1e}$, we show that it is possible that the closed-loop state trajectory hits the boundary of $\mathcal{S}^o_{1e}$ before it hits the boundary of $\mathcal{S}_1$. If both hitting times $\tau_{\mathbf{R}^n \setminus \mathcal{S}_{1e}}(\Delta)$ and $\tau_{\mathcal{S}_1}(\Delta)$ are longer than a sampling period Δ, Equation (15) is trivially satisfied. However, if one of them or both occur within one sampling period, we show that the probability of Equation (15) holds by first showing that the extreme case that $x(0) \in (\Omega_\rho \setminus \Omega^o_{\rho_e}) \cap (\mathcal{U}_s \setminus \mathcal{U}^o_{s_e})$ (which are the corners where the risk margins $\rho - \rho_e$ and $s - s_e$ overlap in Figure 3) satisfies Equation (15). We first show that the probability of the event $A_T := \{\tau_{\mathbf{R}^n \setminus \Omega^o_{\rho_e}} > \tau_{\Omega_\rho}\}$ can be given as follows $\forall x(0) \in \Omega_\rho \setminus \Omega^o_{\rho_e}$:

$$\mathbf{P}^*(\tau_{\mathbf{R}^n \setminus \Omega^o_{\rho_e}} > \tau_{\Omega_\rho}) \leq \mathbf{P}^*(\frac{V_1(x(\tau_{\Omega_\rho \setminus \Omega^o_{\rho_e}}))}{\rho} \geq 1) \leq \frac{V_1(x(0))}{\rho} \tag{20}$$

The event A_T indicates that the state of the closed-loop system of Equation (1) reaches the boundary of Ω_ρ before it reaches the boundary of Ω_{ρ_e}. The probability of Equation (20) is determined via Equation (17) and the fact that the event $\{\tau_{\mathbf{R}^n \setminus \Omega^o_{\rho_e}} > \tau_{\Omega_\rho}\}$ belongs to the event $\{\frac{V_1(x(\tau_{\Omega_\rho \setminus \Omega^o_{\rho_e}}))}{\rho} \geq 1\}$.

Assuming $x(0) \in \partial\Omega_{\rho_c}$, where $\Omega_{\rho_c} := \{x \in \phi_d \mid V_1(x) \leq \rho_c\}$ and $\rho_c \in [\rho_e, \rho]$, the following probability is derived by bounding Equation (20) by Equation (12c):

$$\sup_{x(0)\in\Omega_{\rho_c}\backslash\Omega^o_{\rho_e}} \mathbf{P}^*(\tau_{\mathbf{R}^n\backslash\Omega^o_{\rho_e}} > \tau_{\Omega_\rho}) \leq \gamma \tag{21}$$

Using the same steps as performed above, we can prove that $\forall x(0) \in \mathcal{U}_s\backslash\mathcal{U}^o_{s_e}$, the probabilities similar to Equations (20) and (21) are derived as follows:

$$\mathbf{P}^*(\tau_{\mathbf{R}^n\backslash\mathcal{U}^o_s} > \tau_{\mathcal{U}_s}) \leq \frac{V_2(x(0))}{s} \tag{22a}$$

$$\sup_{x(0)\in\mathcal{U}_{s_c}\backslash\mathcal{U}^o_{s_e}} \mathbf{P}^*(\tau_{\mathbf{R}^n\backslash\mathcal{U}^o_{s_e}} > \tau_{\mathcal{U}_s}) \leq \gamma' \tag{22b}$$

where $\mathcal{U}_{s_c} := \{x \in \phi_d \mid V_2(x) \leq s_c\}$ and $s_c \in [s_e, s]$. Hence, the probability $\mathbf{P}(\tau_{\mathbf{R}^n\backslash\mathcal{S}_{1e}}(\Delta) \leq \tau_{\mathcal{S}_1}(\Delta))$ (i.e., Equation (15)) for the case where $x(0) \in (\Omega_\rho\backslash\Omega^o_{\rho_e}) \cap (\mathcal{U}_s\backslash\mathcal{U}^o_{s_e}) \subset \mathcal{S}_1\backslash\mathcal{S}^o_{1e}$ is obtained by taking the complementary event of Equations (21) and (22b) and using the definition of conditional probability. We now address the other two possibilities for $x(0) \in \mathcal{S}_1\backslash\mathcal{S}^o_{1e}$ besides $x(0) \in (\Omega_\rho\backslash\Omega^o_{\rho_e}) \cap (\mathcal{U}_s\backslash\mathcal{U}^o_{s_e})$, which are: (1) $x(0) \in (\Omega_\rho\backslash\Omega^o_{\rho_e}) \cap \mathcal{U}^o_{s_e}$ and (2) $x(0) \in (\mathcal{U}_s\backslash\mathcal{U}^o_{s_e}) \cap \Omega^o_{\rho_e}$. Consider the case where $x(0) \in (\Omega_\rho\backslash\Omega^o_{\rho_e}) \cap \mathcal{U}^o_{s_e}$. If $x(t) \in \mathcal{U}^o_{s_e}$, $\forall t \in [0, \Delta)$, then $\mathbf{P}^*(\tau_{\mathbf{R}^n\backslash\mathcal{S}_{1e}}(\Delta) \leq \tau_{\mathcal{S}_1}(\Delta))$=$\mathbf{P}^*(\tau_{\mathbf{R}^n\backslash\Omega^o_{\rho_e}}(\Delta) \leq \tau_{\Omega_\rho}(\Delta))$. If $x(t)$ enters $\mathcal{U}_s\backslash\mathcal{U}^o_{s_e}$ before it leaves $\Omega_\rho\backslash\Omega^o_{\rho_e}$, for some $t \in [0, \Delta)$, then for sure it holds that $\tau_{\mathbf{R}^n\backslash\mathcal{S}_{1e}}(\Delta) < \tau_{\mathcal{S}_1}(\Delta)$ because the closed-loop state trajectory crosses the boundary of $\mathcal{S}_{1e}$ first. Therefore, Equation (15) holds for both cases. The same analysis can be performed for the case where $x(0) \in (\mathcal{U}_s\backslash\mathcal{U}^o_{s_e}) \cap \Omega^o_{\rho_e}$. However, if $x(0) \in \Omega_\rho\backslash\Omega^o_{\rho_e}$ and $x(0) \notin (\mathcal{U}_s \cup \mathcal{D})$, it is readily shown that Equation (21) holds due to the fact that $x(0) \in \Omega_\rho\backslash\Omega^o_{\rho_e}$. Additionally, since it is demonstrated in *Part* 2 that the change of $V_i(x)$ within one sampling period is limited in probability, it follows that $\forall x(0) \in \Omega_\rho\backslash\Omega^o_{\rho_e}$ and $x(0) \notin (\mathcal{U}_s \cup \mathcal{D})$, $x(t)$ does not enter $\mathcal{D}$ in one sampling period with the probability of $1 - \lambda$, which implies that the closed-loop state either stays in $\mathcal{S}_2$ or moves into $\mathcal{S}_1$ in probability. □

Remark 5. *The Safeness Index-based LEMPC of Equation (7) is unable to ensure process operational safety for the closed-loop system of Equation (1) because of stochastic disturbances with unbounded variation. Additionally, in order to achieve process operational safety with higher probability, we should characterize the safe operating region $\mathcal{S}$ well and design large enough risk margins (i.e., $\rho - \rho_e$ and $s - s_e$) to avoid frequent activations of backup safety systems. Specifically, in Theorem 2, it is shown that as ρ_e and s_e decrease, the probabilities of Equations (13)–(15) become larger, which implies that if we want to improve process operational safety, the Safeness Index-based LEMPC design of Equation (7) should be designed with more conservatism (i.e., choosing smaller ρ_e and s_e). However, an operating region with smaller ρ_e and s_e in turn leads to less economic benefits, which is undesired for the Safeness Index-based LEMPC of Equation (7). Therefore, the uncertain process operational safety caused by stochastic disturbances with unbounded variation is essentially a trade-off between economic benefits and probabilistic process operational safety (i.e., in practice, we will choose a conservative operating region to make the process sufficiently safe with respect to the unbounded disturbances, especially considering the other safety systems online and the risks involved, while also optimizing process economics).*

3.7. Feasibility in Probability

Recursive feasibility for the nominal system of Equation (1) with $w(t) \equiv 0$ under the Safeness Index-based LEMPC of Equation (6) is guaranteed since there always exists a solution (e.g., the Lyapunov-based controller $\Phi_n(x)$ in sample-and-hold) that satisfies all the constraints of Equation (6). Now, consider the system of Equation (1) that has disturbance $w(t)$ with unbounded variation. Recursive feasibility under the stochastic Safeness Index-based LEMPC of Equation (7) can only be guaranteed in probability over the operation period $t \in [0, \tau_N\Delta)$. The probability is established as follows, from which it is shown that the probabilistic bounds on recursive feasibility for the remainder of the entire time of operation decrease as the operation period becomes longer (however, this does not necessarily mean the closed-loop system will not remain recursively feasible because at every sampling time, the remaining time of operation decreases and therefore the probability that the LEMPC will remain recursively feasible for the remaining time of operation increases at the next sampling time if the closed-loop state was maintained within $\mathcal{S}$ throughout the prior sampling period).

Theorem 3. *Consider the system of Equation (1) under the stochastic Safeness Index-based LEMPC of Equation (7) applied in a sample-and-hold fashion. Then, if $x(0) \in \mathcal{S}$, let $V_1(x(t+i\Delta)) = \rho_i < \rho$, $V_2(x(t+i\Delta)) = s_i < s$, $i = 0, 1, ..., \tau_N - 1$, and let A_F represent the event that the optimization problem of Equation (7) is solved with the satisfaction of recursive feasibility for time $t \in [0, \tau_N\Delta)$. The probability of A_F can be calculated as follows:*

$$\mathbf{P}(A_F) \geq (1-\lambda)^{\tau_N} \prod_{i=0,1,...,\tau_N-1} (1-\beta_i)(1-\beta_i') \tag{23}$$

where β_i and β_i' are given as follows:

$$\beta_i = \max\{\beta, \frac{\sup_{x\in\partial\Omega_{\rho_i}} V_1(x)}{\inf_{x\in\mathbf{R}^n\setminus\Omega_\rho} V_1(x)}\} \tag{24a}$$

$$\beta_i' = \max\{\beta', \frac{\sup_{x\in\partial\mathcal{U}_{s_i}} V_2(x)}{\inf_{x\in\mathbf{R}^n\setminus\mathcal{U}_s} V_2(x)}\} \tag{24b}$$

Proof. We can derive the probability of Equation (23) following similar arguments to those in [11]. Since the deterministic prediction model of Equation (7b) is used in the stochastic Safeness Index-based LEMPC of Equation (7), it follows that there always exists a solution $u(t) = \Phi_s(\tilde{x}(t_q)) \in U$, $\forall t \in [t_q, t_{q+1})$, $q = k, \ldots, k+\tau_P - 1$ that satisfies the constraints of Equation (7d–g) over the prediction horizon provided that $x(t_k), t_k \geq 0$ is inside the safe operating region $\mathcal{S}$. Therefore, this implies that the probability of recursive feasibility (i.e., Equation (23)) is equal to the probability of closed-loop process operational safety over $t \in [0, \tau_N\Delta)$, which can be obtained via the recursive application of Equation (13) with β_i and β_i' of Equation (24) and the definition of conditional probability. Additionally, it should be noted that if $x(0) \in \mathcal{S}_{2e}$, the state is not in $\partial\mathcal{U}_{s_i}$ in Equation (24). In this case, β_i' simply takes the value of β', and the probability of Equation (23) still holds since it is shown in the proof of Theorem 2 that the state either stays in $\mathcal{S}_2$ or moves into $\mathcal{S}_1$ with the probability of $1-\lambda$ (i.e., Equation (23) gives a conservative result in this case). □

Remark 6. *In Theorem 3, probabilistic process operational safety and probabilistic recursive feasibility over the operation period $t \in [0, \tau_N\Delta)$ are established for the closed-loop system of Equation (1) under the Safeness Index-based LEMPC of Equation (7). Due to the disturbance $w(t)$ with unbounded variation, the closed-loop state $x(t)$ may leave $\mathcal{S}$ at any sampling step, and thus, closed-loop process operational safety and recursive feasibility of the Safeness Index-based LEMPC of Equation (7) can only be derived in a probabilistic manner (i.e., $\forall t \in [0, \tau_N\Delta)$, these properties hold with the probability of Equation (23)). Since the existence of a feasible control action is only guaranteed in the safe operating region $\mathcal{S}$, backup safety systems should be designed to handle the process if the state exits the safe operating region. Additionally, since the probabilities of Equations (13)–(15) are less than one if $\rho_e < \rho$ and $s_e < s$, the probabilities of recursive feasibility and process operational safety for $t \in [0, \tau_N\Delta)$ decrease as the operation period $\tau_N\Delta$ becomes longer. However, it should be noted that this dependence is not unique to the MPC, but to all control designs that try to keep the process state within a specific region in state-space in the presence of stochastic disturbances with unbounded variation (i.e., the probability to keep the closed-loop state in $\mathcal{S}$ for all the remaining time of operation goes to zero at t_0 as the process operation time $\tau_N \to \infty$).*

4. Application to a Chemical Process Example

A chemical process example is used to illustrate the application of the stochastic Safeness Index-based LEMPC of Equation (7) to maintain the closed-loop state within a safe operating region in state-space in probability. Specifically, a well-mixed, non-isothermal continuous stirred tank reactor (CSTR) where an irreversible second-order exothermic reaction takes place is considered. The reaction transforms a reactant A to a product B ($A \rightarrow B$). The inlet concentration of A, the inlet temperature and the feed volumetric flow rate of the reactor are C_{A0}, T_0 and F, respectively. The CSTR is equipped with a heating jacket that supplies/removes heat at a rate Q. The CSTR dynamic model is described by the following material and energy balance equations:

$$\begin{aligned} dC_A =& \frac{F}{V_L}(C_{A0} - C_A)dt - k_0 e^{-E/RT} C_A^2 dt \\ &+ \sigma_1 (C_A - C_{As}) dw_1(t) \end{aligned} \tag{25a}$$

$$\begin{aligned} dT =& \frac{F}{V_L}(T_0 - T)dt - \frac{\Delta H k_0}{\rho_L C_p} e^{-E/RT} C_A^2 dt + \frac{Q}{\rho_L C_p V_L} dt \\ &+ \sigma_2 (T - T_s) dw_2(t) \end{aligned} \tag{25b}$$

where C_A is the concentration of reactant A in the reactor, V_L is the volume of the reacting liquid in the reactor, T is the temperature of the reactor and Q denotes the heat input rate. The concentration of reactant A in the feed is C_{A0}. The feed temperature and the volumetric flow rate are T_0 and F, respectively. The reacting liquid has a constant density of ρ_L and a heat capacity of C_p. ΔH, k_0, E and R represent the enthalpy of reaction, pre-exponential constant, activation energy and ideal gas constant, respectively. Process parameter values are given in Table 1. The disturbance terms dw_1 and dw_2 in Equation (25) are independent standard Gaussian white noise with the standard deviations $\sigma_1 = 2.5 \times 10^{-3}$ and $\sigma_2 = 0.15$, respectively. It is noted that the disturbance terms of Equation (25) vanish at the steady state.

Table 1. Parameter values of the continuous stirred tank reactor (CSTR).

$T_0 = 300$ K	$F = 5$ m^3/h
$V_L = 1$ m^3	$E = 5 \times 10^4$ kJ/kmol
$k_0 = 8.46 \times 10^6$ m^3/kmol h	$\Delta H = -1.15 \times 10^4$ kJ/kmol
$C_p = 0.231$ kJ/kg K	$R = 8.314$ kJ/kmol K
$\rho = 1000$ kg/m^3	$C_{A0_s} = 4$ kmol/m^3
$Q_s = 0.0$ kJ/h	$C_{A_s} = 1.22$ kmol/m^3
$T_s = 438$ K	

The initial steady-state of the CSTR is at $(C_{As}, T_s) = (1.22 \text{ kmol/m}^3, 438 \text{ K})$, and $(C_{A0_s}\ Q_s) = (4 \text{ kmol/m}^3, 0 \text{ kJ/h})$. The manipulated inputs are the inlet concentration of species A and the heat input rate, which are represented by the deviation variables $u_1 = \Delta C_{A0} = C_{A0} - C_{A0_s}$ and $u_2 = \Delta Q = Q - Q_s$, respectively. The manipulated inputs are bounded as follows: $|\Delta C_{A0}| \leq 3.5$ kmol/m^3 and $|\Delta Q| \leq 5 \times 10^5$ kJ/h. Therefore, the states and the inputs of the closed-loop system are represented by $x^T = [C_A - C_{As}\ T - T_s]$ and $u^T = [\Delta C_{A0}\ \Delta Q]$, respectively.

The control objective of the stochastic Safeness Index-based LEMPC of Equation (7) is to maximize the production rate of B, while maintaining the closed-loop state trajectories in the safe operating region $\mathcal{S}$ in probability. The objective function of Equation (7a) is the production rate of B: $L_e(\tilde{x}, u) = k_0 e^{-E/RT} C_A^2$. The Lyapunov functions are designed using the standard quadratic form $V_i(x) = x^T P_i x$, $i = 1, 2$, where the positive definite matrices $P_1 = \begin{bmatrix} 1060 & 22 \\ 22 & 0.52 \end{bmatrix}$ and $P_2 = \begin{bmatrix} 1060 & 10 \\ 10 & 5 \end{bmatrix}$ are chosen to characterize the set ϕ_d for the stochastic system of Equation (25). The nonlinear feedback controllers in [13,18] are utilized as $\Phi_n(x)$ and $\Phi_s(x)$, respectively. The level sets of the Lyapunov functions $V_1(x)$ and $V_2(x)$ are chosen as $\rho = 368$ and $s = 8100$ to create a safe operating region $\mathcal{S}$. The explicit Euler method with an integration time step of $h_c = 10^{-4}$ h is applied to numerically simulate the dynamic model of Equation (25). The nonlinear optimization problem of the stochastic Safeness Index-based LEMPC of Equation (7) is solved using the IPOPT software package [19] with the sampling period $\Delta = 10^{-2}$ h. With the fixed sampling period $\Delta = 10^{-2}$ h, $\rho = 368$ and $s = 8100$, we focus on the impact of ρ_e and s_e on probabilistic process operational safety in the following simulations.

It is first shown in Figure 4 that under the Safeness Index-based LEMPC of Equation (6) designed for the nominal system of Equation (25), the closed-loop state of the nominal system of Equation (25) stays in the safe operating region $\mathcal{S}$ within the entire operation period $t_s = 1$ h. Additionally, the Safeness Index-based LEMPC of Equation (6) is solved successfully in each iteration to obtain a feasible control action $u(t)$ that is applied in the next sampling period.

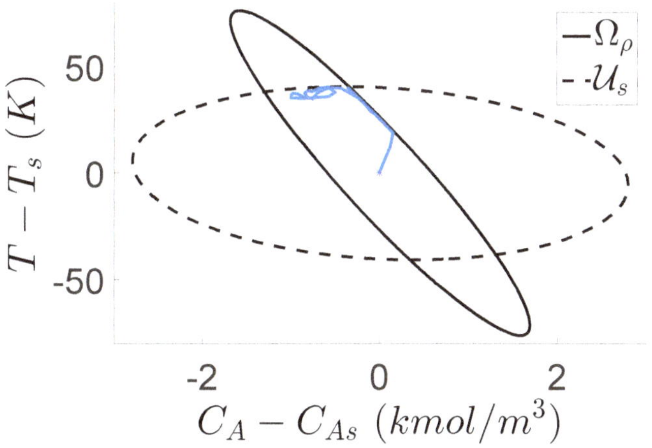

Figure 4. Closed-loop trajectory under the Safeness Index-based LEMPC of Equation (6) for the initial condition (0, 0) (in deviation variable form) with the additional material constraint: $\frac{1}{t_s}\int_0^{t_s} u_1(\tau)d\tau = 0\ \text{kmol/m}^3$.

It follows that under the stochastic Safeness Index-based LEMPC of Equation (7), the state of the closed-loop system of Equation (25) stays in $\mathcal{S}$ with different probabilities for different ρ_e and s_e. To better understand the relationship between probabilistic process operational safety and the choices of ρ_e and s_e, we derived the experimental probabilities via 500 simulation runs for the same initial condition $(\Delta C_{As}, \Delta T_s) = (0\ \text{kmol/m}^3, 0\ \text{K})$ and different choices of ρ_e and s_e (without the material constraint applied for the nominal system). Let A_V denote the event that the closed-loop state stays in $\mathcal{S}$ over the operation period $t_s = 1$ h. The results are reported in Table 2.

Table 2. Experimental probability for different values of ρ_e and s_e.

ρ_e/ρ	s_e/s	$P(A_V)$
0.98	0.99	14.0%
0.95	0.99	63.1%
0.92	0.99	82.0%
0.92	0.97	82.8%
0.92	0.95	83.6%
0.92	0.92	85.8%

From Table 2, it is observed that with fixed s_e, $\mathbf{P}(A_V)$ becomes larger as ρ_e decreases. Likewise, with fixed ρ_e, $\mathbf{P}(A_V)$ increases as s_e decreases. It is demonstrated that a higher probability of closed-loop process operational safety of the system of Equation (25) is achieved when ρ_e and s_e are more conservative. Let $\rho_e = 320$ and $s_e = 6800$. It is obtained that the probability of the states of the closed-loop system of Equation (25) remaining in the safe operating region $\mathcal{S}$ reaches 97.4%. Additionally, the averaged total economic benefit (i.e., the time integral of the stage cost L_e over the operation period $t_s = 1$ h) is 24.3 under the Safeness Index-based LEMPC of Equation (7), which has an improvement of 81% compared to 13.4 under steady-state operation. Therefore, in this example, the closed-loop system of Equation (25) under the Safeness Index-based LEMPC achieves a relatively high probability of process safety and a satisfactory process economic performance simultaneously with $\rho_e = 320$ and $s_e = 6800$. For an actual process, additional work should likely be performed, which can use techniques like those demonstrated here, to increase the probability of the states of the closed-loop system remaining within the safe operating region to higher values considered acceptable for the process at hand given its design, hazards and the backup measures (alarms/operators, safety systems, relief systems) in place.

On the other hand, it is observed from Table 2 that decreasing ρ_e increases the probability $\mathbf{P}(A_V)$. By looking at unsafe closed-loop trajectories (i.e., trajectories that leave the safe operating region $\mathcal{S}$ under the Safeness Index-based LEMPC of Equation (7) during the operation period t_s) in 500 simulation runs (one of them is shown in Figure 5), it is observed that almost all of the unsafe trajectories leave $\mathcal{S}$ through the boundary of Ω_ρ (i.e., the right edge of Ω_ρ in Figure 5). The reason for this behavior is that the local optimum value of L_e is calculated to be at the right edge of Ω_ρ, which is shown as the yellow region in Figure 6. Therefore, under the Safeness Index-based LEMPC of Equation (7), the closed-loop trajectory is optimized to approach this high production rate region and begin circling back due to the disturbances, which leads to a higher probability of leaving the safe operating region $\mathcal{S}$ from Ω_ρ. Additionally, it is observed in Figure 6 that the production rate decreases as the safe operating region shrinks (i.e., the color becomes darker), which is consistent with the fact that smaller ρ_e and s_e lead to safer process operation, at the cost of lower economic performance.

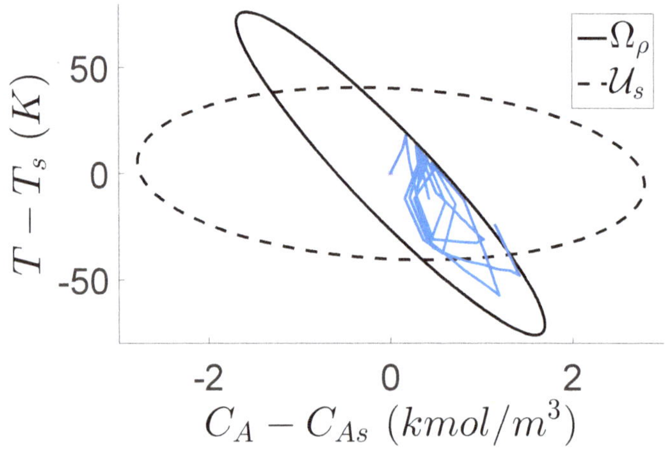

Figure 5. An example closed-loop trajectory under the Safeness Index-based LEMPC of Equation (7) for the initial condition $(0, 0)$ that leaves the safe operating region $\mathcal{S}$, in which $\rho_e = 320$ and $s_e = 6800$.

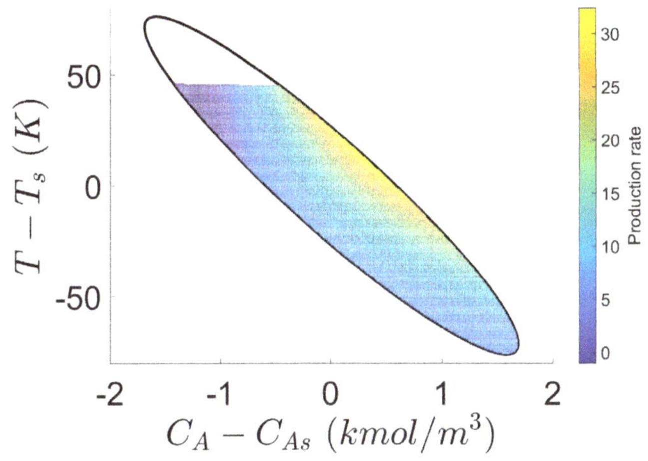

Figure 6. The production rate $L_e = k_0 e^{-E/RT} C_A^2$ within the safe operating region $\mathcal{S}$.

5. Conclusions

In this work, a Safeness Index-based LEMPC design was developed for stochastic nonlinear systems. Under the assumption of stabilizability of the origin of the stochastic nonlinear system via a stochastic Lyapunov-based control law, an economic model predictive controller was developed to account for process operational safety by utilizing Lyapunov-based constraints to maintain the closed-loop state in a safe operating region defined by a Safeness Index function. Under the stochastic Safeness Index-based LEMPC, economic optimality may be achieved with respect to the objective function and sampling period. Additionally, recursive feasibility and process operational safety of the closed-loop stochastic nonlinear system were derived in probability for a well-characterized safe operating region. A chemical reactor example was used to demonstrate the effectiveness of the proposed control method.

Author Contributions: Zhe Wu developed the main results, performed the simulation studies and prepared the initial draft of the paper. Helen Durand contributed to the theory of probabilistic process operational safety and revised this manuscript. Panagiotis D. Christofides developed the idea of Safeness Index-based LEMPC for stochastic nonlinear systems, oversaw all aspects of the research and revised this manuscript.

Acknowledgments: Financial support from the National Science Foundation and the Department of Energy is gratefully acknowledged.

Conflicts of Interest: The authors declare no conflict of interest.

References

1. Sanders, R.E. *Chemical Process Safety: Learning from Case Histories*; Butterworth-Heinemann: Oxford, UK, 2015.
2. Albalawi, F.; Durand, H.; Christofides, P.D. Process operational safety using model predictive control based on a process Safeness Index. *Comput. Chem. Eng.* **2017**, *104*, 76–88. [CrossRef]
3. Albalawi, F.; Durand, H.; Alanqar, A.; Christofides, P.D. Achieving operational process safety via model predictive control. *J. Loss Prev. Process Ind.* **2018**, *53*, 74–88. [CrossRef]
4. Heidarinejad, M.; Liu, J.; Christofides, P.D. Economic model predictive control of nonlinear process systems using Lyapunov techniques. *AIChE J.* **2012**, *58*, 855–870. [CrossRef]
5. Angeli, D.; Amrit, R.; Rawlings, J.B. On average performance and stability of economic model predictive control. *IEEE Trans. Autom. Control* **2012**, *57*, 1615–1626. [CrossRef]

6. Müller, M.A.; Angeli, D.; Allgöwer, F. Economic model predictive control with self-tuning terminal cost. *Eur. J. Control* **2013**, *19*, 408–416. [CrossRef]
7. Ellis, M.; Durand, H.; Christofides, P.D. A tutorial review of economic model predictive control methods. *J. Process Control* **2014**, *24*, 1156–1178. [CrossRef]
8. Van Hessem, D.; Bosgra, O. Stochastic closed-loop model predictive control of continuous nonlinear chemical processes. *J. Process Control* **2006**, *16*, 225–241. [CrossRef]
9. Mahmood, M.; Mhaskar, P. Lyapunov-based model predictive control of stochastic nonlinear systems. *Automatica* **2012**, *48*, 2271–2276. [CrossRef]
10. Maciejowski, J.M.; Visintini, A.L.; Lygeros, J. NMPC for complex stochastic systems using a Markov chain Monte Carlo approach. In *Assessment and Future Directions of Nonlinear Model Predictive Control*; Springer: Berlin/Heidelberg, Germany, **2007**, 269–281.
11. Wu, Z.; Zhang, J.; Zhang, Z.; Albalawi, F.; Durand, H.; Mahmood, M.; Mhaskar, P.; Christofides, P.D. Economic Model Predictive Control of Stochastic Nonlinear Systems. *AIChE J.* **2018**. [CrossRef]
12. Khasminskii, R. *Stochastic Stability of Differential Equations*; Springer Science & Business Media: Berlin/Heidelberg, Germany, 2011; Volume 66.
13. Florchinger, P. A universal formula for the stabilization of control stochastic differential equations. *Stoch. Anal. Appl.* **1993**, *11*, 155–162. [CrossRef]
14. Deng, H.; Krstic, M.; Williams, R. Stabilization of stochastic nonlinear systems driven by noise of unknown covariance. *IEEE Trans. Autom. Control* **2001**, *46*, 1237–1253. [CrossRef]
15. Wu, Z.; Albalawi, F.; Zhang, Z.; Zhang, J.; Durand, H.; Christofides, P.D. Contro Lyapunov-Barrier Function-Based Model Predictive Control of Nonlinear Systems. In Proceedings of the American Control Conference, Milwaukee, WI, USA, 27–29 June 2018; in press.
16. Ciesielski, Z.; Taylor, S.J. First passage times and sojourn times for Brownian motion in space and the exact Hausdorff measure of the sample path. *Trans. Am. Math. Soc.* **1962**, *103*, 434–450. [CrossRef]
17. Øksendal, B. Stochastic differential equations. In *Stochastic Differential Equations*; Springer: Berlin/Heidelberg, Germany, 2003; pp. 65–84.
18. Sontag, E.D. A 'universal' construction of Artstein's theorem on nonlinear stabilization. *Syst. Control Lett.* **1989**, *13*, 117–123. [CrossRef]
19. Wächter, A.; Biegler, L.T. On the implementation of an interior-point filter line-search algorithm for large-scale nonlinear programming. *Math. Progr.* **2006**, *106*, 25–57. [CrossRef]

mathematics

MDPI

Article

Enhancing Strong Neighbor-Based Optimization for Distributed Model Predictive Control Systems

Shan Gao [1,2], Yi Zheng [1,2,*] and Shaoyuan Li [1,2,*]

1 Department of Automation, Shanghai Jiao Tong University, Shanghai 200240, China; gaoshan1117@sjtu.edu.cn

2 Key Laboratory of System Control and Information Processing, Ministry of Education of China, Shanghai 200240, China

* Correspondence: yizheng@sjtu.edu.cn (Y.Z.); syli@sjtu.edu.cn (S.L.)

Received: 1 April 2018; Accepted: 8 May 2018; Published: 22 May 2018

Abstract: This paper considers a class of large-scale systems which is composed of many interacting subsystems, and each of them is controlled by an individual controller. For this type of system, to improve the optimization performance of the entire closed-loop system in a distributed framework without the entire system's information or too-complicated network information, connectivity is always an important topic. To achieve this purpose, a distributed model predictive control (DMPC) design method is proposed in this paper, where each local model predictive control (MPC) considers the optimization performance of its strong coupling subsystems and communicates with them. A method to determine the strength of the coupling relationship based on the closed-loop system's performance and subsystem network connectivity is proposed for the selection of each subsystem's neighbors. Finally, through integrating the steady-state calculation, the designed DMPC is able to guarantee the recursive feasibility and asymptotic stability of the closed-loop system in the cases of both tracking set point and stabilizing system to zeroes. Simulation results show the efficiency of the proposed DMPC.

Keywords: model predictive control; distributed model predictive control; large-scale systems; neighborhood optimization

1. Introduction

There is a class of complex large-scale industrial control systems which are composed of many interacting and spatially distributed subsystems, and each subsystem is controlled by an individual controller (e.g., large-scale chemical process [1], smart micro-grid [2,3] systems, distributed generation systems [4]), where the controllers exchange information with each other through a communication network. The objective is to achieve a good global performance of the entire closed-loop system or a common goal of all subsystems by the controller network. This objective is usually to track setpoints with minimized total error or to stabilize the entire system to zeroes in the dynamic control layer.

Distributed model predictive control (DMPC) controls each subsystem by an individual local model predictive control (MPC), and is one of the most important distributed control or optimization algorithms [1,5–8], since it not only inherits MPC's ability to get good optimization performance and explicitly accommodate constraints [9,10], but also has the advantages of a distributed framework of fault-tolerance, less computation, and being flexible to system structure [7,11–14]. However, compared with the centralized control scheme, its performance is still not as good as that of centralized MPC for coupling systems in a peer-to-peer distributed control framework.

Many algorithms and design methods have appeared in the literature for different types of systems and for different problems in the design of DMPCs. For example, the design of DMPC for nonlinear systems [15,16], DMPC for uncertain systems [15,17], DMPC for networked systems with time delay [18], a decentralized optimization algorithm for solving DMPC [19], the design of cooperative strategies for improving the performance of DMPC [20], the design of an event-based communication DMPC for reducing the load on the communication network [21], as well as the design of a DPMC control structure [22]. Among these algorithms, several DMPC algorithms relate to the purpose of improving the closed-loop optimization performance while considering the information connectivity [5,21,23–26]. Information connectivity is considered because it directly affects the structural flexibility and error tolerance ability. Reference [27] proposed a DMPC where each subsystem-based MPC only communicates with its directly-impacted neighbors and uses an iterative algorithm to obtain the "Nash optimality". References [20,28,29] proposed cooperative DMPC, where each MPC considers the cost of the entire system and communicates with all the other MPCs to obtain "Pareto optimality". To reduce the information connectivity and increase the structural flexibility, Reference [30] proposed that each subsystem optimize all the subsystems impacted by it over the optimization horizon. The solution of this method is equal to the cooperative DMPC, while its communication efforts are less than the cooperative DMPC, especially for sparse systems. References [31,32] gave a strategy to dynamically adjust the weighting of performance index in cooperative MPC to avoid bad performance occurring in some subsystems. In an effort to achieve a trade-off between the optimization performance of the entire system and the information connectivity, an intuitively appealing strategy, called impacted-region cost optimization-based DMPC, is proposed in [33–35], where each subsystem-based MPC only considers the cost of its own subsystem and those of the subsystems directly impacted by it. Consequently, each MPC only communicates with its neighboring MPCs. In addition, in some papers, the control flexibility and information connectivity are paid more attention by researchers. References [14,36] provide a tube-based DMPC where all interactions are considered as disturbances and each subsystem-based MPC is solved independently. It does not exchange the state and input trajectory, but the interaction constraints, to avoid the interaction consistency problem. This method is able to improve the flexibility and fault tolerance ability of the control network [37]. References [25,37] proposed reconfigurable DMPC and plug-and-play DMPC based on dissipative theory, which focus on the problem of how to design a DMPC which allows the addition or deletion of subsystems without any change in existing controllers. It can be seen that the optimization performance of the entire system and structural flexibility are two conflicting key points in DMPC design. The selection of the range of each subsystem's neighbors to be optimized in each subsystem-based MPC is important in the design of DMPC in order to obtain good optimization performance without unnecessary information connections. Thus, the aim of this paper is to design an algorithm to determine the range of each subsystem optimized from the point of view of enlarging each subsystem MPC's feasible region, then to improve the entire system's optimization performance without too-complicated network connectivity. Then, based on the result of this algorithm, we aim to design a stabilized neighborhood optimization-based DMPC that handles state constraints and is able to be used in target tracking.

As for target tracking, the difficulty in DMPC is to guarantee the recursive feasibility. References [38–40] provide a tracking algorithm for a series of MPC systems, where a steady-state target optimizer (SSTO) is integrated in the design of the cost function. The proposed controller is able to drive the whole system to any admissible setpoint in an admissible way, ensuring feasibility under any change of setpoint. As for distributed systems, [38] gives a DMPC for tracking based on the method introduced in [39] and a cooperative DMPC strategy. Reference [41] proposes another method based on global calculations of targeting tracking. It does not require a feasible starting point of each distributed predictive controller. These methods provide good references and possible methods for designing a tracking DMPC that considers optimization performance improvement and network connectivity.

In this paper, strong coupling neighbor-based optimization DMPC is proposed. With this method, each local MPC coordinates and communicates with its strong coupling neighbors. It takes its strongly-coupling downstream subsystems cost function into account in its cost function to improve the performance of the entire closed-loop system. To reduce the unnecessary network connectivity, the interaction terms of weak coupling upstream neighbors are ignored in its predictive model and are considered as bounded disturbances. In addition, the closed-loop optimization performance is used to determine which interaction should be regarded as strong coupling and be considered in DMPC. The strategy proposed in [38] is used to guarantee the recursive feasibility and stability in the target tracking problem. An asymptotically-stable closed-loop system with state constraints is guaranteed.

The remainder of this paper is organized as follows. Section 2 describes the problem to be solved. Section 3 describes the design of the proposed DMPC. Section 4 analyzes the stability of the closed-loop system. Section 5 presents the simulation results to demonstrate the effectiveness of the proposed algorithm. Finally, a brief conclusion to the paper is drawn in Section 6.

2. Problem Description

Considering a large-scale discrete-time linear system which is composed of many interacting subsystems, the overall system model is:

$$\begin{cases} x^+ = Ax + Bu, \\ y = Cx, \end{cases} \tag{1}$$

where $x \in \mathcal{R}^{n_x}$ is the system state, $u \in \mathcal{R}^{n_u}$ is the system current control input, $y \in \mathcal{R}^{n_y}$ is the controlled output, and x^+ is the successor state. The state of the system and control input applied at sample time t are denoted as $x(t)$,$u(t)$, respectively. Moreover, there are hard constraints in the system state and control input. That is, for $\forall t \geq 0$:

$$x(t) \in \mathcal{X}, u(t) \in \mathcal{U}, \tag{2}$$

where $\mathcal{X} \subset \mathcal{R}^{n_x}$ and $\mathcal{U} \subset \mathcal{R}^{n_u}$ are compact convex polyhedra containing the origin in their interior.

Given Model (1), without loss of generality, the overall system is divided in to m subsystems, denoted as $\mathcal{S}_i, i \in \mathcal{I}_{0:m}$. Thus, $u = (u_1, u_2, ..., u_m)$ and $x = (x_1, x_2, ..., x_m)$, then the subsystem model for $\mathcal{S}_i, \forall i \in \mathcal{I}_{0:m}$ is:

$$x_i^+ = A_{ii}x_i + B_{ii}u_i + \sum\nolimits_{j \in \mathcal{N}_i} B_{ij}u_j, \tag{3}$$

where $\mathcal{N}_i$ is the set of subsystems that send inputs to the current subsystem $\mathcal{S}_i$. For subsystem $\mathcal{S}_j, j \in \mathcal{N}_i$, $\mathcal{S}_j$ couples with $\mathcal{S}_i$ by sending control input u_i to $\mathcal{S}_i$. In particular, $j \in \mathcal{N}_i$ if $B_{ij} \neq 0$. Given the overall system constraints set $\mathcal{X}, \mathcal{U}$, x_i, u_i fit hard constraints $x_i(t) \in \mathcal{X}_i, u_i(t) \in \mathcal{U}_i$.

In this paper, for ease of analysis, here the definitions of neighbor (upstream-neighbor) and downstream neighbor are given.

Definition 1. *Given subsystem $\mathcal{S}_i$ with state evolution Equation (3), define $\mathcal{S}_j, \mathcal{S}_j \in \mathcal{N}_i$, which send input information to $\mathcal{S}_i$ as the neighbor (upstream neighbor) of $\mathcal{S}_i$. Moreover, for arbitrary $\mathcal{S}_j, \mathcal{S}_j \in \mathcal{N}_i$, since $\mathcal{S}_i$ receives input information from $\mathcal{S}_j$, $\mathcal{S}_i$ is defined as a downstream neighbor of $\mathcal{S}_j$.*

Denote the tracking target as y_t. Assume that (A, B) is stabilizable and the state is measurable. The aim of a tracking problem given a target y_t is to design a controller which enables $y(t) \to y_t$ in an admissible way when $t \to \infty$. Hence, the origin control objective function of the overall system is:

$$V_{N_{origin}}(x, y_t; u) = \sum_{k=0}^{N-1} (\|Cx(k) - \hat{y}_t\|_{Q_o}^2 + \|u(k) - \hat{u}_t\|_R^2) + \|Cx(N) - \hat{y}_t\|_{P_o}^2, \tag{4}$$

where $P_o > 0$, $Q_o > 0$, and $R > 0$ is the weighting coefficients matrix, and u_t is steady input corresponding to y_t.

The problem considered here is to design a DMPC algorithm to control a physical network, which coordinate with each other considering the following performance indicators:

- to achieve a good optimization performance of the entire closed-loop system.
- to guarantee the feasibility of target tracking.
- to simplify the information connectivity among controllers to guarantee good structural flexibility and error-tolerance of the distributed control framework.

To solve this problem, in this paper, an enhanced strong neighbor-based optimization DMPC is designed, and is detailed in the next section.

3. DMPC Design

In an interacting distributed system, the state evolution of each subsystem is affected by the optimal control decisions of its upstream neighbors. Each subsystem considers if these effects will help to improve the performance of entire closed-loop system. On the other hand, these impacts have different strengths for different downstream subsystems. Some of the effects are too small and can be ignored. If these weakly-coupling downstream subsystems' cost functions are involved in each subsystem's optimization problem, additional information connections arise with little improvement of the performance of the closed-loop system. The increase of information connections will hinder the error tolerance and flexibility of the distributed control system. Thus, each subsystem-based MPC takes the cost functions of its strongly-interacting downstream subsystems into account to improve the closed-loop performance of the entire system and receive information from its strong-coupling neighbors.

3.1. Strong-Coupling Neighbor-Based Optimization for Tracking

Given that the coupling degrees between different subsystems differ substantially, here we enable the subsystem to cooperate with strong-coupling neighbors while treating the weak-coupling ones as disturbance. Define $\mathcal{N}_{i(strong)}$ as a set of strong-coupling neighboring subsystems and $\mathcal{N}_{i(weak)}$ as set of weak-coupling neighbors. The rule for deciding strong-coupling systems is detailed in Section 3.4.

Then, for $\mathcal{S}_i$, we have:

$$x_i^+ = A_{ii}x_i + B_{ii}u_i + \sum\nolimits_{j\in\mathcal{N}_{i(strong)}} B_{ij}u_j + w_i, \tag{5}$$

where

$$\begin{gathered} w_i = \sum\nolimits_{j\in\mathcal{N}_{i(weak)}} B_{ij}u_j, \\ w_i \in \mathcal{W}_i, \mathcal{W}_i = (\oplus B_{ij}\mathcal{U}_j), \\ \mathcal{N}_{i(weak)} \cup \mathcal{N}_{i(strong)} = \mathcal{N}_i = \{j | B_{ij} \neq 0, j \neq i\}. \end{gathered}$$

The deviation w_i represents the influence collection of weak-coupling upstream neighbors in $N_{i,(weak)}$. w_i is contained in a convex and compact set $\mathcal{W}_i$ which contains the origin.

If the weak coupling influence w_i is neglected, a simplified model based on $\mathcal{S}_i$ is acquired. That is:

$$\bar{x}_i^+ = A_{ii}\bar{x}_i + B_{ii}\bar{u}_i + \sum\nolimits_{j\in\mathcal{N}_{i(strong)}} B_{ij}\bar{u}_j. \tag{6}$$

Here $\bar{x}_i$, $\bar{u}_i$, and $\bar{u}_j, j \in \mathcal{N}_{i(strong)}$ represent the state and input of a simplified subsystem model which neglects weak-coupling upstream neighbors' influence w_i.

The simplified overall system model with new coupling relation matrix $\bar{B}$ is:

$$\bar{x}^+ = A\bar{x} + \bar{B}\bar{u}, \tag{7}$$

where $\bar{x} = (\bar{x}_1, \bar{x}_2, ..., \bar{x}_m)$ and $\bar{u} = (\bar{u}_1, \bar{u}_2, ..., \bar{u}_m)$ represent states and inputs in this simplified model.

Considering the target-tracking problem of the simplified model, in order to ensure the output track, given target y_t, constraints are given for terminal state prediction. If the current target y_t is set as the tracking target through the controller optimization, when y_t changes, the terminal constraints need to change immediately. The optimal solution at a previous time may not fit the terminal constraints brought by the changed y_t. This violates the recursive feasibility of the system. Thus, here a steady state optimization is integrated in the MPC for tracking where an artificial feasible tracking goal y_s is proposed as a medium variable. This variable works as an optimized variable. With setting tracking point y_s equal to the previous target, the recursive feasibility will not be violated by the target change.

The medium target y_s and its state $\bar{x}_S$ and input $\bar{u}_s$ should satisfy the simplified system's steady state equations. It has

$$\begin{bmatrix} A - I_{n_x} & \bar{B} & 0 \\ C & 0 & -I \end{bmatrix} \begin{bmatrix} \bar{x}_s \\ \bar{u}_s \\ y_s \end{bmatrix} = \begin{bmatrix} 0 \\ 0 \end{bmatrix}, \tag{8}$$

$$\begin{bmatrix} \bar{x}_s & \bar{u}_s \end{bmatrix} = M_y y_s. \tag{9}$$

Here M_y is a suitable matrix. That is, target y_s's corresponding inputs $\bar{u}_s$ and states $\bar{x}_s$ in the simplified model can be expressed by y_s. The equation is based on the premise of Lemma 1.14 in [42]. If Lemma 1.14 does not hold, a M_θ and θ which fits $\begin{bmatrix} \bar{x}_s & \bar{u}_s \end{bmatrix} = M_\theta \theta$ can be found, which can replace the y_s as a variable to be solved.

For the manual tracking target y_s for the overall system, we have $y_s = \{y_{1,s}, \ldots, y_{i,s}, \ldots, y_{m,s}\}$. That is, given y_s, arbitrary subsystem $\mathcal{S}_i$ gets a subtracking target $y_{s,i}$. Similar to (9), $\bar{x}_{s,i}, \bar{u}_{s,i}$ are solved.

With the simplified model and artificial tracking target $y_{s,i}$, according to (9), in the strong-coupling neighbor-based optimization MPC algorithm, the objective function optimized in subsystem $\mathcal{S}_i, \forall i \in [1, m]$ is set as $V'_{iN}(x_i, y_t; x_i, u_{i,0:N-1}, y_s)$ as follows:

$$\begin{aligned} V'_{iN}(x_i, y_t; x_i, u_{i,0:N-1}, y_s) = &\sum_{k=0}^{N-1} (\|x_i(k) - \bar{x}_{i,s}\|^2_{Q_i} + \|u_i(k) - \bar{u}_{i,s}\|^2_{R_i}) + \|x_i(N) - \bar{x}_{i,s}\|^2_{P_i} + V_O(y_{i,s}, y_{t,i}) \\ &+ \sum_{k=0}^{N-1} \sum_{h \in \mathcal{H}_i} \|x_h(k) - \bar{x}_{s,h}\|^2_{Q_h} + \|x_h(N) - \bar{x}_{s,h}\|_{P_h}, \end{aligned} \tag{10}$$

where x_i, y_t is the given initial state and target, $u_{i,0:N-1}$ are input predictions in $0 : N-1$ sample time ahead. y_s is the admissible target. $Q_i = C_i' Q_{o,i} C_i > 0$ and

$$\mathcal{H}_i = \{h | i \in \mathcal{N}_{h(strong)}, \forall \mathcal{S}_h, h \in [1, m], h \neq i\}. \tag{11}$$

Here, $\mathcal{S}_i$'s controller design takes the strong-coupling downstream neighbors' performances as part of its optimized objective. That is, the current subsystem $\mathcal{S}_i$'s optimal solution is decided by its own and downstream neighbors in set $\mathcal{H}_i$, which is strongly impacted by $\mathcal{S}_i$.

Next, we will use the simplified model in (6) with only strong couplings to solve the tracking problem (10) for each subsystem. To guarantee control feasibility and stability, the following definitions and assumptions are given.

One important issue is to deal with the deviation caused by neglecting weak-coupling neighbor inputs. Here robust positively invariantt sets are adopted to enable the deviation of states to be bounded and the real system's states to be controlled in $\mathcal{X}$.

Definition 2. *(Robust positively invariant set control law) Given* $e = (x - \bar{x})$ *which represents the dynamics of the error between the origin plant and the simplified model:*

$$e^+ = A_k e + w, \tag{12}$$

with $A_k = (A + BK)$. *A set* ϕ *is called a robust positively invariant set for system (12) if* $A_k\phi \oplus \mathcal{W} \subseteq \phi$, *and the control law is called a robust positively invariant set control law.*

The definition of a robust positively invariant set illustrates that for system $x = Ax + Bu + w$ if ϕ and robust positively invariant set control law K exist, then for $e(0) = x(0) - \bar{x}(0)$, the trajectories of the original system at arbitrary time t denoted as $x(t)$ can be controlled in $x(t) = \bar{x}(t) \oplus \phi$.

Based on this definition, in this paper the dynamics of deviation $(x_i - \bar{x}_i)$ introduced by neglecting weak-coupling neighbors can be solved. For subsystem $\mathcal{S}_i$ proposed as (5), the deviation is written as:

$$e_i^+ = A_{ii}e_i + B_{ii}u_{i,e} + w_i,$$

where $e_i = x_i - \bar{x}_i$ is the deviation from the simplified model to the original model and $u_{i,e}$ is the control law. There exists the set ϕ_i as a robust positively invariant set for $\mathcal{S}_i$ if $(A_{ii} + B_{ii}K_i)e_i \in \phi_i$ for all $e_i \in \phi_i$ and all $w_i \in \mathcal{W}_i$. Here $u_{i,e} = K_i e_i$ is a feedback control input and we denote K_i as the robust positively invariant set control law for $\mathcal{S}_i$. Then, it is easy to obtain $x_i(t) = \bar{x}_i(t) \oplus \phi_i$ for time t. Let $(\bar{x}_i(t), \bar{u}_i(t)) \in \mathcal{F}_i$, where $\mathcal{F}_i = (\mathcal{X}_i \times \mathcal{U}_i) \ominus (\phi_i \times K_i\phi_i)$, the origin system state and input satisfy $(x_i(t), K_i(x_i(t) - \bar{x}_i(t)) + \bar{u}_i(t)) \in \mathcal{X}_i \times \mathcal{U}_i$. Thus, with the help of a robust positively invariant set, the original system optimization is transferred to a simplified model. For the overall system, we have $K = diag(K_1, K_2, ..., K_m)$.

With Definition 2, if the deviation brought by omitting weak-coupling neighbors is controlled in a robust positively invariant (RPI) set ϕ_i with control law K_i and simplified model in (7) has control law and state $\bar{u}_i, \bar{x}_i$ confined in $\mathcal{U}_i \ominus K_i\phi_i, \mathcal{X}_i \ominus \phi_i$, respectively, the local subsystem will have a feasible solution for the optimization.

As for the manually-selected tracking target y_s, based on the overall simplified model in (7), the following definition is given:

Definition 3. *(Tracking invariant set control law). Consider that overall system (7) is controlled by the following control law:*

$$\bar{u} = \bar{K}(\bar{x} - \bar{x}_s) + \bar{u}_s = \bar{K}\bar{x} + Ly_s. \tag{13}$$

Let $A + \bar{B}\bar{K}$ *be Hurwitz, then this control law steers system (7) to the steady state and input* $(\bar{x}_s, \bar{u}_s) = M_y y_s$. $\bar{K}$ *is denoted as the tracking invariant set control law.*

Denote the set of initial state and steady output that can be stabilized by control law (13) while fulfilling the system constraints throughout its evolution as an invariant set for tracking $\Omega_{\bar{K}}$. For any $(x(0), y_s) \in \Omega_{\bar{K}}$, the trajectory of the system $\bar{x}^+ = A\bar{x} + B\bar{u}$ controlled by $\bar{u} = \bar{K}x + Ly_s$ is confined in $\Omega_{\bar{K}}$ and tends to $(x_s, u_s) = M_y y_s$.

Under Definitions 2 and 3, before introducing the enhancing strong neighbor-based optimization DMPC, some assumptions for the closed-loop system feasibility and stability are given as follows. The concrete theorem and an analysis of stability and feasibility are given in Section 4.

Assumption 1. *The eigenvalues of* $A_{ii} + B_{ii}K_i$ *are in the interior of the unitary circle.* ϕ_i *is an admissible robust positively invariant set for* $\mathcal{S}_i$*'s deviation* $(x_i - \bar{x}_i)$ *subject to constraints* $\mathcal{F}_i$, *and the corresponding feedback control law is* $u_{i,e} = K_i e_i$.

Assumption 2. *Let* $\Omega_{\bar{K}}$ *be a tracking invariant set for the simplified system (7) subject to constraints* $\mathcal{F} = \{\{(\bar{x}_1, \bar{u}_1), ..., (\bar{x}_m, \bar{u}_m)\} | \forall i, (\bar{x}_i, \bar{u}_i) \in (\mathcal{X}_i \times \mathcal{U}_i) \ominus (\phi_i \times \bar{K}_i\phi_i)\}$ *, and the corresponding feedback gain matrix is* $\bar{K} = \{\bar{K}_1, \bar{K}_2, \ldots, \bar{K}_m\}$.

Assumption 3. *For* $Q = \text{block-diag}\{Q_1, Q_2, \ldots, Q_m\}$, $R = \text{block-diag}\{R_1, R_2, \ldots, R_m\}$ *and* $P = \text{block-diag}(P_1, P_2, \ldots, P_m)$, *it has:*

$$(A + \bar{B}\bar{K})'P(A + \bar{B}\bar{K}) - P = -(Q + \bar{K}'R\bar{K}). \tag{14}$$

Assumption 1 ensures that with the feedback control law $u_{i,e} = K_i e_i$, $i \in \mathcal{I}_{0:m}$, the state estimated by the simplified model (7) is near to the real system's trajectory before the system reaches the target. In Assumption 2, $\Omega_{\bar{k}}$ is set as a terminal constraint of DMPC. Assumption 3 is used in the proof of the convergence of system presented in the Appendix A.

So far, the strong-coupling neighbor-based optimization DMPC algorithm, which is solved iteratively, can be defined as follows:

Firstly, denote the optimal objective of subsystem $\mathcal{S}_i$ as V_{iN}. According to (10), at iterating step p, V_{iN} fits:

$$\begin{aligned}
&V_{iN}(x_i, y_t, p; \bar{x}_i, \bar{u}_{i,0:N-1}, y_{i,s}) \\
&= \sum_{k=0}^{N-1} (\|\bar{x}_i(k) - \bar{x}_{i,s}\|^2_{Q_i} + \|\bar{u}_i(k) - \bar{u}_{i,s}\|^2_{R_i}) + \|\bar{x}_i(N) - \bar{x}_{i,s}\|^2_{P_i} + V_0(y_{i,s}, y_{i,t}) \\
&+ \sum_{k=0}^{N-1} \sum_{h \in \mathcal{H}_i} \left\| \bar{x}_h(k) - \bar{x}_{h,s}^{[p-1]} \right\|^2_{Q_h} + \left\| \bar{x}_h(N) - \bar{x}_{h,s}^{[p-1]} \right\|_{P_h}.
\end{aligned} \tag{15}$$

Compute the optimization solution

$$(\bar{x}_i'(0), \bar{u}_{i,0:N-1}{}', y_{i,s}{}') = \arg\min V_{iN}(x_i, y_t, p; \bar{x}_i, \bar{u}_{i,0:N-1}, y_{i,s}), \tag{16}$$

Subject to constraints:

$$\bar{x}_{h_i}(k+1) = A_{h_i h_i}\bar{x}_{h_i}(k) + \sum\nolimits_{h_j \in \mathcal{N}_{h(strong)}} B_{h_j}\bar{u}_{h_j}^{[p]}(k) + B_{h_i h_i}\bar{u}_{h_i}(k),$$
$$(\bar{x}_{h_i}(k)\ \bar{u}_{h_i}(k)) \in \mathcal{F}, \mathcal{F} : (\mathcal{X}_{h_i}, \mathcal{U}_{h_i}) \ominus (\mathcal{W}_{h_i}, K_{h_i}\mathcal{W}_i), \tag{17a}$$
$$(\bar{x}(N), y_s) \in \Omega_{\bar{K}}, \tag{17b}$$
$$\bar{x}_i(0) \in x_i - \phi_i, \tag{17c}$$
$$M_y y_{i,s} = (\bar{x}_{i,s}, \bar{u}_{i,s}), \tag{17d}$$

with $h_i \in \mathcal{H}_i \cup \{i\}$, and $\phi_i, \Omega_{\bar{k}}$ defined in Assumptions 2 and 3, respectively. The optimization function (16) updates $\mathcal{S}_i$'s initial state, inputs in N steps $\bar{u}_{i,0:N-1}$ and current tracking target $y_{i,s}$ based on the information from subsystems in $\mathcal{H}$.

Secondly, set

$$\bar{u}_{i,0:N-1}^{[p]} = \gamma_i \bar{u}_{i,0:N-1}' + (1 - \gamma_i)\bar{u}_{i,0:N-1}^{[p-1]}, \tag{18}$$
$$y_{i,s}^{[p]} = \gamma_i y_{i,s}' + (1 - \gamma_i) y_{i,s}^{[p-1]}, \tag{19}$$
$$\bar{x}_i^{[p]}(0) = \gamma_i x_i'(0) + (1 - \gamma_i)\bar{x}_i^{[p-1]}(0), \tag{20}$$
$$\sum_{i=1}^{m} \gamma_i = 1, \gamma_i > 0. \tag{21}$$

$\gamma_i \in \mathcal{R}, 0 < \gamma_i < 1$ is to guarantee the consistency of the optimization problem. That is, at the end of the current sample time, all shared variables converge.

After that, we take

$$p = p + 1$$

to iterate until the solutions convergence. Then, we have $\bar{x}_i^* = \bar{x}_i^{[p]}$, $\bar{u}_i^* = \bar{u}_{i,0:N-1}^{[p]}$, $y_{i,s}^* = y_{i,s}^{[p]}$.

Finally, when the solution converges, according to Assumption 1, take the control law of $\mathcal{S}_i$ as

$$u_{i,0}^* = \bar{u}_{i,0}^* + K_i(x_i - \bar{x}_i^*), \tag{22}$$

where K_i is the robust positively invariant set control law. $\bar{u}_{i,0}^*$ is the first element of $\bar{u}_i^*$. For better understanding, the algorithm is also presented in Algorithm 1.

Algorithm 1: Enhancing Strong Neighbor-Based Optimization DMPC

Data: initial time t_0, inital state $x_i(t_0)$, and tracking target y_{target} (target can be changed with time according to production demand) for subsystem $\mathcal{S}_i$

Result: the control law $u_i^*(t)$ for $t = t_0 : +\infty$

Firstly, determine the strong-coupling neighbors set $N_{i(strong)}$ by solving $C_{i,(d^*)}$'s optimization in 3.4.

Secondly, confirm $\mathcal{S}_i$'s downstream neighbor set $\mathcal{H}_i$ which is the set of subsystems that SS_i has control influence on.

Set $t = t_0$.

while *True* **do**

  Select Warm Start (reference solution at iteration $p = 0$, details in Section 3.2) :

$$v_i^{[0]}(t) = (\bar{x}_i^{[0]}(0|t), \bar{u}_{i,0:N-1}^{[0]}(t), \bar{y}_{i,s}^{[0]}(t))$$

  Set $x_i = x_i(t), y_t = y_{target}, p = 1$.

  while *True* **do**

    Set $(\bar{x}_i'(0), \bar{u}_{i,0:N-1}{'}, y_{i,s}{'}) = \arg\min\{V_{iN}(x_i, y_t, p; \bar{x}_i, \bar{u}_{i,0:N-1}, y_{i,s}) : s.t(17)\}$

    Get optimization solution at p:

$$\begin{aligned}
\bar{u}_{i,0:N-1}^{[p]} &= \gamma_i \bar{u}_{i,0:N-1}' + (1 - \gamma_i)\bar{u}_{i,0:N-1}^{[p-1]} \\
y_{i,s}^{[p]} &= \gamma_i y_{i,s}' + (1 - \gamma_i) y_{i,s}^{[p-1]} \\
\bar{x}_i^{[p]}(0) &= \gamma_i x_i'(0) + (1 - \gamma_i)\bar{x}_i^{[p-1]}(0) \\
\sum_{i=1}^{m} \gamma_i &= 1, \gamma_i > 0
\end{aligned}$$

    if $||\bar{u}_i^{[p]} - \bar{u}_i^{[p-1]}|| \le 1e^{-6}$ **then**

      Break

    end

    Set $p = p + 1$

  end

  Set $\bar{x}_i^* = \bar{x}_i^{[p]}(0)$, $\bar{u}_i^* = \bar{u}_{i,0:N-1}^{[p]}$, $y_{i,s}^* = y_{i,s}^{[p]}$. $u_{i,0}^* = \bar{u}_{i,0}^* + K_i(x_i - \bar{x}_i^*)$

  Get $u_i^*(t) = u_{i,0}^*$. **Set** $t = t + 1$.

end

In this algorithm, we use an iterative strategy to guarantee the distributed control solution $(\bar{x}(0), \bar{u}_{0:N-1}, y_s)$ is consistent. Next, the selection of warm start, the given solution for each subsystem at initial iterative step 0, is proposed in the next section.

3.2. Warm Start

Considering a new sample time, with updated system states, the choice of a warm start is based on the principle that it fits the simplified system's constraints in (17), so that real subsystem solution's feasibility is guaranteed. The warm start is designed as the following algorithm:

Algorithm 2: Warm Start for Iterative Algorithm

Data: $x(t+1), y_t, \bar{u}^*_{i,0:N-1}(t), y^*_{i,s}(t)$

Result: the warm start $v_i^{[0]}(t+1) = (\bar{x}_i^{[0]}(0|t+1), \bar{u}^{[0]}_{i,0:N-1}(t+1), \bar{y}^{[0]}_{i,s}(t+1))$

For control inputs, two options are given:

Option 1: $\tilde{u}_{i,0:N-1}(t+1) = (\bar{u}^*_{i,1}(t), ..., \bar{u}^*_{i,N-1}(t), \bar{u}_i(N))$

where $\bar{u}(N) = (\bar{u}_1(N), ..., \bar{u}_m(N)) = \bar{K}\bar{x}^*(N) + Ly^*_s(t)$.

Option 2:

$\hat{u}_{i,0:N-1}(t+1) = (\hat{u}_i(0), ..., \hat{u}_i(N-1))$, where

$$\hat{x}_i(0) = \bar{x}_i^*(1,t), \text{ and } \hat{x}_i(k+1) \text{ fits}$$
$$\hat{x}(k+1) = (\hat{x}_1(k+1), ..., \hat{x}_m(k+1)) = (A+B\bar{K})\hat{x}(k) + BLy^*_s(k), k \in I_{0:N-1}$$
$$\hat{u}_i(k) = \bar{K}\hat{x}(k) + Ly^*_s(t), k \in I_{0:N-1}$$

As for $\bar{x}_i^{[0]}(0|t+1), y^{[0]}_{i,s}(t+1)$. To ensure feasibility, denote:

$$\bar{x}_i^{[0]}(0|t+1) = \bar{x}_i^*(1|t)$$

$$y^{[0]}_{i,s}(t+1) = y^*_{i,s}(t)$$

Warm Start

if $(\bar{x}_i^{[0]}(t+1), y^{[0]}_{i,s}(t+1))$ *in tracking invariant set, and*

$$V_{iN}(x_i(t+1), y_t, 0; \bar{x}_i^{[0]}(0|t+1), \hat{u}_{i,0:N-1}(t+1), y^{[0]}_{i,s}(t+1)$$
$$\leq V_{iN}(x_i(t+1), y_t, 0; \bar{x}_i^{[0]}(0|t+1), \tilde{u}_{i,0:N-1}(t+1), y^{[0]}_{i,s}(t+1)$$

then

  $v_i(t+1)^{[0]} = (\bar{x}_i^{[0]}(0|t+1), \hat{u}_{i,0:N-1}(t+1), y^{[0]}_{i,s}(t+1))$

end

else

  $v_i(t+1)^{[0]} = (\bar{x}_i^{[0]}(0|t+1), \tilde{u}_{i,0:N-1}(t+1), y^{[0]}_{i,s}(t+1))$

end

The algorithm illustrates that two choices are provided for the warm start. One is acquiring a solution from the tracking invariant set control law $\bar{K}$, with the simplified model prediction $(\bar{x}_i^*(1|t), y^*_{i,s}(t))$ as initial state and tracking target, respectively. The other is taking a solution from the simplified model prediction at time t. Both of them fit the constraints of (17). Note that the second option will only be considered when the subsystem enters the tracking invariant set.

3.3. RPI Control Law and RPI Set

Here one constraint coupling subsystem is considered. Given that for $\mathcal{S}_i$ we have $x_i \in \mathcal{X}_i$ and $u_i \in \mathcal{U}_i$, express the constraints in inequalities: $\mathcal{X}_i = \{x_i | l_i^T x_i| \leq 1\}$ and $\mathcal{U}_i = \{u_i | h_i^T u_i| \leq 1\}$. The robust positively invariant set ϕ_i is denoted as $\phi_i = \{x_i : x_i^T P_i x_i \leq 1\}$.

With the definition a of robust positively invariant set in Definition 2, ϕ_i should ensure that $\forall x_i \in \phi_i, x_i \in \mathcal{X}_i$. That is:

$$|h_i^T x_i| \leq 1, \forall x_i \in \phi_i. \tag{23}$$

Based on definitions of $\mathcal{N}_{i(strong)}$ and $\mathcal{N}_{i(weak)}$, $\mathcal{W}_i$ is decided according to the constraints of $\mathcal{N}_{i(weak)}$. For deviation caused by neglecting the subsystem in $\mathcal{N}_{i(weak)}$, a minimization of robust positive invariant set ϕ_i by introducing a parameter $\gamma_i \in [0, 1]$ can be obtained.

The parameter γ_i controls the size of the robust positive invariant set ϕ_i by further minimizing ϕ_i in $\phi_i \subseteq \sqrt{\gamma_i}\mathcal{X}$. That is:

$$\begin{aligned} &\min \gamma_i \\ s.t. \quad &|h_i^T x_i| \leq \sqrt{\gamma_i}, \forall x_i \in \phi_i. \end{aligned} \tag{24}$$

We should also consider the input constraint $\mathcal{U}_i$:

$$|l_i^T K_i x_i| \leq 1, \forall x_i \in \phi_i, \tag{25}$$

and the constraint brought by the property of robust positive invariant set ϕ_i itself should be considered.

Based on the above analysis, referring to [43], we can obtain γ_i and K_i by solving the following linear matrix inequality optimization problem:

$$\min_{W_i, Y_i, \gamma_i} \gamma_i, \tag{26}$$

$$\begin{bmatrix} \lambda_i W_i & * & * \\ 0 & 1 - \lambda_i & * \\ A_{ii} W_i + B_i Y_i & w_i & W_i \end{bmatrix} > 0, \forall w_i \in vert(\mathcal{W}_i), \tag{27}$$

$$\begin{bmatrix} 1 & * \\ Y_i^T l_i & W_i \end{bmatrix} > 0, \tag{28}$$

$$\begin{bmatrix} \gamma_i & * \\ W_i h_i & W_i \end{bmatrix} > 0, \tag{29}$$

and $K_i = Y_i W_i^{-1}$. Thus, we get RPI control law K_i and γ_i, which illustrates the size of ϕ_i. To get ϕ_i, we use the procedure in Reference [43].

3.4. Determination of Strong Coupling

There are many measurements to measure the strength of interactions among subsystems. Different measurements lead to different optimization performance. This paper focuses on the performance and connectivity of subsystems. Thus, the determination of strong-coupling neighbors is based on the influence on the size of current subsystem's robust positively invariant (RPI) set and subsystem connectivity.

On the one hand, as defined in Definition 2, ϕ_i is a robust positively invariant set for subsystem $\mathcal{S}_i$ described as $x_i^+ = A_{ii}x_i + B_{ii}u_i + \sum_{j \in \mathcal{N}_i} B_{ij}u_j$ when u_j is set to zero. Given that ϕ_i deals with deviation caused by neglecting some of the inputs $u_j, j \in \mathcal{N}_i$, the size of ϕ_i is expected to be sufficiently small. The benefit is that the solution in (15) can get a larger feasible domain. Here we consider that a sufficiently large domain means the solution has more degrees of freedom and brings better subsystem performance. Based on the idea above, to decide to omit the weak-coupling neighbor set $N_{i(weak)}$, we choose a neighbor collection which results in a small size of robust positively invariant set ϕ_i. The basis of measuring the robust positively invariant set ϕ_i by introducing γ_i is mentioned in the previous section. On the other hand, connectivity, as the measurement of subsystem topology complexity, is easy to obtain. Next, we give the numerical analysis.

Denote an arbitrary option for deciding the strong-, weak-coupling neighbors as $\mathcal{C}_{i,(d)}, d \in D_i$. $D_i = \{1, ..., d_{max}\} \in I$ is the label set of ways of $\mathcal{S}_i$'s neighbors' distribution. d_{max} represents the size of feasible distribution methods which fits $d_{max} \leq 2^{size(\mathcal{N}_i)}$. For better understanding of $\mathcal{C}_{i,(d)}$, here we take an arbitrary neighbor set $\mathcal{N}_i = \{j_1, j_2, j_3\}$ as an example. If we treat j_i as a strong-coupling neighbor and j_2, j_2 as weak, we have $\exists d \in D_i, \mathcal{C}_{i,(d)}$, satisfying:

$$\mathcal{C}_{i,(d)} = \{(\mathcal{N}_{i(strong)}, \mathcal{N}_{i(weak)}) | \mathcal{N}_{i(strong)} = \{j_1\}, \mathcal{N}_{i(weak)} = \{j_2, j_3\}\}.$$

Option $\mathcal{C}_{i,(d)}$ results in a specified connectivity amount (normalized) $c_{i,(d)} \in [0,1]$ and an RPI set denoted as $\phi_{(i,d)} \subseteq \sqrt{\gamma_{i,(d)}}\mathcal{X}_i$. Here $c_{i,(d)} \in [0,1]$ are defined as:

$$c_{i,(d)} = \frac{size(\mathcal{N}_{i(strong)})}{size(\mathcal{N}_i)} \in [0,1]. \tag{30}$$

To find the optimal distribution $\mathcal{C}_{i,(d^*)}$ of strong- and weak-coupling neighbors, here we take:

$$\mathcal{C}_{i,(d^*)} = \underset{\mathcal{C}_{i,(d)}, W_{i,(d)}, Y_{i,(d)}, \gamma_{i,(d)}}{\operatorname{argmin}} ((\gamma_{i,(1)} + \mu_i c_{i,(1)}), ..., (\gamma_{i,(d)} + \mu_i c_{i,(d)}), ..., (\gamma_{i,(d_{max})} + \mu_i c_{i,(d_{max})})),$$

where for $d \in D_i$,

$$\begin{bmatrix} \lambda_i W_{i,(d)} & * & * \\ 0 & 1-\lambda_i & * \\ A_{ii}W_{i,(d)} + B_i Y_{i,(d)} & w_i & W_{i,(d)} \end{bmatrix} > 0, \forall w_i \in vert(\mathcal{W}_{i,(d)}), \tag{31}$$

$$\begin{bmatrix} 1 & * \\ Y_{i,(d)}^T l_i & W_{i,(d)} \end{bmatrix} > 0, \tag{32}$$

$$\begin{bmatrix} \gamma_{i,(d)} & * \\ W_{i,(d)} h_i & W_{i,(d)} \end{bmatrix} > 0, \tag{33}$$

$$0 \leq \gamma_{i,(d)} \leq 1. \tag{34}$$

In this equation, μ_i is a weight coefficient for the optimization. $\gamma_{i,(d)}, W_{i,(d)}, Y_{i,(d)}, X_{i,(d)}$ represent the γ_i, W_i, Y_i, X_i under distribution $\mathcal{C}_{i,(d)}$. Moreover, $\mathcal{C}_{i,(d^*)}$ is the optimal solution.

This optimization means that in order to make the optimal decision on strong-coupling neighbors and weak-coupling neighbors while taking both connectivity and performance into account, the optimization that minimizes the combination of subsystem connectivity and size of ϕ_i should be solved. To decide whether a neighbor $\mathcal{S}_j, j \in \mathcal{N}_i$ is a strong-coupling neighbor or a weak one, the size of ϕ_i is expected to be small so that the solution in (15) can get a larger feasible domain. At the same time, the connectivity is expected to be small to reduce the system's topological complexity.

The optimization achieves the goal of choosing neighbors which result in smaller size of robust positively invariant set ϕ_i and connectivity. Solution $\mathcal{C}_{i,(d^*)}$ reflects the consideration of influence on RPI set ϕ_i and connectivity. With this method, even though "weak-coupling" neighbors are omitted and deviation is brought, the simplified model has a large degree of freedom to design the control law of tracking and reduces the connectivity at the same time. Thus, a good system performance and error tolerance can be obtained.

4. Stability and Convergence

In this section, the feasibility and stability theorem of strong-coupling neighbor-based DMPC are given. Denote

$$\mathcal{X}_N = \{x \in \mathcal{X} | \exists v = (x, u_{0:N-1}, y_s), u(k) \in \mathcal{U}, k \in \mathcal{I}_{0:N-1}, y_s \in \mathcal{Y}_s, s.t. v \in \mathcal{Z}_N\},$$
$$\mathcal{Z}_N = \{v | u(k) \in \mathcal{U}, k \in \mathcal{I}_{0:N-1}, y_s \in \mathcal{Y}_s, x(k; x, u) \in \mathcal{X}, k \in \mathcal{I}_{0:N-1}, x(N; x, u) \in \Omega_{\tilde{K}}\}.$$

$x(k; x, u)$ represents the current time's state prediction after k sample time. $\mathcal{Y}_s$ is the feasible tracking set based on hard constraints of x and u.

Theorem 1. *Assume that Assumptions 1–3 hold. Then, for all initial state $x(0)$ with tracking target y_t if $v(0) \in \mathcal{Z}_N$, the closed-loop system based on a strong-coupling neighbor-based DMPC algorithm is feasible and asymptotically stable and converges to $\hat{y}_s \oplus C\phi_k$, where $\hat{y}_s = (\hat{y}_{1,s}, ..., \hat{y}_{m,s})$, $\hat{y}_{i,s} = argmin V_0(y_{i,s}, y_{i,t})$ among feasible targets.*

Proof. Feasibility is proved by Lemmas A1, A2. Stability's proofs are in Lemmas A3, A4 in the Appendix A. □

5. Simulation

The simulation takes an industrial system model with five subsystems interacting with each other as an example. Between different subsystems, the coupling degrees vary substantially. The relationships of subsystems and the designed MPC are shown in Figure 1.

In Figure 1, dotted lines are used to represent weak coupling, while solid lines are used to represent strong coupling. With the strategy we have defined in our paper, weak couplings are neglected. As a result, it can be seen in Figure 1 that only parts of the subsystems are joint in cooperation. Subsystem models are also given as follows:

$$\mathcal{S}_1 : x_{1,t+1} = \begin{bmatrix} 0.5 & 0.6 \\ 0 & 0.66 \end{bmatrix} x_{1,t} + \begin{bmatrix} 0.1 \\ 0.7 \end{bmatrix} u_{1,t} + \begin{bmatrix} 0 \\ 0.04 \end{bmatrix} u_{2,t},$$
$$y_{1,t} = \begin{bmatrix} 0 & 1 \end{bmatrix} x_{1,t}, \tag{35a}$$

$$\mathcal{S}_2 : x_{2,t+1} = \begin{bmatrix} 0.6 & 0.1 \\ 0 & 0.71 \end{bmatrix} x_{2,t} + \begin{bmatrix} 0.5 \\ 1 \end{bmatrix} u_{2,t} + \begin{bmatrix} 0 \\ 0.3 \end{bmatrix} u_{1,t} + \begin{bmatrix} 0 \\ 0.01 \end{bmatrix} u_{3,t},$$
$$y_{1,t} = \begin{bmatrix} 0 & 1 \end{bmatrix} x_{1,t}, \tag{35b}$$

$$\mathcal{S}_3 : x_{3,t+1} = \begin{bmatrix} 0.7 & 0.2 \\ 0.1 & 0.4 \end{bmatrix} x_{3,t} + \begin{bmatrix} 0.9 \\ 1 \end{bmatrix} u_{3,t} + \begin{bmatrix} 0 \\ 0.4 \end{bmatrix} u_{2,t} + \begin{bmatrix} 0 \\ 0.05 \end{bmatrix} u_{4,t},$$
$$y_{1,t} = \begin{bmatrix} 0 & 1 \end{bmatrix} x_{1,t}, \tag{35c}$$

$$\mathcal{S}_4 : x_{4,t+1} = \begin{bmatrix} 0.9 & 0.7 \\ 0 & 0.6 \end{bmatrix} x_{4,t} + \begin{bmatrix} 0.4 \\ 0.4 \end{bmatrix} u_{4,t} + \begin{bmatrix} 0.3 \\ 0.6 \end{bmatrix} u_{3,t} + \begin{bmatrix} 0 \\ 0.01 \end{bmatrix} u_{5,t},$$
$$y_{1,t} = \begin{bmatrix} 0 & 1 \end{bmatrix} x_{1,t}, \tag{35d}$$

$$\mathcal{S}_5 : x_{5,t+1} = \begin{bmatrix} 0.8 & 0 \\ 0.5 & 0.78 \end{bmatrix} x_{5,t} + \begin{bmatrix} 0 \\ 1 \end{bmatrix} u_{5,t} + \begin{bmatrix} 0.4 \\ 0.2 \end{bmatrix} u_{4,t},$$
$$y_{1,t} = \begin{bmatrix} 0 & 1 \end{bmatrix} x_{1,t}. \tag{35e}$$

By the strong-coupling neighbor-based DMPC, connections including $\mathcal{S}_2 \rightarrow \mathcal{S}_1$, $\mathcal{S}_3 \rightarrow \mathcal{S}_2$, $\mathcal{S}_4 \rightarrow \mathcal{S}_3$, $\mathcal{S}_5 \rightarrow \mathcal{S}_4$ are neglected. For the five subsystems in the given model, $\gamma_1, \gamma_2, \gamma_3, \gamma_4$, and γ_5, which evaluate the system performance, are obtained by optimization in Section 3.4, they are:

$$(\gamma_1, \gamma_2, \gamma_3, \gamma_4, \gamma_5) = (0.54, 0.66, 0.72, 0.53, 0). \tag{36}$$

Among them, $\gamma_5 = 0$ illustrates that subsystem $\mathcal{S}_5$ has no weak-coupling upstream neighbors. Additionally, the robust positively invariant set feedback control laws are

$$\{K_1, K_2, K_3, K_4\} = \{[-0.119 - 0.762]^{\mathrm{T}}, [-0.171 - 0.434]^{\mathrm{T}}, [-0.316 - 0.251]^{\mathrm{T}}, [-0.724 - 0.966]^{\mathrm{T}}\}.$$

The optimization horizon N is 10 sample time. Take $Q = I_{10\times 10}$ and $R = I_{5\times 5}$. To accelerate the iterative process, in both of these iterative algorithms, the terminal conditions of iteration are $||u_i^{[p]} - u_i^{[p-1]}||_2 \leq 10^{-3}$ or $p > 100$. If either of these two conditions is satisfied, iteration terminates.

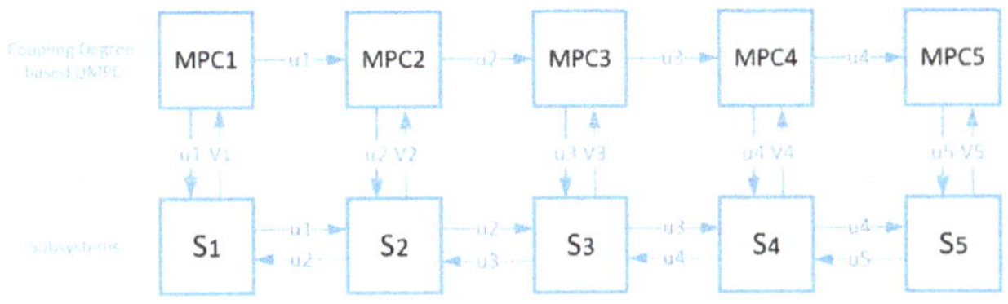

Figure 1. An illustration of the structure of a distributed system and its distributed control framework. MPC: model predictive control; DMPC: distributed MPC.

The following shows the system performance when the strong-coupling neighbor-based DMPC algorithm is applied. Here we chose different set-points to detect the system stability. Three groups of setpoints were given to verify the system's feasibility and stability. For a better understanding, cooperative DMPC strategy control results which cooperate with all neighbors are also introduced to make a comparison. The simulation took a total of 74.3 seconds for 90 sampling times. The performance comparison of strong-coupling neighbor-based DMPC (SCN-DMPC) with cooperative DMPC where each subsystem used the full system's information in their controller is shown in Figures 2–4.

Figure 2 shows the state evolution of each subsystem. The two curves of SCN-DMPC and cooperative are close to each other. This is because the weak couplings in the given example are tiny compared with the strong couplings and thus do not have much of an impact on system dynamics. Besides, SCN-DMPC optimization algorithm was always feasible and was able to keep stable with a changing tracking target. Figure 3 shows the input difference between these two algorithms. The control laws of these two algorithms are almost the same. Tracking results are shown in Figure 4. There was a small off-set in subsystem $\mathcal{S}_1, \mathcal{S}_3$, which could be eliminated by adding an observer. All other subsystems could track the steady-state target without steady-state off-set. From the simulation results of Figures 2–4, the stability and good optimization performance of the closed-loop system using SCN-DMPC is verified.

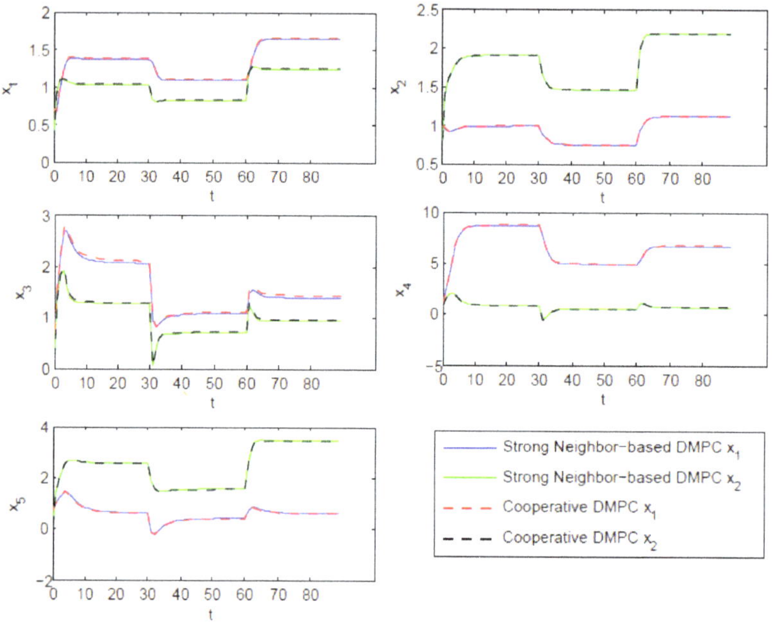

Figure 2. States of each subsystem under the control of strong-coupling neighbor-based DMPC (SCN-DMPC) and cooperative DMPC.

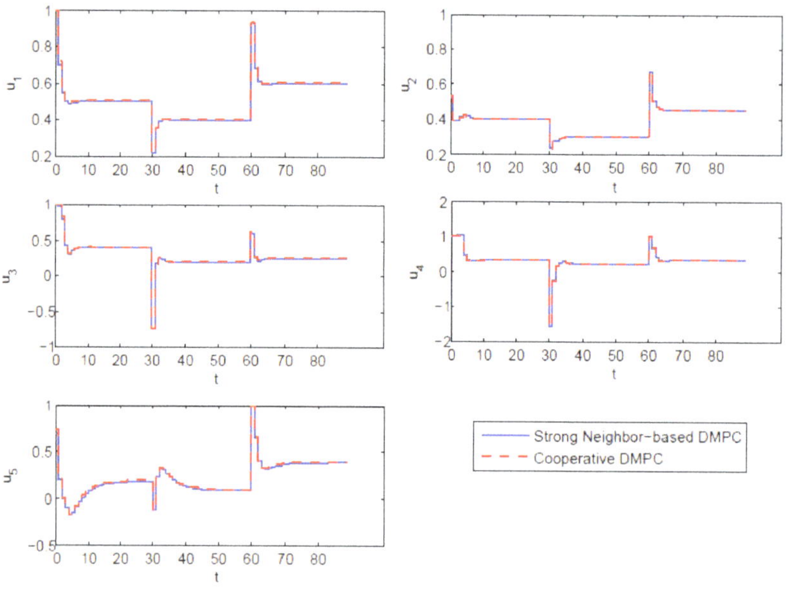

Figure 3. Inputs of each subsystem under the control of SCN-DMPC and cooperative DMPC.

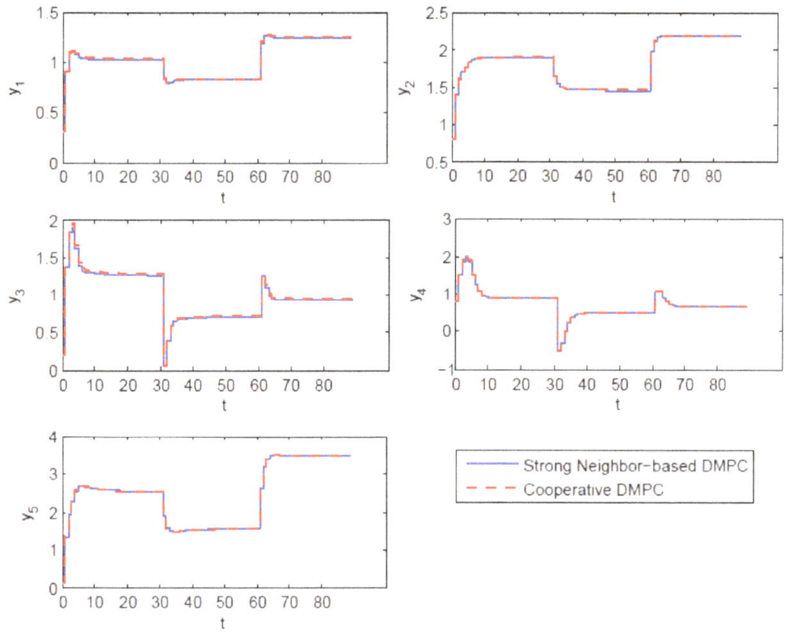

Figure 4. Output of each subsystem under the control of SCN-DMPC and cooperative DMPC.

In Figures 2–4, the curves of SCN-DMPC and cooperative DMPC are close to each other. The reason is that the weak couplings in the given example were tiny compared with the strong couplings, and thus they did not have much of an impact on system dynamics, even though a small difference existed. Specifically, given the state equation input weight coefficients in each subsystem, the deviations of the five subsystems fit $\|w_1\| \leq 0.04$, $\|w_2\| \leq 0.01$, $\|w_3\| \leq 0.05$, $\|w_4\| \leq 0.01$. The effects of these disturbances were very small compared with those of each subsystem's inputs. Under robust feedback control law, they do not have much of an influence on system dynamics. Besides, the performance of the simplified model under SCN-DMPC equals to that under the control law optimizing the global performance of simplified system. As a result, the system performance under SCN-DMPC was close to that in cooperative DMPC. Under circumstances where the weak interactions are close to the impact of each subsystem's inputs (which sacrifices part of the performance to achieve less network connectivity), omitting weak couplings may result in a greater influence on system dynamics, and the simulation results can differ.

Moreover, mean square errors between the closed-loop systems with strong-coupling neighbor-based optimization DMPC and cooperative DMPC outputs are listed in Table 1. The total error of five subsystems was only 3.5, which illustrates the good optimization performance of SCN-DMPC.

Table 1. Mean square error (MSE) of outputs between SCN-DMPC and cooperative DMPC.

Item	$\mathcal{S}_1$	$\mathcal{S}_2$	$\mathcal{S}_3$	$\mathcal{S}_4$	$\mathcal{S}_5$
MSE	0.5771	1.1512	0.7111	0.1375	0.9162

Connectivities are compared in the following table.

Table 2 shows that when strong-coupling neighbor-based DMPC was applied, the total information connections reduced to eight, which means that five connections were avoided compared with cooperative DMPC.

Table 2. Comparison of system connectivity with different control methods.

System	SCN-DMPC	Cooperative DMPC
$\mathcal{S}_1$	1	2
$\mathcal{S}_2$	2	3
$\mathcal{S}_3$	2	4
$\mathcal{S}_4$	2	3
$\mathcal{S}_5$	1	1
$\mathcal{S}$	8	13

Above all, the simulation results show that the proposed SCN-DMPC achieved a good performance close to the cooperative DMPC with a significant reduction of information connectivity.

6. Conclusions

In this paper, a strong-coupling neighbor-based optimization DMPC method is proposed to decide the cooperation among subsystems, where each subsystem's MPC considers the optimization performance and evolution of its strong-coupling downstream subsystems and communicates with them. For strongly-coupled subsystems, the influence on state and objective function are considered. For weakly-coupled subsystems, influence is neglected in the cooperative design. A closed-loop system's performance and network connectivity-based method is proposed to determine the strength of coupling relationships among subsystems. The feasibility and stability of the closed-loop system in the case of target-tracking are analyzed. Simulation results show that the proposed SCN-DMPC was able to achieve similar performance in comparison to the DMPC which did not neglect the information or influence of weakly coupling subsystems. At the same time, connectivity was significantly decreased.

Author Contributions: Shan Gao developed the main algorithm, contributed to the stability analysis, designed the simulation and prepared the draft of the paper. Yi Zheng and Shaoyuan Li proposed the idea of Enhancing Strong Neighbor-based coordination strategy. They contributed to the main theory of the work and gave the inspiration and guidance of the strong-coupling neighbors' determination, the algorithm design and stability analysis.

Acknowledgments: This work is supported by the National Nature Science Foundation of China (61673273, 61590924).

Conflicts of Interest: The authors declare no conflict of interest.

Appendix

In the strong-coupling neighbor-based optimization DMPC algorithm proposed in this paper, the optimal solution of each subsystem equals to the solution of optimizing the overall system objective function. That is:

$$\arg\min V_N(x_i, y_t, \infty; \bar{x}_i, \bar{u}_{i,0:N-1}, y_s) = \arg\min V_{iN}(x_i, y_t; \bar{x}_i, \bar{u}_{i,0:N-1}, y_s), \tag{A1}$$

where

$$V_N(x_i, y_t; \bar{x}_i, \bar{u}_{i,0:N-1}, y_s) = \sum_{k=0}^{N-1} (\|x(k) - \bar{x}_s\|_Q^2 + \|u(k) - \bar{u}_s\|_R^2) + \|x(N) - \bar{x}_s\|_P^2 + V_0(y_s, y_t). \tag{A2}$$

Thus, for easy analysis, here we take the overall objective function V_N to prove the feasibility and stability, and define:

$$v^{[p]} = \{\bar{x}_1^{[p]}, ..., \bar{x}_m^{[p]}, \bar{u}_{1,0:N-1}^{[p]}, ..., \bar{u}_{m,0:N-1}^{[p]}, y_{1,s}^{[p]}, ..., y_{1,m}^{[p]}\}, \tag{A3}$$

$$v^* = \{\bar{x}_1^*, ..., \bar{x}_m^*, \bar{u}_{1,0:N-1}^*, ..., \bar{u}_{m,0:N-1}^*, y_{1,s}^*, ..., y_{1,m}^*\}. \tag{A4}$$

Denote $v_i^{\prime[p]}$ as

$$v_i^{\prime[p]} = (\bar{x}_1^{[p-1]}, ..., \bar{x}_i^{\prime[p]}, ..., \bar{x}_m^{[p-1]}, \bar{u}_{1,0:N}^{[p-1]}, ..., \bar{u}_{i,0:N}^{\prime[p]}, ..., \bar{u}_{m,0:N}^{[p-1]}, y_{1,s}^{[p-1]}, ..., y_{i,s}^{\prime[p]}, ..., y_{m,s}^{[p-1]}). \tag{A5}$$

Lemma A1. *Feasibility. Feasibility can be proved by assuming $v(t) \in \mathcal{Z}_n$. Then, we get $v(t+1) \in \mathcal{Z}_n$.*

Proof. The feasibility of the model is proved by analyzing the simplified model.

Refer to the warm start algorithm. At time $t+1$, the warm start for arbitrary $\mathcal{S}_i : \bar{v}_i^{[0]}(t+1) = (\bar{x}_i^{[0]}(0|t+1), \bar{u}_{i,0:N}^{[0]}(t+1), y_{i,s}^{[0]}(t+1))$. With the warm start, all constraints in (17) are satisfied.

From p^{th} to $(p+1)^{th}$ iterations, given that

$$(v_1^{\prime[p+1]}, v_2^{\prime[p+1]}, ..., v_m^{\prime[p+1]})$$

are feasible, obviously their convex sum $v^{[p+1]}$ is feasible. As a result, the converged simplified solution $v^*(t+1)$ is feasible. Based on this, the real system solution is around the invariant set of $v^*(t+1)$, and fits the system constraints. That is, $(x(t+1), v(t+1)) \in \mathcal{Z}_n$. □

Lemma A2. *Convergence. Here we have*

$$V_N(x(t), y_t; v^{[p+1]}(t)) \leq V_N(x(t), y_t; v^{[p]}(t)).$$

Proof. Since it has

$$\begin{aligned} V_N(x, y_t, v^{[p+1]}) &\leq \gamma_1 V_N(x, y_t; v_1^{\prime[p+1]}) + ... + \gamma_m V_N(x, y_t; v_m^{\prime[p+1]}) \\ &\leq \gamma_1 V_N(x, y_t; v^{[p]}) + ... + \gamma_m V_N(x, y_t; v^{[p]}) \\ &= V_N(x, y_t, v^{[p]}), \end{aligned} \tag{A6}$$

convergence is proved. □

According to the Lemma above, we have:

$$V_N(x(t), y_t; v^{[p+1]}(t)) \leq V_N(x(t), y_t; v^{[0]}(t)). \tag{A7}$$

Lemma A3. *Local Bounded. When $(x(t), y_s^{[0]}(t) \in \Omega_{\bar{K}}$, then*

$$V_N(x(t), y_t; v^{[p]}(t)) \leq \left\| x(t) - \bar{x}_s^{[0]}(t) \right\|_P^2 + V_0(y_s^{[0]}(t), y_t).$$

Proof. Firstly,

$$V_N(x(t), y_t; v^{[0]}(t)) \leq \left\| x(t) - \bar{x}_s^{[0]}(t) \right\|_P^2 + V_0(y_s^{[0]}(t), y_t)$$

will be proved.

According to the definition of warm start and Assumption 3, here we have:

$$\begin{aligned}
&V_N(x(t), y_t; v^{[0]}(t)) \\
&\leq \sum_{k=0}^{N-1} \left\| x(k;t) - \bar{x}_s^{[0]}(t) \right\|_Q + \left\| u(k) - \bar{u}_s^{[0]}(t) \right\|_R + \left\| x(N) - \bar{x}_s^{[0]}(t) \right\|_P^2 + V_0(y_s^{[0]}(t), y_t) \\
&= \sum_{k=0}^{N-1} \left\| x(k;t) - \bar{x}_s^{[0]}(t) \right\|_{Q+\bar{K}'R\bar{K}} + \left\| x(N) - \bar{x}_s^{[0]}(t) \right\|_P^2 + V_0(y_s^{[0]}(t), y_t) \qquad \text{(A8)} \\
&= \left\| x(N) - \bar{x}_s^{[0]}(t) \right\|_P^2 + V_0(y_s^{[0]}(t), y_t).
\end{aligned}$$

Thus, we have:

$$\begin{aligned}
V_N(x(t), y_t; v^{[p]}(t)) &\leq V_N(x(t), y_t; v^{[0]}(t)) \\
&\leq \left\| x(t) - \bar{x}_s^{[0]}(t) \right\|_P^2 + V_0(\bar{y}_s^{[0]}(t), y_t). \qquad \text{(A9)}
\end{aligned}$$

□

Lemma A4. *Convergence. Let Assumption 3 hold, for any feasible solution* $z(0) = (x(0), v(0)) \in \mathcal{Z}_N$*, the system converges to equilibrium point* z_s*. That is,*

$$V_N(x(t+1), y_t; \bar{v}^*(t+1)) - V_N(x(t), y_t; \bar{v}^*(t)) \leq \|x(t) - \bar{x}_s(t)\|_Q^2. \qquad \text{(A10)}$$

The final tracking points of the simplified system (the optimal solution of V_N*) are* $(\bar{x}^*(x_s, y_t), \bar{u}^*(x_s, y_t)) = (x_s, u_s)$*, which are the centralized optimal solution.*

Proof. For simplified system optimization, we have

$$V_N(x(t+1), y_t, v^*(t+1)) \leq V_N(x(t+1), y_t; v(0; t+1)), \qquad \text{(A11)}$$

and also

$$V_N(x(t+1), y_t; v(0; t+1)) \leq V_N(x(t), y_t, v^*(t)) - \|x(t) - \bar{x}_s(t)\|_Q^2 - \|u(t) - \bar{u}_s(t)\|_R. \qquad \text{(A12)}$$

According to (A11) and (A12), we have

$$\begin{aligned}
&V_N(x(t+1), y_t; v^*(t+1)) - V_N(x(t), y_t; v^*(t)) \qquad \text{(A13)} \\
&\leq -\|x(t) - \bar{x}_s(t)\|_Q^2 - \|u(t) - \bar{u}_s(t)\|_R \\
&\leq -\|x(t) - \bar{x}_s(t)\|_Q^2.
\end{aligned}$$

Since the robust positively invariant set feedback control law $K = diag(K_1, K_2, ..., K_m)$ ensures the real states in the invariant set of the simplified model, the real system's stability is proved. □

References

1. Christofides, P.D.; Scattolini, R.; Muñoz de la Peña, D.; Liu, J. Distributed model predictive control: A tutorial review and future research directions. *Comput. Chem. Eng.* **2013**, *51*, 21–41. [CrossRef]
2. Del Real, A.J.; Arce, A.; Bordons, C. Combined environmental and economic dispatch of smart grids using distributed model predictive control. *Int. J. Electr. Power Energy Syst.* **2014**, *54*, 65–76. [CrossRef]
3. Zheng, Y.; Li, S.; Tan, R. Distributed Model Predictive Control for On-Connected Microgrid Power Management. *IEEE Trans. Control Syst. Technol.* **2018**, *26*, 1028–1039. [CrossRef]

4. Yu, W.; Liu, D.; Huang, Y. Operation optimization based on the power supply and storage capacity of an active distribution network. *Energies* **2013**, *6*, 6423–6438. [CrossRef]
5. Scattolini, R. Architectures for distributed and hierarchical model predictive control-a review. *J. Process Control* **2009**, *19*, 723–731. [CrossRef]
6. Du, X.; Xi, Y.; Li, S. Distributed model predictive control for large-scale systems. In Proceedings of the 2001 American Control Conference, Arlington, VA, USA, 25–27 June 2001; Volume 4, pp. 3142–3143.
7. Li, S.; Yi, Z. *Distributed Model Predictive Control for Plant-Wide Systems*; John Wiley & Sons: Hoboken, NJ, USA, 2015.
8. Mota, J.F.C.; Xavier, J.M.F.; Aguiar, P.M.Q.; Püschel, M. Distributed Optimization With Local Domains: Applications in MPC and Network Flows. *IEEE Trans. Autom. Control* **2015**, *60*, 2004–2009. [CrossRef]
9. Qin, S.; Badgwell, T. A survey of industrial model predictive control technology. *Control Eng. Pract.* **2003**, *11*, 733–764. [CrossRef]
10. Maciejowski, J. *Predictive Control: With Constraints*; Pearson Education: London, UK, 2002.
11. Vaccarini, M.; Longhi, S.; Katebi, M. Unconstrained networked decentralized model predictive control. *J. Process Control* **2009**, *19*, 328–339. [CrossRef]
12. Leirens, S.; Zamora, C.; Negenborn, R.; De Schutter, B. Coordination in urban water supply networks using distributed model predictive control. In Proceedings of the American Control Conference (ACC), Baltimore, MD, USA, 30 June–2 July 2010; pp. 3957–3962.
13. Wang, Z.; Ong, C.J. Distributed Model Predictive Control of linear discrete-time systems with local and global constraints. *Automatica* **2017**, *81*, 184–195. [CrossRef]
14. Trodden, P.A.; Maestre, J. Distributed predictive control with minimization of mutual disturbances. *Automatica* **2017**, *77*, 31–43. [CrossRef]
15. Al-Gherwi, W.; Budman, H.; Elkamel, A. A robust distributed model predictive control algorithm. *J. Process Control* **2011**, *21*, 1127–1137. [CrossRef]
16. Kirubakaran, V.; Radhakrishnan, T.; Sivakumaran, N. Distributed multiparametric model predictive control design for a quadruple tank process. *Measurement* **2014**, *47*, 841–854. [CrossRef]
17. Zhang, L.; Wang, J.; Li, C. Distributed model predictive control for polytopic uncertain systems subject to actuator saturation. *J. Process Control* **2013**, *23*, 1075–1089. [CrossRef]
18. Liu, J.; Muñoz de la Peña, D.; Christofides, P.D. Distributed model predictive control of nonlinear systems subject to asynchronous and delayed measurements. *Automatica* **2010**, *46*, 52–61. [CrossRef]
19. Cheng, R.; Fraser Forbes, J.; Yip, W.S. Dantzig–Wolfe decomposition and plant-wide MPC coordination. *Comput. Chem. Eng.* **2008**, *32*, 1507–1522. [CrossRef]
20. Zheng, Y.; Li, S.; Qiu, H. Networked coordination-based distributed model predictive control for large-scale system. *IEEE Trans. Control Syst. Technol.* **2013**, *21*, 991–998. [CrossRef]
21. Groß, D.; Stursberg, O. A Cooperative Distributed MPC Algorithm With Event-Based Communication and Parallel Optimization. *IEEE Trans. Control Netw. Syst.* **2016**, *3*, 275–285. [CrossRef]
22. Walid Al-Gherwi, H.B.; Elkamel, A. Selection of control structure for distributed model predictive control in the presence of model errors. *J. Process Control* **2010**, *20*, 270–284. [CrossRef]
23. Camponogara, E.; Jia, D.; Krogh, B.; Talukdar, S. Distributed model predictive control. *IEEE Control Syst. Mag.* **2002**, *22*, 44–52.
24. Conte, C.; Jones, C.N.; Morari, M.; Zeilinger, M.N. Distributed synthesis and stability of cooperative distributed model predictive control for linear systems. *Automatica* **2016**, *69*, 117–125. [CrossRef]
25. Tippett, M.J.; Bao, J. Reconfigurable distributed model predictive control. *Chem. Eng. Sci.* **2015**, *136*, 2–19. [CrossRef]
26. Zheng, Y.; Wei, Y.; Li, S. Coupling Degree Clustering-Based Distributed Model Predictive Control Network Design. *IEEE Trans. Autom. Sci. Eng.* **2018**, 1–10. doi:10.1109/TASE.2017.2780444. [CrossRef]
27. Li, S.; Zhang, Y.; Zhu, Q. Nash-optimization enhanced distributed model predictive control applied to the Shell benchmark problem. *Inf. Sci.* **2005**, *170*, 329–349. [CrossRef]
28. Venkat, A.N.; Hiskens, I.A.; Rawlings, J.B.; Wright, S.J. Distributed MPC Strategies with Application to Power System Automatic Generation Control. *IEEE Trans. Control Syst. Technol.* **2008**, *16*, 1192–1206. [CrossRef]
29. Stewart, B.T.; Wright, S.J.; Rawlings, J.B. Cooperative distributed model predictive control for nonlinear systems. *J. Process Control* **2011**, *21*, 698–704. [CrossRef]

30. Zheng, Y.; Li, S. N-Step Impacted-Region Optimization based Distributed Model Predictive Control. *IFAC-PapersOnLine* **2015**, *48*, 831–836. [CrossRef]
31. de Lima, M.L.; Camponogara, E.; Marruedo, D.L.; de la Peña, D.M. Distributed Satisficing MPC. *IEEE Trans. Control Syst. Technol.* **2015**, *23*, 305–312. [CrossRef]
32. de Lima, M.L.; Limon, D.; de la Pena, D.M.; Camponogara, E. Distributed Satisficing MPC With Guarantee of Stability. *IEEE Trans. Autom. Control* **2016**, *61*, 532–537. [CrossRef]
33. Zheng, Y.; Li, S.; Wang, X. Distributed model predictive control for plant-wide hot-rolled strip laminar cooling process. *J. Process Control* **2009**, *19*, 1427–1437. [CrossRef]
34. Zheng, Y.; Li, S.; Li, N. Distributed model predictive control over network information exchange for large-scale systems. *Control Eng. Pract.* **2011**, *19*, 757–769. [CrossRef]
35. Li, S.; Zheng, Y.; Lin, Z. Impacted-Region Optimization for Distributed Model Predictive Control Systems with Constraints. *IEEE Trans. Autom. Sci. Eng.* **2015**, *12*, 1447–1460. [CrossRef]
36. Riverso, S.; Ferrari-Trecate, G. Tube-based distributed control of linear constrained systems. *Automatica* **2012**, *48*, 2860–2865. [CrossRef]
37. Riverso, S.; Boem, F.; Ferrari-Trecate, G.; Parisini, T. Plug-and-Play Fault Detection and Control-Reconfiguration for a Class of Nonlinear Large-Scale Constrained Systems. *IEEE Trans. Autom. Control* **2016**, *61*, 3963–3978. [CrossRef]
38. Ferramosca, A.; Limon, D.; Alvarado, I.; Camacho, E. Cooperative distributed MPC for tracking. *Automatica* **2013**, *49*, 906–914. [CrossRef]
39. Limon, D. MPC for tracking of piece-wise constant references for constrained linear systems. In Proceedings of the 16th IFAC World Congress, Prague, Czech Republic, 3–8 July 2005; p. 882.
40. Limon, D.; Alvarado, I.; Alamo, T.; Camacho, E.F. Robust tube-based MPC for tracking of constrained linear systems with additive disturbances. *J. Process Control* **2010**, *20*, 248–260. [CrossRef]
41. Shao, Q.M.; Cinar, A. Coordination scheme and target tracking for distributed model predictive control. *Chem. Eng. Sci.* **2015**, *136*, 20–26. [CrossRef]
42. Rawlings, J.B.; Mayne, D.Q. *Model Predictive Control: Theory and Design*; Nob Hill Publishing: Madison, WI, USA, 2009; pp. 3430–3433.
43. Alvarado, I. On the Design of Robust Tube-Based MPC for Tracking. In Proceedings of the 17th World Congress The International Federation of Automatic Control, Seoul, Korea, 6–11 July 2008; pp. 15333–15338.

mathematics

Article

Model Predictive Control of Mineral Column Flotation Process

Yahui Tian [1,2], Xiaoli Luan [1], Fei Liu [1] and Stevan Dubljevic [2,*]

1 Key Laboratory of Advanced Process Control for Light Industry (Ministry of Education), Institute of Automation, Jiangnan University, Wuxi 214122, China; tian894@hotmail.com (Y.T.); xlluan@jiangnan.edu.cn (X.L.); fliu@jiangnan.edu.cn (F.L.)
2 Department of Chemical and Materials Engineering, University of Alberta, Edmonton, AB T6G 2V4, Canada
* Correspondence: stevan.dubljevic@ualberta.ca; Tel.: +1-(780)-248-1596

Received: 28 April 2018; Accepted: 4 June 2018; Published: 13 June 2018

Abstract: Column flotation is an efficient method commonly used in the mineral industry to separate useful minerals from ores of low grade and complex mineral composition. Its main purpose is to achieve maximum recovery while ensuring desired product grade. This work addresses a model predictive control design for a mineral column flotation process modeled by a set of nonlinear coupled heterodirectional hyperbolic partial differential equations (PDEs) and ordinary differential equations (ODEs), which accounts for the interconnection of well-stirred regions represented by continuous stirred tank reactors (CSTRs) and transport systems given by heterodirectional hyperbolic PDEs, with these two regions combined through the PDEs' boundaries. The model predictive control considers both optimality of the process operations and naturally present input and state/output constraints. For the discrete controller design, spatially varying steady-state profiles are obtained by linearizing the coupled ODE–PDE model, and then the discrete system is obtained by using the Cayley–Tustin time discretization transformation without any spatial discretization and/or without model reduction. The model predictive controller is designed by solving an optimization problem with input and state/output constraints as well as input disturbance to minimize the objective function, which leads to an online-solvable finite constrained quadratic regulator problem. Finally, the controller performance to keep the output at the steady state within the constraint range is demonstrated by simulation studies, and it is concluded that the optimal control scheme presented in this work makes this flotation process more efficient.

Keywords: model predictive control; column flotation; coupled PDE–ODE; Cayley–Tustin discretization; input/state constraints

1. Introduction

Since its first commercial application in the 1980s [1], column flotation has attracted the attention of many researchers. As a result of its industrial relevance and importance, the modeling and control of mineral column flotation has gradually become a popular research field. In general, column flotation methods are based on different physical properties of the particle surface and the flotability for mineral separation [2,3]. Column flotation has many advantages compared to conventional mechanical flotation processes, such as simplicity of construction, low energy consumption, higher recovery and product grade, and so forth [4]. It has been claimed that appropriate process regulation could improve recovery and product grades or process operations for greater benefits [5,6].

The column flotation process uses a complex distributed parameter system (DPS) that is highly nonlinear, and various important parameters are highly interrelated. The whole process consists of water, solid, and gas three-phase flows with multiple inputs as well as the occurrence of various sub-processes, such as particle–bubble attachment, detachment, and bubble coalescence, making the process more complex and difficult to predict. After several decades of research and development, the process is still not fully understood; the process control of column flotation has proven to be a great challenge and remains a very important topic for the research community.

The process control for column flotation consists of three to four interconnected levels [6–8], but according to the control effects, it can be divided into stability control and optimal control [9,10]. At present, most flotation control systems are based on stability control, and the traditional control method uses PID control to achieve automatic control of the froth depth as well as other easily measurable variables to keep the flotation process as close as possible to the steady state [11,12]. A growing number of scholars have begun to apply advanced control methods, such as model predictive control, fuzzy control, expert systems, and neural network control, to regulate the column flotation process and/or combine these novel control methods to achieve better flotation column regulation [13–16].

Model predictive control is the most widely used multivariable control algorithm in current industrial practice. One of its major advantages is that it can explicitly handle constraints while dealing with multiple-input multiple-output process setting [17]. This ability comes from its model-based prediction of the future dynamic behaviour of the system. By adding constraints to future inputs, outputs, or state variables, constraints can be explicitly accounted for in an online quadratic programming problem realization. This paper proposes and develops a model predictive control design for the column flotation process, considering the process state/output and input control constraints.

For the model predictive controller design, a three-phase column flotation dynamic model was developed. A typical column flotation process can be divided into two regions [3]: the collection region and the froth region (a schematic representation of the column flotation process is given in Figure 1). This work considers the collection region as a model given by the continuous stirred tank reactor (CSTR) and considers the froth region as a plug flow reactor (PFR) model. According to the mass balance laws, the overall column flotation system is described by a set of nonlinear heterodirectional coupled hyperbolic partial differential equations (PDEs) and ordinary differential equations (ODEs) that are connected through the PDEs' boundaries. The steady-state profiles are utilized to linearize the original nonlinear system, and then the discrete model is realized by the Cayley–Tustin time discretization transformation [18–20]. By using this method, the continuous linear infinite-dimensional PDE system can be mapped into a discrete infinite-dimensional system without spatial discretization; the discretized model is structure-preserving and does not imply any model reduction [21]. Finally, the model predictive controller is designed on the basis of the infinite-dimensional discrete model. The paper is organized as follows: Section 2 develops the model of the column flotation process, and the discrete version of the system is obtained by using the Cayley–Tustin time discretization transformation. Section 3 addresses the model predictive controller design for the coupled PDE–ODE model with the consideration of input disturbance rejection and input and state/output constraints. Simulation results are shown in Section 4 to demonstrate the controller performance. Finally, Section 5 provides the conclusions.

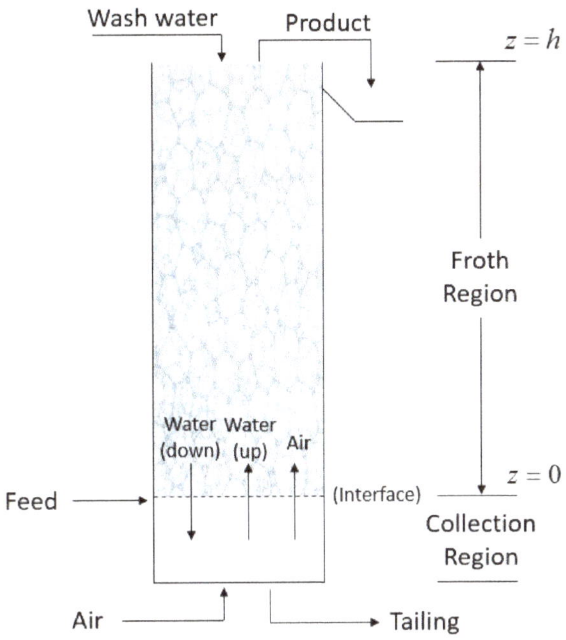

Figure 1. Schematic representation of a flotation column.

2. Model Formulation of Column Flotation

2.1. Model Description

Column flotation utilizes the principle of countercurrent flow, in which air is introduced into the column at the bottom through a sparger or in the form of externally generated bubbles and rises through the downward-flowing slurry that contains mineral, locked, and gangue particles. By countercurrent flow, contact, and collision, hydrophobic particles (minerals) attach to the bubbles forming bubble–particle aggregates and reach the top of the column; they are subsequently removed at the top as a valuable product. Above the overflowing froth, there is a fine spray of water, washing down the undesired particles that could have been entrained by the bubbles from the froth region [22,23]. Meanwhile, rising bubbles entrain some water flow together through bubble coalescence, and the interaction of wash water and particles also simultaneously occurs. Therefore, the essential process step in column flotation is the transfer of particles between the water phase and air phase as well as between the upward and downward water phases.

On the basis of the above description and by making appropriate mass balances, the following equations are obtained to describe the column flotation process.

2.1.1. Model for Collection Region

Under the assumption of perfect mixing, the collection region can be considered as a CSTR, which means that the material properties are uniform throughout the reactor. Therefore, the model for the collection region is described by coupled ODEs. The state variables for the process of the collection region are mass concentrations of solid particles (mineral, locked, and gangue) with the air phase (C_a) and water phase (C_w):

$$\begin{aligned}\frac{d(H_a V C_a(t))}{dt} &= \alpha Av f H_w V C_w(t) - \beta H_a V C_a(t) - Q_a C_a(t) &(1)\\ \frac{d(H_w V C_w(t))}{dt} &= -\alpha Av f H_w V C_w(t) + \beta H_a V C_a(t) + Q_F C_F - Q_T C_w(t)\\ &+ Q_{w_d} C_{w_d}(0,t) - Q_{w_u} C_w(t), &(2)\end{aligned}$$

where $f = 1 - \frac{C_a}{C_a^*}$ is the fractional free surface area of the bubbles; $Q_i = U_i Ac$ is the flow rate of i-th phase; Q_a is the air flow rate; Q_{w_u} is the upward water flow rate; Q_{w_d} is the downward water flow rate; Q_F is the feed flow rate; Q_T is the tailing flow rate; U_a, U_{w_u}, U_{w_d}, U_F, U_T are the velocities of particles with air, upward water, downward water, feed, and tailing phase, respectively. Ac is the cross-sectional area of the column, V is the volume of the collection region, H_a is the holdup of the air phase, H_w is the holdup of the water phase, α is the particle–bubble attachment-rate parameter, β is the particle–bubble detachment rate parameter, and Av is the air–water interfacial area per unit volume of the column. The initial conditions for the ODE model of the collection region are given by

$$C_a(0) = C_{a0}, \quad C_w(0) = C_{w0}. \tag{3}$$

2.1.2. Model for Froth Region

The froth region can be considered as a PFR, which means that the material is perfectly mixed perpendicular to the direction of flow but is not mixed along the flow direction. This region is modeled by a set of transport hyperbolic PDEs. An upward water phase is added in this region because of the bubble entrainment. The state variables for the process of the froth region are mass concentrations of solid particles (mineral, locked, and gangue) with the air phase (C_a^F), downward water phase (C_{w_d}), and upward water phase (C_{w_u}).

$$\begin{aligned}\frac{\partial(H_a C_a^F(z,t))}{\partial t} &= -\frac{\partial(U_a C_a^F(z,t))}{\partial z} + \alpha Av f C_{w_d}(z,t) + \sigma Av f C_{w_u}(z,t) - \beta C_a^F(z,t), &(4)\\ \frac{\partial(H_{w_d} C_{w_d}(z,t))}{\partial t} &= \frac{\partial[(U_{w_d} + H_{w_d} U_s) C_{w_d}(z,t)]}{\partial z} - \alpha Av f C_{w_d}(z,t) + \rho C_{w_u}(z,t) + k\beta C_a^F(z,t), &(5)\\ \frac{\partial(H_{w_u} C_{w_u}(z,t))}{\partial t} &= -\frac{\partial(U_{w_u} C_{w_u}(z,t))}{\partial z} - \sigma Av f C_{w_u}(z,t) - \rho C_{w_u}(z,t) + (1-k)\beta C_a^F(z,t). &(6)\end{aligned}$$

The letter F is marked in the upper right corner of the parameter C_a to indicate that it is a froth region parameter that can be easily distinguished from the collection region parameter. The term $\alpha Av f C_{w_d}$ represents the transfer of particles from the downward water flow to the bubble, $\sigma Av f C_{w_u}$ represents the transfer of particles from the upward water flow to the bubble, βC_a represents the particles' detachment from the bubble, and ρC_{w_u} represents the transfer of particles from the upward water flow to the downward water flow. This transport-reaction model belongs to the class of fully state-coupled heterodirectional transport systems (transporting velocities have opposite signs).
The boundary and initial conditions for the PED model of the froth region are given by the following:

$$\begin{aligned}C_a^F(0,t) &= C_a(t), \quad C_{w_u}(0,t) = C_w(t), \quad C_{w_d}(h,t) = 0. &(7)\\ C_a^F(z,0) &= f_a(z), \quad C_{w_u}(z,0) = f_{w_u}(z), \quad C_{w_d}(z,0) = f_{w_d}(z). &(8)\end{aligned}$$

In the system given by Equations (1)–(8), the ODE system provides boundary conditions for the PDE system. The froth overflow (production) is controlled by the velocity of the feed; that is, the velocity of the feed U_F is the control input and the mass concentration of solid particles with the air phase of froth overflow $C_a^F(h,t)$ is the output.

2.2. Linearized Model

The system described by Equations (1)–(8) is nonlinear, and it is essential for it to be linearized for further analysis (taking the mineral, e.g., the steady-state profiles of the mineral within three phases are illustrated in Figure 2). C_{as}, $C_{w_d s}$, $C_{w_u s}$, and U_{F0} are defined at steady state. With the consideration of steady-state conditions, defining the variables $C_a(t) = C_{as}(0) + x_{ab}(t)$, $C_w(t) = C_{w_u s}(0) + x_{wb}(t)$, $C_a^F(z,t) = C_{as}(z) + x_a(z,t)$, $C_{w_d}(z,t) = C_{w_d s}(z) + x_{w_d}(z,t)$, $C_{w_u}(z,t) = C_{w_u s}(z) + x_{w_u}(z,t)$, and $U_F = U_{F0} + u(t)$, one can obtain the following linear coupled PDE–ODEs:

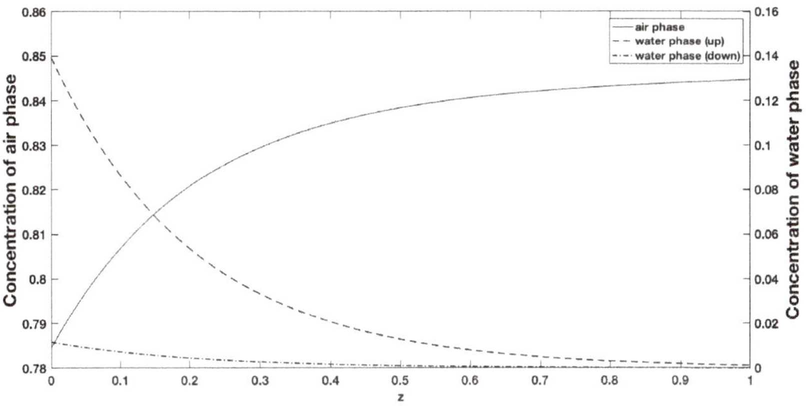

Figure 2. Steady-state profile of mineral.

$$\frac{\partial}{\partial t}\begin{bmatrix} x_a(z,t) \\ x_{w_d}(z,t) \\ x_{w_u}(z,t) \end{bmatrix} = \begin{bmatrix} -m_1\frac{\partial}{\partial z}+J_{11}(z) & J_{12}(z) & J_{13}(z) \\ J_{21}(z) & m_2\frac{\partial}{\partial z}+J_{22}(z) & J_{23}(z) \\ J_{31}(z) & J_{32}(z) & -m_3\frac{\partial}{\partial z}+J_{33}(z) \end{bmatrix}\begin{bmatrix} x_a(z,t) \\ x_{w_d}(z,t) \\ x_{w_u}(z,t) \end{bmatrix}, \tag{9}$$

$$\frac{d}{dt}\begin{bmatrix} x_{ab}(t) \\ x_{wb}(t) \end{bmatrix} = \begin{bmatrix} b_{11} & b_{12} \\ b_{21} & b_{22} \end{bmatrix}\begin{bmatrix} x_{ab}(t) \\ x_{wb}(t) \end{bmatrix} + \begin{bmatrix} 0 \\ \frac{C_F}{H_w l} \end{bmatrix} u(t) + \begin{bmatrix} 0 \\ \frac{U_{w_d}}{H_w l} \end{bmatrix} x_{w_d}(0,t), \tag{10}$$

with the following boundary conditions and initial conditions:

$$x_a(0,t) = x_{ab}(t), \quad x_{w_u}(0,t) = x_{wb}(t), \quad x_{w_d}(h,t) = 0; \tag{11}$$

$$x_{ab}(0) = x_{ab0}, \quad x_{wb}(0) = x_{wb0}, \quad x_a(z,0) = x_{a0}, \quad x_{w_u}(z,0) = x_{w_u 0}, \quad x_{w_d}(z,0) = x_{w_d 0}; \tag{12}$$

where $l = \frac{V}{Ac}$, $m_1 = \frac{U_a}{H_a}$, $m_2 = \frac{U_{w_d}+H_{w_d}U_s}{H_{w_d}}$, $m_3 = \frac{U_{w_u}}{H_{w_u}}$, $b_{11} = -\beta - \frac{U_{a_b}}{H_{a_b} l} - \frac{\alpha A v H_w}{H_{a_b} C_a^*} C_{w_u s}(0)$, $b_{12} = \frac{\alpha A v H_w}{H_{a_b}} - \frac{\alpha A v H_w}{H_{a_b} C_a^*} C_{as}(0)$, $b_{21} = \frac{\beta H_{a_b}}{H_w} + \frac{\alpha A v}{C_a^*} C_{w_u s}(0)$, and $b_{22} = -\alpha A v - \frac{U_T}{H_w l} - \frac{U_{w_u}}{H_w l} + \frac{\alpha A v}{C_a^*} C_{as}(0)$. $J_{ij}(z)$ $(i = 1,2,3;\ j = 1,2,3)$ is the Jacobian of the nonlinear term evaluated at steady state.

$$\begin{aligned} J :=& \begin{bmatrix} J_{11}(z) & J_{12}(z) & J_{13}(z) \\ J_{21}(z) & J_{22}(z) & J_{23}(z) \\ J_{31}(z) & J_{32}(z) & J_{33}(z) \end{bmatrix} \\ =& \begin{bmatrix} -\frac{\beta}{H_a} - \frac{\alpha A v}{H_a C_a^*} C_{w_d s}(z) - \frac{\sigma A v}{H_a C_a^*} C_{w_u s}(z) & \frac{\alpha A v}{H_a} - \frac{\alpha A v}{H_a C_a^*} C_{as}(z) & \frac{\sigma A v}{H_a} - \frac{\sigma A v}{H_a C_a^*} C_{as}(z) \\ \frac{k\beta}{H_{w_d}} + \frac{\alpha A v}{H_{w_d} C_a^*} C_{w_d s}(z) & -\frac{\alpha A v}{H_{w_d}} + \frac{\alpha A v}{H_{w_d} C_a^*} C_{as}(z) & \frac{\rho}{H_{w_d}} \\ \frac{(1-k)\beta}{H_{w_u}} + \frac{\sigma A v}{H_{w_u} C_a^*} C_{w_u s}(z) & 0 & -\frac{\sigma A v}{H_{w_u}} + \frac{\sigma A v}{H_{w_u} C_a^*} C_{as}(z) - \frac{\rho}{H_{w_u}} \end{bmatrix} \end{aligned} \tag{13}$$

The interconnection of the hyperbolic PDE system and ODE system can be considered as the boundary-controlled hyperbolic PDE system (see Figure 3). The state transformation that transfers the boundary actuation to in-domain actuation is given as $x_a(z,t) = \bar{x}_a(z,t) + B_1(z)x_a(0,t)$, and $x_{w_u}(z,t) = \bar{x}_{w_u}(z,t) + B_2(z)x_{w_u}(0,t)$, with $\bar{x}_a(0,t) = 0$, $\bar{x}_{w_u}(0,t) = 0$, $B_1(0) = 1$, and $B_2(0) = 1$. Equations (9) and (10) become

$$\begin{aligned}
\frac{\partial \bar{x}_a(z,t)}{\partial t} &= -m_1 \frac{\partial \bar{x}_a(z,t)}{\partial z} + J_{11}(z)\bar{x}_a(z,t) + J_{12}(z)x_{w_d}(z,t) + J_{13}(z)\bar{x}_{w_u}(z,t) \\
&\quad + [J_{13}(z)B_2(z) - b_{12}B_1(z)]x_{w_u}(0,t) + [-m_1\frac{dB_1(z)}{dz} - b_{11}B_1(z) + J_{11}(z)B_1(z)]x_a(0,t), \quad (14)\\
\frac{\partial x_{w_d}(z,t)}{\partial t} &= m_2 \frac{\partial x_{w_d}(z,t)}{\partial z} + J_{21}(z)\bar{x}_a(z,t) + J_{22}(z)x_{w_d}(z,t) + J_{23}(z)\bar{x}_{w_u}(z,t) \\
&\quad + [J_{21}(z)B_1(z)x_a(0,t) + J_{23}(z)B_2(z)x_{w_u}(0,t)], \quad (15)\\
\frac{\partial \bar{x}_{w_u}(z,t)}{\partial t} &= -m_3 \frac{\partial \bar{x}_{w_u}(z,t)}{\partial z} + J_{31}(z)\bar{x}_a(z,t) + J_{32}(z)x_{w_d}(z,t) + J_{33}(z)\bar{x}_{w_u}(z,t) \\
&\quad + [J_{31}(z)B_1(z) - b_{21}B_2(z)]x_a(0,t) + [-m_3\frac{dB_2(z)}{dz} - b_{22}B_2(z) + J_{33}(z)B_2(z)]x_{w_u}(0,t) \\
&\quad - B_2(z)\frac{U_{w_d}}{H_{wl}}x_{w_d}(0,t) - B_2(z)\frac{C_F}{H_{wl}}u(t), \quad (16)\\
\frac{dx_{ab}(t)}{dt} &= b_{11}x_{ab}(t) + b_{12}x_{wb}(t), \quad (17)\\
\frac{dx_{wb}(t)}{dt} &= b_{21}x_{ab}(t) + b_{22}x_{wb}(t) + \frac{U_{w_d}}{H_{wl}}x_{w_d}(0,t) + \frac{C_F}{H_{wl}}u(t), \quad (18)
\end{aligned}$$

with the following boundary conditions and initial conditions:

$$\bar{x}_a(0,t) = 0, \quad \bar{x}_{w_u}(0,t) = 0, \quad x_{w_d}(h,t) = 0; \quad (19)$$

$$x_{ab}(0) = x_{ab0}, \quad x_{wb}(0) = x_{wb0}, \quad \bar{x}_a(z,0) = \bar{x}_{a0}, \quad \bar{x}_{w_u}(z,0) = \bar{x}_{w_u0}, \quad x_{w_d}(z,0) = x_{w_d0}; \quad (20)$$

with $\bar{x}_{a0} = x_a(z,0) - B_1(z)x_{ab}(0)$ and $\bar{x}_{w_u0} = x_{w_u}(z,0) - B_2(z)x_{wb}(0)$. We consider the conditions $-m_1\frac{dB_1(z)}{dz} - b_{11}B_1(z) + J_{11}(z)B_1(z) = 0$ and $-m_3\frac{dB_2(z)}{dz} - b_{22}B_2(z) + J_{33}(z)B_2(z) = 0$ and solve for the $B_1(z)$ and $B_2(z)$ expressions, which simplifies the system given by Equations (14)–(18) as follows:

$$\begin{aligned}
\frac{\partial \bar{x}_a(z,t)}{\partial t} &= -m_1 \frac{\partial \bar{x}_a(z,t)}{\partial z} + J_{11}(z)\bar{x}_a(z,t) + J_{12}(z)x_{w_d}(z,t) + J_{13}(z)\bar{x}_{w_u}(z,t) \\
&\quad + [J_{13}(z)B_2(z) - b_{12}B_1(z)]x_{wb}(t), \quad (21)\\
\frac{\partial x_{w_d}(z,t)}{\partial t} &= m_2 \frac{\partial x_{w_d}(z,t)}{\partial z} + J_{21}(z)\bar{x}_a(z,t) + J_{22}(z)x_{w_d}(z,t) + J_{23}(z)\bar{x}_{w_u}(z,t) \\
&\quad + J_{21}(z)B_1(z)x_{ab}(t) + J_{23}(z)B_2(z)x_{wb}(t), \quad (22)\\
\frac{\partial \bar{x}_{w_u}(z,t)}{\partial t} &= -m_3 \frac{\partial \bar{x}_{w_u}(z,t)}{\partial z} + J_{31}(z)\bar{x}_a(z,t) + J_{32}(z)x_{w_d}(z,t) + J_{33}(z)\bar{x}_{w_u}(z,t) \\
&\quad + [J_{31}(z)B_1(z) - b_{21}B_2(z)]x_{ab}(t) - B_2(z)\frac{U_{w_d}}{H_{wl}}x_{w_d}(0,t) - B_2(z)\frac{C_F}{H_{wl}}u(t), \quad (23)\\
\frac{dx_{ab}(t)}{dt} &= b_{11}x_{ab}(t) + b_{12}x_{wb}(t), \quad (24)\\
\frac{dx_{wb}(t)}{dt} &= b_{21}x_{ab}(t) + b_{22}x_{wb}(t) + \frac{U_{w_d}}{H_{wl}}x_{w_d}(0,t) + \frac{C_F}{H_{wl}}u(t). \quad (25)
\end{aligned}$$

We consider the extended state $x \in H \bigoplus R^n$, where H is a real Hilbert space $L_2(0,1)$ with the inner product $< \cdot, \cdot >$ and R^n is a real space. The input $u(t) \in U$; the disturbance $g(t) \in G$; and the output $y(t) \in Y$, U, G, and Y are real Hilbert spaces. The system given by Equations (21)–(25) can be expressed as the equivalent state-space description:

$$\dot{x}(t) \quad = \quad Ax(t) + Bu(t) + Eg(t), \tag{26}$$

where the system state is $x(\cdot,t) = \left[\bar{x}_a(z,t), x_{w_d}(z,t), \bar{x}_{w_u}(z,t), x_{ab}(t), x_{wb}(t)\right]^T$ and the disturbance is $g(t) = x_{w_d}(0,t)$. The system operator A is defined on its domain as $D(A) = \left\{x = [x_1, x_2, x_3, x_4, x_5]^T \in H : x \text{ is a.c.}, \frac{dx}{dz} \in H, x_1(0) = 0, x_2(h) = 0, x_3(0) = 0\right\}$ by

$$A = \begin{bmatrix} A_F & A_O \\ 0 & A_C \end{bmatrix} \tag{27}$$

$$= \begin{bmatrix} -m_1\frac{\partial}{\partial z} + J_{11}(z) & J_{12}(z) & J_{13}(z) & 0 & J_{13}(z)B_2(z) - b_{12}B_1(z) \\ J_{21}(z) & m_2\frac{\partial}{\partial z} + J_{22}(z) & J_{23}(z) & J_{21}(z)B_1(z) & J_{23}(z)B_2(z) \\ J_{31}(z) & J_{32}(z) & -m_3\frac{\partial}{\partial z} + J_{33}(z) & J_{31}(z)B_1(z) - b_{21}B_2(z) & 0 \\ 0 & 0 & 0 & b_{11} & b_{12} \\ 0 & 0 & 0 & b_{21} & b_{22} \end{bmatrix}.$$

The input, disturbance, and output operators are given by $B = \begin{bmatrix} 0 \\ 0 \\ -B_2(z)\frac{C_F}{H_w l} \\ 0 \\ \frac{C_F}{H_w l} \end{bmatrix}$ and $E = \begin{bmatrix} 0 \\ 0 \\ -B_2(z)\frac{U_{w_d}}{H_w l} \\ 0 \\ \frac{U_{w_d}}{H_w l} \end{bmatrix}$.

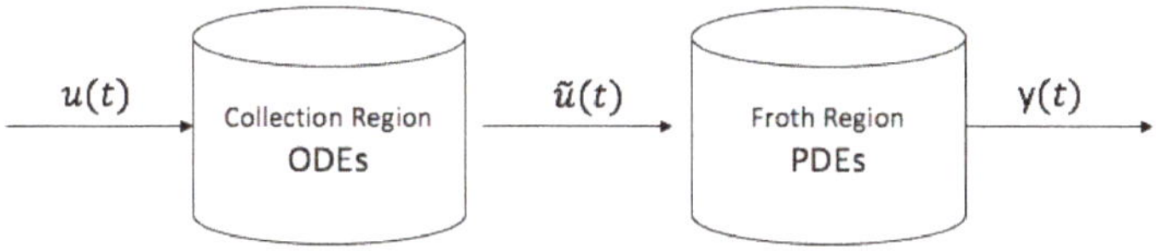

Figure 3. Schematic representation of coupled partial differential equations (PDEs) and ordinary differential equations (ODEs) system connected through boundary.

2.3. Discretized Model

For the discrete controller design, a discretized model is required. Cayley–Tustin time discretization transformation is used to obtain discrete models without consideration of spatial discretization and/or without any other type of model spatial approximation.

2.3.1. Time Discretization for Linear PDE

The linear infinite-dimensional system is described by the following state-space system:

$$\begin{aligned} \dot{x}(z,t) &= Ax(z,t) + Bu(t), \quad x(z,0) = x_0; \\ y(t) &= Cx(z,t) + Du(t). \end{aligned} \tag{28}$$

The operator $A : D(A) \subset H \to H$ is a generator of a C_0-semigroup on H and has a Yoshida extension operator A_{-1}. B, C, and D are linear operators associated with the actuation and output measurement or direct feed-forward element; that is, $B \subset L(U,H)$, $C \subset L(H,Y)$, and $D \subset L(U,Y)$. Given a time discretization parameter $d > 0$, the Tustin time discretization is given by the following [24]:

$$\begin{aligned} \frac{x(jd) - x((j-1)d)}{d} &\approx A\frac{x(jd) + x((j-1)d)}{2} + Bu(jd), \\ y(jd) &\approx C\frac{x(jd) + x((j-1)d)}{2} + Du(jd), \\ x(0) &= x_0. \end{aligned} \tag{29}$$

We let $u_j^d/\sqrt{d}$ be an approximation to $u(jd)$, under the assumptions that $y_j^d/\sqrt{d}$ converges to $y(jd)$ as $d \to 0$. Then, one obtains discrete time dynamics as follows:

$$\begin{aligned} \frac{x_j^d - x_{j-1}^d}{d} &= A\frac{x_j^d + x_{j-1}^d}{2} + Bu_j^d/\sqrt{d}, \\ y_j^d/\sqrt{d} &= C\frac{x_j^d + x_{j-1}^d}{2} + Du_j^d/\sqrt{d}, \\ x_0^d &= x_0. \end{aligned} \tag{30}$$

After some calculations, one obtains the form of a discrete system:

$$\begin{aligned} x(z,k) &= A_d x(z,k-1) + B_d u(k), \quad x(z,0) = x_0; \\ y(k) &= C_d x(z,k-1) + D_d u(k). \end{aligned} \tag{31}$$

Here $\delta = 2/d$, and the operators A_d, B_d, C_d, and D_d are given by

$$\begin{bmatrix} A_d & B_d \\ C_d & D_d \end{bmatrix} = \begin{bmatrix} [\delta - A]^{-1}[\delta + A] & \sqrt{2\delta}[\delta - A]^{-1}B \\ \sqrt{2\delta}C[\delta - A]^{-1} & G(\delta) \end{bmatrix}, \tag{32}$$

where $G(\delta)$ denotes the transfer function of the system, $G(\delta) = C[\delta - A]^{-1}B + D$. Operator A_d can be expressed as $A_d = -I + 2\delta[\delta - A]^{-1}$, where I is the identity operator. By introducing the affine disturbance input, the time discretization for the most general form:

$$\begin{aligned} \dot{x}(z,t) &= Ax(z,t) + Bu(t) + Eg(t), \quad x(z,0) = x_0, \\ y(t) &= Cx(z,t) + Du(t) + Fg(t), \end{aligned} \tag{33}$$

is given by [25]; the corresponding discrete operator of the linear operators $E \subset L(R^n, H)$ and $F \subset L(R^n, Y)$ are $E_d = \sqrt{2\delta}[\delta - A]^{-1}E$ and $F_d = C[\delta - A]^{-1}E + F$.

2.3.2. Time Discretization of Column Flotation System

From the previous section, one can find resolvent $R(\delta, A) = [\delta - A]^{-1}$ of the column operator A (Equation (27)), which provides discrete operators (A_d, B_d, C_d, and D_d) to be easily realized. Returning to the system given by Equation (26), the resolvent operator can be obtained by taking a Laplace transform. After the Laplace transform, one obtains

$$\frac{\partial \bar{x}_a(z,s)}{\partial z} = \frac{1}{m_1}(J_{11}(z)-s)\bar{x}_a(z,s)+\frac{1}{m_1}J_{12}(z)x_{w_d}(z,s)+\frac{1}{m_1}J_{13}(z)\bar{x}_{w_u}(z,s) + \frac{1}{m_1}[J_{13}(z)B_2(z)-b_{12}B_1(z)]x_{w_u}(0,s)+\frac{1}{m_1}\bar{x}_a(z,0), \quad (34)$$

$$\frac{\partial x_{w_d}(z,s)}{\partial z} = -\frac{1}{m_2}J_{21}(z)\bar{x}_a(z,s)-\frac{1}{m_2}(J_{22}(z)-s)x_{w_d}(z,s)-\frac{1}{m_2}J_{23}(z)\bar{x}_{w_u}(z,s) - \frac{1}{m_2}J_{21}(z)B_1(z)x_a(0,s)-\frac{1}{m_2}J_{23}(z)B_2(z)x_{w_u}(0,s)-\frac{1}{m_2}x_{w_d}(z,0), \quad (35)$$

$$\frac{\partial \bar{x}_{w_u}(z,s)}{\partial z} = \frac{1}{m_3}J_{31}(z)\bar{x}_a(z,s)+\frac{1}{m_3}J_{32}(z)x_{w_d}(z,s)+\frac{1}{m_3}(J_{33}(z)-s)\bar{x}_{w_u}(z,s) + \frac{1}{m_3}[J_{31}(z)B_1(z)-b_{21}B_2(z)]x_a(0,s)+\frac{1}{m_3}\bar{x}_{w_u}(z,0), \quad (36)$$

$$x_{ab}(s) = \frac{s-b_{22}}{b}x_{ab}(0)+\frac{b_{12}}{b}x_{wb}(0), \quad (37)$$

$$x_{wb}(s) = \frac{b_{21}}{b}x_{ab}(0)+\frac{s-b_{11}}{b}x_{wb}(0), \quad (38)$$

where $b=(s-b_{11})(s-b_{22})-b_{21}b_{12}$.

By solving this ODE, one obtains

$$\begin{bmatrix}\bar{x}_a(z,s)\\ x_{w_d}(z,s)\\ \bar{x}_{w_u}(z,s)\end{bmatrix}=e^{\bar{A}z}\begin{bmatrix}\bar{x}_a(0,s)\\ x_{w_d}(0,s)\\ \bar{x}_{w_u}(0,s)\end{bmatrix}+\int_0^z e^{\bar{A}(z-\eta)}\begin{bmatrix}\frac{1}{m_1}\bar{x}_a(\eta,0)+\frac{1}{m_1}J_{a1}(\eta)x_{ab}(0)+\frac{1}{m_1}J_{a2}(\eta)x_{wb}(0)\\ -\frac{1}{m_2}x_{wd}(\eta,0)-\frac{1}{m_2}J_{d1}(\eta)x_{ab}(0)-\frac{1}{m_2}J_{d2}(\eta)x_{wb}(0)\\ \frac{1}{m_3}\bar{x}_{w_u}(\eta,0)+\frac{1}{m_3}J_{u1}(\eta)x_{ab}(0)+\frac{1}{m_3}J_{u2}(\eta)x_{wb}(0)\end{bmatrix}d\eta, \quad (39)$$

where $J_{a1}(z) = \frac{b_{21}}{b}[J_{13}(z)B_2(z)-b_{12}B_1(z)]$, $J_{a2}(z) = \frac{s-b_{11}}{b}[J_{13}(z)B_2(z)-b_{12}B_1(z)]$, $J_{d1}(z) = \frac{s-b_{22}}{b}J_{21}(z)B_1(z)+\frac{b_{21}}{b}J_{23}(z)B_2(z)$, $J_{d2}(z) = \frac{b_{12}}{b}J_{21}(z)B_1(z)+\frac{s-b_{11}}{b}J_{23}(z)B_2(z)$, $J_{u1}(z) = \frac{s-b_{22}}{b}[J_{31}(z)B_1(z)-b_{21}B_2(z)]$, and $J_{u2}(z) = \frac{b_{12}}{b}[J_{31}(z)B_1(z)-b_{21}B_2(z)]$. For simplicity, we let J_{ij}, which is defined as a spatial average of $J_{ij}(z)$, replace $J_{ij}(z)$, so that the matrix exponential of $\bar{A}$ can be computed directly. Then, $\bar{A}=\begin{bmatrix}\frac{1}{m_1}(J_{11}-s) & \frac{1}{m_1}J_{12} & \frac{1}{m_1}J_{13}\\ -\frac{1}{m_2}J_{21} & -\frac{1}{m_2}(J_{22}-s) & -\frac{1}{m_2}J_{23}\\ \frac{1}{m_3}J_{31} & \frac{1}{m_3}J_{32} & \frac{1}{m_3}(J_{33}-s)\end{bmatrix}$.

With the boundary conditions given by Equation (19), the resolvent of operator A can be expressed as follows:

$$R(s,A) = [sI-A]^{-1}x(z,0) = \begin{bmatrix}R_{11} & R_{12} & R_{13} & R_{14} & R_{15}\\ R_{21} & R_{22} & R_{23} & R_{24} & R_{25}\\ R_{31} & R_{32} & R_{33} & R_{34} & R_{35}\\ 0 & 0 & 0 & R_{44} & R_{45}\\ 0 & 0 & 0 & R_{54} & R_{55}\end{bmatrix}x(z,0), \quad (40)$$

where R_{11}, R_{12}, ..., R_{55} are shown in Appendix A. Then, the discretized model of the column flotation process takes the following form:

$$x(z,k) = A_d x(z,k-1)+B_d u(k)+E_d g(k). \quad (41)$$

3. Model Predictive Control Design

In this work, the model for the column flotation process uses coupled PDEs and ODEs with input disturbance $g(t)$; the continuous linearized model described in Equation (26) can be rewritten as

$$\dot{x}(t) \quad = \quad Ax(t) + B\bar{u}(t), \tag{42}$$

where $\bar{u}(t) = u(t) + \frac{U_{w_d}}{C_F} g(t)$. The discrete version is obtained by applying Cayley–Tustin discretization as follows:

$$x(z,k) \quad = \quad A_d x(z,k-1) + B_d \bar{u}(k), \tag{43}$$

where we have adopted $x(z,k)$ notation to denote spatial characteristics of the extended state $x(t)$. The model predictive controller was developed as a solution of the optimization problem by minimizing the following open-loop performance objective function across the length of the infinite horizon at sampling time k [26] on the basis of the above system given by Equation (43) without disturbances being present.

$$\begin{aligned} \min_{\bar{u}^N} \quad & \sum_{j=0}^{\infty} [< x(z,k+j|k), Qx(z,k+j|k) > \\ & + < \bar{u}(k+j+1|k), R\bar{u}(k+j+1|k) >] \\ s.t. \quad & x(z,k+j|k) = A_d x(z,k+j-1|k) + B_d \bar{u}(k+j|k)] \\ & \bar{u}^{min} \leq \bar{u}(k+j|k) \leq \bar{u}^{max} \\ & x^{min} \leq x(z,k+j|k) \leq x^{max}, \end{aligned} \tag{44}$$

where Q is a symmetric positive semidefinite matrix and R is a symmetric positive definite matrix. The infinite-horizon open-loop objective function in Equation (44) can be expressed as the finite-horizon open-loop objective function with $u(k+N+1|k) = 0$, as below:

$$\begin{aligned} \min_{\bar{u}^N} \quad & J = \sum_{j=0}^{N-1} [< x(z,k+j|k), Qx(z,k+j|k) > \\ & + < \bar{u}(k+j+1|k), R\bar{u}(k+j+1|k) >] \\ & + < x(z,k+N|k), \bar{Q}x(z,k+N|k) > \\ s.t. \quad & x(z,k+j|k) = A_d x(z,k+j-1|k) + B_d \bar{u}(k+j|k) \\ & \bar{u}^{min} \leq \bar{u}(k+j|k) \leq \bar{u}^{max} \\ & x^{min} \leq x(z,k+j|k) \leq x^{max}, \end{aligned} \tag{45}$$

where $\bar{Q}$ is defined as the infinite sum $\bar{Q} = \sum_{i=0}^{\infty} A_d^{*i} C_d^* Q C_d A_d^i$. This terminal state penalty operator $\bar{Q}$ can be calculated from the solution of the following discrete Lyapunov function:

$$A_d^* \bar{Q} A_d - \bar{Q} = -C_d^* Q C_d. \tag{46}$$

It can be noticed that operator A_d in the equation is applied to some function and that the same holds for C_d; thus to derive $\bar{Q}$ directly from Equation (46) is not a feasible task. However, the unique solution of the discrete Lyapunov function can be related to the solution of the continuous Lyapunov function, which can be solved uniquely. Therefore, $\bar{Q}$ can be obtained by solving the continuous Lyapunov function:

$$A^* \bar{Q} + \bar{Q} A = -C^* Q C, \quad \bar{Q} \in D(A^*). \tag{47}$$

Proof. We establish a link between the continuous and discrete Lyapunov functions. If the continuous Lyapunov function holds, defining $A_d := -I + 2\delta[\delta - A]^{-1}$ and $A_d^* := [-I + 2\delta[\delta - A]^{-1}]^*$, then

$$\begin{aligned} A_d^*\bar{Q}A_d - \bar{Q} &= \left[-I+2\delta(\delta - A)^{-1}\right]^* Q\left[-I+2\delta(\delta - A)^{-1}\right] - \bar{Q} \\ &= (\delta - A)^{-1*}[[-(\delta - A)+2\delta]^* Q[-(\delta - A)+2\delta] \\ &- (\delta - A)^* Q(\delta - A)](\delta - A)^{-1} \\ &= (\delta - A)^{-1*}\,[2A^*\bar{Q}\delta + 2\delta\bar{Q}A]\,(\delta - A)^{-1} \\ &= (\delta - A)^{-1*}\,[2\delta(-C^*QC)]\,(\delta - A)^{-1} \\ &= -(\sqrt{2\delta}C\,[\delta - A]^{-1})^* Q(\sqrt{2\delta}C\,[\delta - A]^{-1}) \\ &= -C_d^* Q C_d, \end{aligned}$$

such that the unique solution of the continuous Lyapunov function (Equation (51)) is also a solution of the discrete Lyapunov function ((Equation 46)). □

Because of the fact that $\bar{u}(t) = u(t) + \frac{U_{w_d}}{C_F} g(t)$, straightforward algebraic manipulation of the objective function presented in Equation (45) results in the following program:

$$\begin{aligned} \min_U J &= U^T < I, H > U + 2U^T[< I, Px(z,k|k) > + < I, R\frac{U_{w_d}}{C_F}G >] \\ &+ < x(z,k|k), \bar{Q}x(z,k|k) > + < \frac{U_{w_d}}{C_F}G, R\frac{U_{w_d}}{C_F}G >, \end{aligned} \quad (48)$$

where $U = [u(k+1|k), u(k+2|k), ..., u(k+N|k)]^T$, $G = [g(k+1|k), g(k+2|k), ..., g(k+N|k)]^T$, and

$$H = \begin{bmatrix} B_d^*\bar{Q}B_d + R & B_d^*A_d^*\bar{Q}B_d & \cdots & B_d^*A_d^{*N-1}\bar{Q}B_d \\ B_d^*\bar{Q}A_dB_d & B_d^*\bar{Q}B_d + R & \cdots & B_d^*A_d^{*N-2}\bar{Q}B_d \\ \vdots & \vdots & \ddots & \vdots \\ B_d^*\bar{Q}A_d^{N-1}B_d & B_d^*\bar{Q}A_d^{N-2}B_d & \cdots & B_d^*\bar{Q}B_d + R \end{bmatrix}, P = \begin{bmatrix} B_d^*\bar{Q}A_d \\ B_d^*\bar{Q}A_d^2 \\ \vdots \\ B_d^*\bar{Q}A_d^N \end{bmatrix}.$$

The objective function is subjected to the following constraints:

$$\begin{aligned} &\bar{U}^{min} \le U + \frac{U_{w_d}}{C_F}G \le \bar{U}^{max}, \\ &X^{min} \le S(U + \frac{U_{w_d}}{C_F}G) + Tx(z,k|k) \le X^{max}. \end{aligned} \quad (49)$$

That is,

$$\begin{bmatrix} I \\ -I \\ S \\ -S \end{bmatrix} U \le \begin{bmatrix} \bar{U}^{max} - \frac{U_{w_d}}{C_F}G \\ -\bar{U}^{min} + \frac{U_{w_d}}{C_F}G \\ X^{max} - Tx(z,k|k) - S\frac{U_{w_d}}{C_F}G \\ -X^{min} + Tx(z,k|k) + S\frac{U_{w_d}}{C_F}G \end{bmatrix}, \quad (50)$$

where $S = \begin{bmatrix} B_d & 0 & \cdots & 0 \\ A_dB_d & B_d & \cdots & 0 \\ \vdots & \vdots & \ddots & \vdots \\ A_d^{N-1}B_d & A_d^{N-2}B_d & \cdots & B_d \end{bmatrix}$ and $T = \begin{bmatrix} A_d \\ A_d^2 \\ \vdots \\ A_d^N \end{bmatrix}$.

The optimization problem described in Equation (48) is a standard finite-dimensional quadratic optimization problem, as inner products in Equation (48) are integrations over the spatial components in the cost function.

Remark 1. *The model predictive controller design in this paper uses the system state $x(z,t)$; therefore, it is necessary to design a discrete observer to reconstruct the system state. At present, the design of the continuous system observer is very mature, and it is feasible to design a discrete observer on this basis of the discrete infinite-dimensional model.*

4. Simulation Results

In this section, the performance of the proposed model predictive control to keep the output at the steady state within the constraint range by adjusting the input is demonstrated by a comparison between high-fidelity numerical simulations of open-loop and controlled system responses.

In the simulations, the values of the system parameters were as given in Table 1. The time discretization parameter was chosen as $d = 0.2$, which implies that $\delta = 10$ and $\Delta z = 0.01$ were chosen for the numerical integration.

Table 1. Model parameters for both collection region (C) and froth region (F).

Symbol	Description	Value
h	Height of froth region	1.0
l	Height of collection region	1.0
Ca^*	Bubble saturation parameter	5
Av	Air–water interfacial area	C: 1.0; F: 0.3
U_w	Water velocity of collection region	0.8
U_s	Settling/slip velocity	1
H_a	Air holdup	C: 0.3; F: 0.7
H_{w_d}	Downward water holdup	0.1
H_{w_u}	Upward water holdup	C: 0.7; F: 0.2
U_a	Air velocity	C: 0.1; F: 0.2
U_{w_d}	Downward water velocity	0.1
U_{w_u}	Upward water velocity	C: 0.08; F: 0.1
α	Attachment-rate parameter for downward water	C: 1.2; F: 1.0
σ	Attachment-rate parameter for upward water	1.5
β	Detachment rate parameter	C: 0.1; F: 0
ρ	Transfer rate from upward water to downward water	0.1
k	Transfer rate from air to downward water	0.01
a_1, a_2, a_3	Initial-condition coefficients	$a_1 = 0.8, a_2 = 0.1, a_3 = 0.2$

In this work, C was considered as $C = I$, which implies that the full state was available for the controller realization, and thus $C^* = C$. The above framework allows for easy extension to the discrete observer design with boundary measurements applied; that is, $C(\cdot) := \int_0^1 \delta(z-1)(\cdot)dz$. The $\bar{Q}$ can be obtained by solving the following continuous Lyapunov equation corresponding to the discrete Lyapunov Equation (46):

$$A^*\bar{Q} + \bar{Q}A = -Q. \tag{51}$$

We consider $\bar{Q} = \begin{bmatrix} \bar{Q}_1 & 0 \\ 0 & \bar{Q}_2 \end{bmatrix}$, where $\bar{Q}_1 \in H$, $\bar{Q}_2 \in R^n$, and assume $Q = \begin{bmatrix} Q_{11} & Q_{12} \\ Q_{21} & Q_{22} \end{bmatrix}$; we can obtain a set of equivalent Lyapunov equations as follows:

$$\begin{aligned} &A_F^*\bar{Q}_1 + \bar{Q}_1 A_F = -Q_{11}, \quad \bar{Q}_1 \in D(A_F^*); \\ &\bar{Q}_1 A_O = -Q_{12}; \\ &A_O^*\bar{Q}_1 = -Q_{21}; \\ &A_C^*\bar{Q}_2 + \bar{Q}_2 A_C = -Q_{22}. \end{aligned} \tag{52}$$

With the assumptions that $\bar{Q}_1 = \begin{bmatrix} \bar{q}_{1-11} & 0 & 0 \\ 0 & \bar{q}_{1-22} & 0 \\ 0 & 0 & \bar{q}_{1-33} \end{bmatrix}$ and $\bar{Q}_2 = \begin{bmatrix} \bar{q}_{2-11} & 0 \\ 0 & \bar{q}_{2-22} \end{bmatrix}$, and by choosing $Q_{11} = \begin{bmatrix} 1 & 0 & 0 \\ 0 & 1 & 0 \\ 0 & 0 & 1 \end{bmatrix}$ and $Q_{22} = \begin{bmatrix} q_{11} & q_{12} \\ q_{12} & q_{22} \end{bmatrix}$, where $q_{11} = q_{22} = 1$, one can obtain $\bar{Q}_2 = \begin{bmatrix} 0.25 & -0.688 \\ -0.688 & 1.15 \end{bmatrix}$ and $\bar{Q}_1$, as shown in Figure 4.

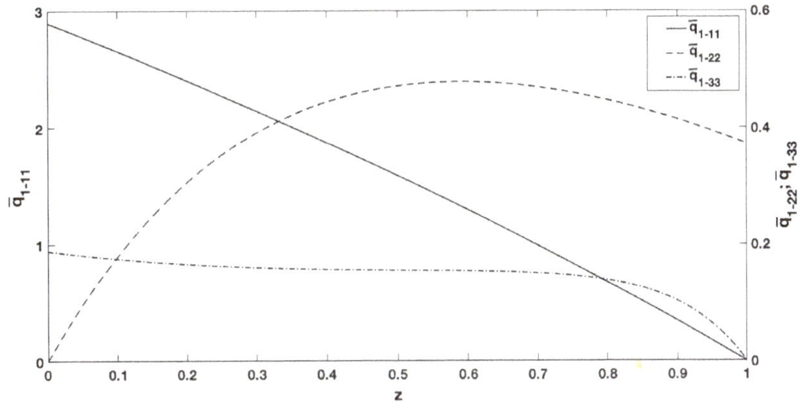

Figure 4. $\bar{Q}_1$ obtained by solving Equation (52).

In order to demonstrate the controller performance, the MPC horizon was $N = 3$, and $R = 0.1$. The input and state constraints were given as $-0.5 \leq u(t) \leq 1$ and $0 \leq x_a(h_2, t) \leq 0.83$. The initial conditions of the system given by Equation (26) were given as follows:

$$\begin{aligned} x_a(z,0) &= a_1(cos(\frac{\pi}{2}z) + sin(\pi z)), \\ x_{w_d}(z,0) &= a_2(cos(\frac{\pi}{2}z) + sin(\pi z)), \\ x_{w_u}(z,0) &= a_3(cos(\frac{\pi}{2}z) + sin(\pi z)). \end{aligned} \tag{53}$$

The performance of the proposed model predictive control can be evaluated from Figures 5–8, and the corresponding control input is given in Figure 9. From Figures 5–7, it can be seen that under the model predictive control, the system reached steady state. Figure 8 compares the output $x_a(h,k)$ evolutions under the model predictive control with state/output constraints to the model predictive control without constraints and using the open-loop system. It is clear from the figure that the application of model predictive control allows the system to reach steady state faster and that the output profile under the model predictive control law satisfies the constraints. It is important to emphasize that the system is stable and that performance under input and state constraints are of interest in this study; however, the extension to the unstable system dynamics case is easily realizable for more general classes of dynamic plants.

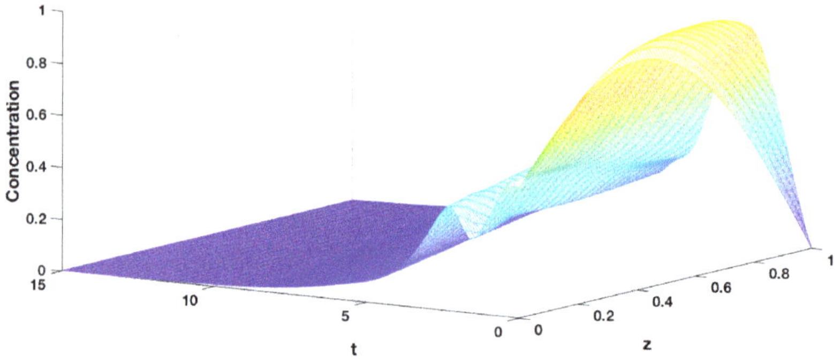

Figure 5. Profile of concentration for mineral particles with air phase under model predictive control law Equations (48) and (49).

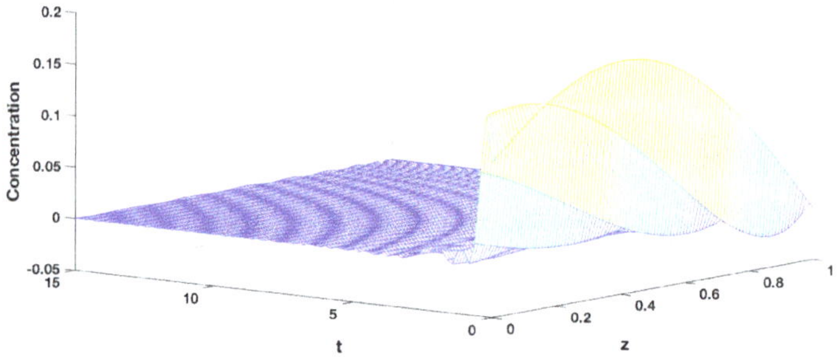

Figure 6. Profile of concentration for mineral particles with downward water phase under model predictive control law Equations (48) and (49).

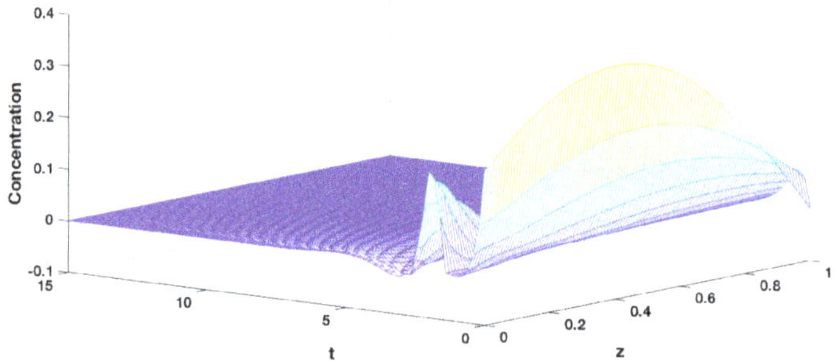

Figure 7. Profile of concentration for mineral particles with upward water phase under model predictive control law Equations (48) and (49).

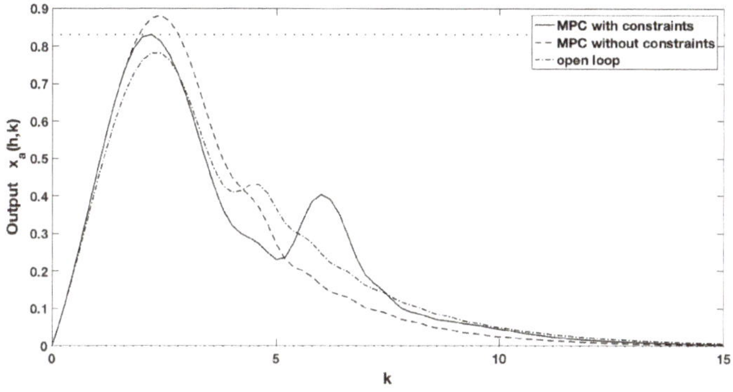

Figure 8. Profile of the state $x_a(h,k)$ (output) under model predictive control law Equations (48) and (49) (dash-dotted line) without constraints (dashed line), and the profile of an open-loop system (dash-dotted line), as well as state/output constraints (dotted line).

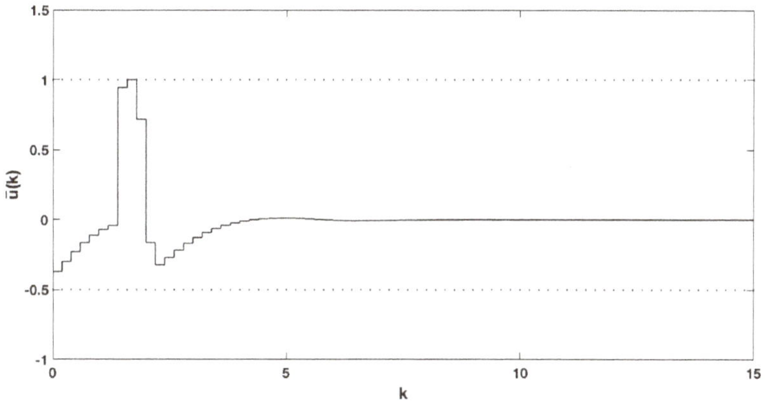

Figure 9. Input profile obtained by model predictive control law Equations (48) and (49) (solid line); input constraints are given by dotted line.

5. Conclusions

In conclusion, model predictive control algorithms are developed for the column flotation process that take into account the input and state/output constraints as well as the input disturbance. The underlying model is described by coupled nonlinear heterodirectional hyperbolic transport PDE–ODEs, and the steady-state profiles are utilized in the linearization of a nonlinear system. By using the Cayley–Tustin time discretization method, the continuous infinite-dimensional system is mapped into the discrete infinite-dimensional system without model reduction or spatial discretization. Finally, the performance of the proposed model predictive control development was demonstrated by applying it to the column flotation system, and the simulation results show that the output (the mass concentration of solid minerals with the air phase of froth overflow) is stabilized at the steady state within the constraints' physical range by adjusting the feed velocity. This optimal control realization improves the column flotation process, enabling it to operate more efficiently.

Author Contributions: Yahui Tian conducted research calculations and simulation work and wrote the first draft of the paper. Xiaoli Luan and Fei Liu supervised this work and proposed feasibility suggestions. Stevan Dubljevic provided a framework for research and guidance.

Acknowledgments: This work is supported by the National Natural Science Foundation of China (NSFC: 61773183 and 61722306), the China Scholarship Council, and the 111 Project (B12018).

Conflicts of Interest: The authors declare no conflict of interest.

Appendix A. Resolution of Operator A for Discretized Model

$$R_{11} = -\frac{\varphi_1(1,2)}{E_1(2,2)}\frac{1}{m_1}\int_0^h E_2(2,1)(\cdot)d\eta + \frac{1}{m_1}\int_0^z \varphi_2(1,1)(\cdot)d\eta, \quad \text{(A1)}$$

$$R_{12} = \frac{\varphi_1(1,2)}{E_1(2,2)}\frac{1}{m_2}\int_0^h E_2(2,2)(\cdot)d\eta - \frac{1}{m_2}\int_0^z \varphi_2(1,2)(\cdot)d\eta, \quad \text{(A2)}$$

$$R_{13} = -\frac{\varphi_1(1,2)}{E_1(2,2)}\frac{1}{m_3}\int_0^h E_2(2,3)(\cdot)d\eta + \frac{1}{m_3}\int_0^z \varphi_2(1,3)(\cdot)d\eta, \quad \text{(A3)}$$

$$\begin{aligned} R_{14} = & -\frac{\varphi_1(1,2)}{E_1(2,2)}[\frac{1}{m_1}\int_0^h E_2(2,1)J_{a1}(\eta)d\eta - \frac{1}{m_2}\int_0^h E_2(2,2)J_{d1}(\eta)d\eta + \frac{1}{m_3}\int_0^h E_2(2,3)J_{u1}(\eta)d\eta](\cdot) \\ & + [\frac{1}{m_1}\int_0^z \varphi_2(1,1)J_{a1}(\eta)d\eta - \frac{1}{m_2}\int_0^z \varphi_2(1,2)J_{d1}(\eta)d\eta + \frac{1}{m_3}\int_0^z \varphi_2(1,3)J_{u1}(\eta)d\eta)](\cdot), \end{aligned} \quad \text{(A4)}$$

$$\begin{aligned} R_{15} = & -\frac{\varphi_1(1,2)}{E_1(2,2)}[\frac{1}{m_1}\int_0^h E_2(2,1)J_{a2}(\eta)d\eta - \frac{1}{m_2}\int_0^h E_2(2,2)J_{d2}(\eta)d\eta + \frac{1}{m_3}\int_0^h E_2(2,3)J_{u2}(\eta)d\eta](\cdot) \\ & + [\frac{1}{m_1}\int_0^z \varphi_2(1,1)J_{a2}(\eta)d\eta - \frac{1}{m_2}\int_0^z \varphi_2(1,2)J_{d2}(\eta)d\eta + \frac{1}{m_3}\int_0^z \varphi_2(1,3)J_{u2}(\eta)d\eta)](\cdot), \end{aligned} \quad \text{(A5)}$$

$$R_{44} = \frac{s - b_{22}}{b}(\cdot), \quad \text{(A6)}$$

$$R_{45} = \frac{b_{12}}{b}(\cdot), \quad \text{(A7)}$$

$$R_{54} = \frac{b_{21}}{b}(\cdot), \quad \text{(A8)}$$

$$R_{55} = \frac{s - b_{11}}{b}(\cdot), \quad \text{(A9)}$$

where $\varphi_1 = e^{\bar{A}z}$, $\varphi_2 = e^{\bar{A}(z-\eta)}$, $E_1 = e^{\bar{A}h}$, and $E_2 = e^{\bar{A}(h-\eta)}$. The other R_{21}, R_{22}, ..., R_{35} values can be obtained in a similar way as for R_{11}, ..., R_{15}.

References

1. Finch, J.; Dobby, G. *Column Flotation*; Pergamon Press: Oxford, UK, 1990.
2. Kawatra, S.K. Fundamental principles of froth flotation. In *SME Mining Engineering Handbook*; Society for Mining, Metallurgy, and Exploration: Englewood, IL, USA, 2009; pp. 1517–1531.
3. Sbárbaro, D.; Del Villar, R. *Advanced Control and Supervision of Mineral Processing Plants*; Springer Science & Business Media: Berlin/Heidelberg, Germany, 2010.
4. Mathieu, G. Comparison of flotation column with conventional flotation for concentration of a mo ore. *Can. Min. Metall. Bull.* **1972**, *65*, 41–45.
5. Bergh, L.; Yianatos, J. The long way toward multivariate predictive control of flotation processes. *J. Process Control* **2011**, *21*, 226–234. [CrossRef]
6. McKee, D. Automatic flotation control—A review of 20 years of effort. *Miner. Eng.* **1991**, *4*, 653–666. [CrossRef]
7. Laurila, H.; Karesvuori, J.; Tiili, O. Strategies for instrumentation and control of flotation circuits. *Miner. Process. Plant Des. Pract. Control* **2002**, *2*, 2174–2195.
8. Bouchard, J.; Desbiens, A.; del Villar, R.; Nunez, E. Column flotation simulation and control: An overview. *Miner. Eng.* **2009**, *22*, 519–529. [CrossRef]

9. Bu, X.; Xie, G.; Peng, Y.; Chen, Y. Kinetic modeling and optimization of flotation process in a cyclonic microbubble flotation column using composite central design methodology. *Int. J. Miner. Process.* **2016**, *157*, 175–183. [CrossRef]
10. Mavros, P.; Matis, K.A. *Innovations in Flotation Technology*; Springer Science & Business Media: Berlin/Heidelberg, Germany, 2013; Volume 208.
11. Del Villar, R.; Grégoire, M.; Pomerleau, A. Automatic control of a laboratory flotation column. *Miner. Eng.* **1999**, *12*, 291–308. [CrossRef]
12. Persechini, M.A.M.; Peres, A.E.C.; Jota, F.G. Control strategy for a column flotation process. *Control Eng. Pract.* **2004**, *12*, 963–976. [CrossRef]
13. Bergh, L.; Yianatos, J.; Leiva, C. Fuzzy supervisory control of flotation columns. *Miner. Eng.* **1998**, *11*, 739–748. [CrossRef]
14. Bergh, L.; Yianatos, J.; Acuna, C.; Perez, H.; Lopez, F. Supervisory control at Salvador flotation columns. *Miner. Eng.* **1999**, *12*, 733–744. [CrossRef]
15. Mohanty, S. Artificial neural network based system identification and model predictive control of a flotation column. *J. Process Control* **2009**, *19*, 991–999. [CrossRef]
16. Maldonado, M.; Desbiens, A.; Del Villar, R. Potential use of model predictive control for optimizing the column flotation process. *Int. J. Miner. Process.* **2009**, *93*, 26–33. [CrossRef]
17. Mayne, D.Q.; Rawlings, J.B.; Rao, C.V.; Scokaert, P.O. Constrained model predictive control: Stability and optimality. *Automatica* **2000**, *36*, 789–814. [CrossRef]
18. Malinen, J. Tustin's method for final state approximation of conservative dynamical systems. *IFAC Proc. Vol.* **2011**, *44*, 4564–4569. [CrossRef]
19. Havu, V.; Malinen, J. Laplace and Cayley transforms—An approximation point of view. In Proceedings of the 44th IEEE Conference on Decision and Control, 2005 and 2005 European Control Conference (CDC-ECC'05), Seville, Spain, 15 December 2005; pp. 5971–5976.
20. Havu, V.; Malinen, J. The Cayley transform as a time discretization scheme. *Numer. Funct. Anal. Opt.* **2007**, *28*, 825–851. [CrossRef]
21. Xu, Q.; Dubljevic, S. Linear model predictive control for transport-reaction processes. *AIChE J.* **2017**, *63*, 2644–2659. [CrossRef]
22. Dobby, G.; Finch, J. Mixing characteristics of industrial flotation columns. *Chem. Eng. Sci.* **1985**, *40*, 1061–1068. [CrossRef]
23. Yianatos, J.; Finch, J.; Laplante, A. Cleaning action in column flotation froths. *Trans. Inst. Min. Metall. Sect. C-Miner. Process. Extract. Metall.* **1987**, *96*, C199–C205.
24. Franklin, G.F.; Powell, J.D.; Workman, M.L. *Digital Control of Dynamic Systems*; Addison-Wesley: Menlo Park, CA, USA, 1998; Volume 3.
25. Xu, Q.; Dubljevic, S. Modelling and control of solar thermal system with borehole seasonal storage. *Renew. Energy* **2017**, *100*, 114–128. [CrossRef]
26. Muske, K.R.; Rawlings, J.B. Model predictive control with linear models. *AIChE J.* **1993**, *39*, 262–287. [CrossRef]

mathematics

MDPI

Article

Forecast-Triggered Model Predictive Control of Constrained Nonlinear Processes with Control Actuator Faults

Da Xue and Nael H. El-Farra *

Department of Chemical Engineering, University of California, Davis, CA 95616, USA; xdxue@ucdavis.edu
* Correspondence: nhelfarra@ucdavis.edu; Tel.: +1-530-754-6919

Received: 21 May 2018; Accepted: 12 June 2018; Published: 19 June 2018

Abstract: This paper addresses the problem of fault-tolerant stabilization of nonlinear processes subject to input constraints, control actuator faults and limited sensor–controller communication. A fault-tolerant Lyapunov-based model predictive control (MPC) formulation that enforces the fault-tolerant stabilization objective with reduced sensor–controller communication needs is developed. In the proposed formulation, the control action is obtained through the online solution of a finite-horizon optimal control problem based on an uncertain model of the plant. The optimization problem is solved in a receding horizon fashion subject to appropriate Lyapunov-based stability constraints which are designed to ensure that the desired stability and performance properties of the closed-loop system are met in the presence of faults. The state-space region where fault-tolerant stabilization is guaranteed is explicitly characterized in terms of the fault magnitude, the size of the plant-model mismatch and the choice of controller design parameters. To achieve the control objective with minimal sensor–controller communication, a forecast-triggered communication strategy is developed to determine when sensor–controller communication can be suspended and when it should be restored. In this strategy, transmission of the sensor measurement at a given sampling time over the sensor–controller communication channel to update the model state in the predictive controller is triggered only when the Lyapunov function or its time-derivative are forecasted to breach certain thresholds over the next sampling interval. The communication-triggering thresholds are derived from a Lyapunov stability analysis and are explicitly parameterized in terms of the fault size and a suitable fault accommodation parameter. Based on this characterization, fault accommodation strategies that guarantee closed-loop stability while simultaneously optimizing control and communication system resources are devised. Finally, a simulation case study involving a chemical process example is presented to illustrate the implementation and evaluate the efficacy of the developed fault-tolerant MPC formulation.

Keywords: model predictive control (MPC); fault-tolerant control; networked control systems; actuator faults; chemical processes

1. Introduction

Model predictive control (MPC), also known as receding horizon control, refers to a class of optimization-based control algorithms that utilize an explicit process model to predict the future response of the plant. At each sampling time, a finite-horizon optimal control problem with a cost functional that captures the desired performance requirements is solved subject to state and control constraints, and a sequence of control actions over the optimization horizon is generated. The first part of the control inputs in the sequence is implemented on the plant, and the optimization problem is solved repeatedly at every sampling time. While developed originally in response to the specialized control needs of large-scale industrial systems, such as petroleum refineries and power plants, MPC technology now spans a broad range of application areas including chemicals, food processing, automotive, and aerospace applications (see, for example, [1]). Motivated by the advantages of MPC, such as constraint handling capabilities, performance optimization, handling multi-variable interactions and ease of implementation, an extensive and growing body of research has been developed over the past few decades on the analysis, design and implementation of MPC, leading to a plethora of MPC formulations (see, for example, References [2–4] for some recent research directions and references in the field).

With the increasing demand over the past few decades for meeting stringent stability and performance specifications in industrial operations, fault-tolerance capabilities have become an increasingly important requirement in the design and implementation of modern day control systems. This is especially the case for safety-critical applications, such as chemical processes, where malfunctions in the control devices or process equipment can cause instabilities and lead to safety hazards if not appropriately mitigated through the use of fault-tolerant control approaches (see, for example, References [5–7] for some results and references on fault-tolerant control). The need for fault-tolerant control is further underscored by the increasing calls in recent times to achieve zero-incident plant operations as part of enabling the transition to smart plant operations ([8]).

As an advanced controller design methodology, MPC is also faced with the challenges of dealing with faults and handling the resulting degradation in the closed-loop stability and performance properties. Not surprisingly, this problem has been the subject of significant research work, and various methods have been investigated for the design and implementation of fault-tolerant MPC for both linear and nonlinear processes (see, for example, References [9–13] for some results and references in this area). An examination of the available literature on fault-tolerant MPC, however, reveals that the majority of existing methods have been developed within the traditional feedback control setting which assumes that the sensor–controller communication takes place over reliable dedicated links with flawless data transfer. This assumption needs to be re-examined in light of the widespread reliance on networked control systems which are characterized by increased levels of integration of resource-limited communication networks in the feedback loop.

The need to address the control-relevant challenges introduced by the intrinsic limitations on the processing and transmission capabilities of the sensor–controller communication medium has motivated a significant body of research work on networked control systems. Examples of efforts aimed at addressing some of these challenges in the context of MPC include the results in [14,15] where resource-aware MPC formulations that guarantee closed-loop stability with reduced sensor–controller communication requirements have been developed using event-based control techniques. In these studies, however, the problem of integrating fault-tolerance capabilities in the MPC design framework was not addressed.

Motivated by the above considerations, the aim of this work is to present a methodology for the design and implementation of fault-tolerant MPC for nonlinear process systems subject to model uncertainties, input constraints, control actuator faults and sensor–controller communication constraints. The co-presence of faults, control and communication resource constraints creates a conflict in the control design objectives where, on the one hand, increased levels of sensor–controller communication may be needed to mitigate the effects of the faults, and, on the other, such levels may be either unattainable or undesirable due to the sensor–controller communication constraints. To reconcile these conflicting objectives, a resource-aware Lyapunov-based MPC formulation that achieves the fault-tolerant stabilization objective with reduced sensor–controller communication is presented in this work.

The remainder of the paper is organized as follows. Section 2 begins by introducing some preliminaries that define the scope of the work and the class of systems considered. Section 3 then introduces an auxiliary Lyapunov-based fault-tolerant controller synthesized on the basis of an uncertain model of the plant and characterizes its closed-loop stability region. An analysis of the effects of discrete measurement sampling on the stability properties of the closed-loop model is conducted using Lyapuonv techniques and subsequently used in Section 4 to formulate a Lyapunov-based MPC that retains the same closed-loop stability and fault-tolerance properties enforced by the auxiliary model-based controller. The stability properties of the closed-loop system are analyzed and precise conditions that guarantee ultimate boundedness of the closed-loop trajectories in the presence of faults, discretely-sampled measurements and plant-model mismatch are provided. A forecasting scheme is then developed to predict the evolution of the Lyapunov function and its time-derivative over each sampling interval. The forecasts are used to trigger updates of the model states using the actual state measurements whenever certain stability-based thresholds are projected to be breached. Finally, Section 6 presents a simulation study that demonstrates the implementation and efficacy of the developed MPC formulation.

2. Preliminaries

We consider the class of finite-dimensional nonlinear process systems with the following state-space representation:

$$\dot{\mathbf{x}} = \mathbf{f}(\mathbf{x}) + \mathbf{G}(\mathbf{x})\Theta\mathbf{u} \tag{1}$$

where $\mathbf{x} \in \mathbb{R}^{n_\mathbf{x}}$ is the vector of process state variables, and $\mathbf{f}(\cdot)$ and $\mathbf{G}(\cdot)$ are sufficiently smooth nonlinear functions of their arguments on the domain of interest which contains the origin in its interior. Without loss of generality, the origin is assumed to be an equilibrium point of the uncontrolled plant (i.e., $\mathbf{f}(\mathbf{0}) = \mathbf{0}$). The matrix $\Theta = \text{diag}\{\theta_1\ \theta_2\ \cdots\ \theta_m\}$ is a diagonal deterministic (but unknown) fault coefficient matrix, where θ_i is a fault parameter whose value indicates the fault or health status of the i-th control actuator. A value of $\theta_i = 1$ indicates that the i-th actuator is perfectly healthy, whereas a value of $\theta_i = 0$ represents a completely failed (non-functioning) control actuator. Any other value, $\theta_i \in (0, 1)$, represents a certain degree of fault. The parameter θ_i essentially measures the effectiveness (or control authority) of the i-th control actuator, with $\theta_i = 0$ indicating an ineffective failed actuator, $\theta_i = 1$ indicating a fully effective actuator, and any other value indicating a partially effective actuator that implements only a fraction of the required control action prescribed by the controller. The vector of manipulated input variables, $\mathbf{u} \in \mathbb{R}^{n_\mathbf{u}}$, takes values in a nonempty compact convex set $\mathcal{U} \triangleq \{\mathbf{u} \in \mathbb{R}^{n_\mathbf{u}} : \|\mathbf{u}\| \leq u^{\max}\}$ where $u^{\max} > 0$ represents the magnitude of input constraints and $\|\cdot\|$ denotes the Euclidean norm of a vector or matrix.

The control objective is to steer the process state from a given initial condition to the origin in the presence of input constraints, control actuator faults and limited sensor–controller communication. To facilitate controller synthesis, we assume that an uncertain dynamic model of the system of Equation (1) is available and has the following form:

$$\dot{\widehat{\mathbf{x}}} = \widehat{\mathbf{f}}(\widehat{\mathbf{x}}) + \widehat{\mathbf{G}}(\widehat{\mathbf{x}})\widehat{\Theta}\mathbf{u} \tag{2}$$

where $\widehat{\mathbf{x}}$ is the model state, $\widehat{\mathbf{f}}(\cdot)$ and $\widehat{\mathbf{G}}(\cdot)$ are sufficiently smooth nonlinear functions that approximate the functions $\mathbf{f}(\cdot)$ and $\mathbf{G}(\cdot)$, respectively, in Equation (1), and are given by:

$$\widehat{\mathbf{f}}(\mathbf{x}) = \mathbf{f}(\mathbf{x}) + \delta_\mathbf{f}(\mathbf{x}) \tag{3a}$$

$$\widehat{\mathbf{G}}(\mathbf{x}) = \mathbf{G}(\mathbf{x}) - \delta_\mathbf{G}(\mathbf{x}) \tag{3b}$$

where $\delta_\mathbf{f}(\cdot)$ and $\delta_\mathbf{G}(\cdot)$ are smooth nonlinear functions that capture the model uncertainties, and the following Lipschitz conditions hold on a certain region of interest:

$$\|\widehat{\mathbf{f}}(\mathbf{x}_1) - \widehat{\mathbf{f}}(\mathbf{x}_2)\| \leq \widehat{L}_1\|\mathbf{x}_1 - \mathbf{x}_2\| \tag{4a}$$

$$\|\widehat{\mathbf{G}}(\mathbf{x}_1) - \widehat{\mathbf{G}}(\mathbf{x}_2)\| \leq \widehat{L}_2\|\mathbf{x}_1 - \mathbf{x}_2\| \tag{4b}$$

where $\widehat{L}_1$ and $\widehat{L}_2$ are known positive constants. $\widehat{\Theta} = \text{diag}\{\widehat{\theta}_1\ \widehat{\theta}_2\ \cdots \widehat{\theta}_m\}$ is a diagonal matrix, where $\widehat{\theta}_i$ is an estimate of the actual fault coefficient, θ_i. As discussed below, $\widehat{\theta}_i$ can also be viewed as a fault accommodation parameter that can be adjusted within the model to help achieve the fault-tolerant stabilization objective.

Towards our goal of designing a fault-tolerant MPC with well-characterized stability and performance properties, we begin in the next section by introducing an auxiliary bounded Lyapunov-based fault-tolerant controller that has an explicitly-characterized region of stability in the presence of faults. The stability properties of this controller are used as the basis for the development of a Lyapunov-based MPC formulation that retains the same closed-loop stability and fault-tolerance characteristics.

3. An Auxiliary Model-Based Fault-Tolerant Controller

3.1. Controller Synthesis and Analysis under Continuous State Measurements

Based on the dynamic model of Equation (2), we consider the following bounded Lyapunov-based state feedback controller:

$$\mathbf{u}(\widehat{\mathbf{x}}) = -k(\widehat{\mathbf{x}},\widehat{\Theta})(L_{\widehat{\mathbf{G}}}V(\widehat{\mathbf{x}})\widehat{\Theta})^T \doteq \mathbf{k}(\widehat{\mathbf{x}},\widehat{\Theta}) \tag{5a}$$

$$k(\widehat{\mathbf{x}},\widehat{\Theta}) = \frac{\alpha(\widehat{\mathbf{x}}) + \sqrt{(\alpha(\widehat{\mathbf{x}}))^2 + (u^{\max}\|\beta^T(\widehat{\Theta},\widehat{\mathbf{x}})\|)^4}}{\|\beta^T(\widehat{\Theta},\widehat{\mathbf{x}})\|^2[1+\sqrt{1+(u^{\max}\|\beta^T(\widehat{\Theta},\widehat{\mathbf{x}})\|)^2}]} \tag{5b}$$

$$\alpha(\widehat{\mathbf{x}}) \doteq L_{\widehat{\mathbf{f}}}V + \lambda V = \frac{\partial V}{\partial \widehat{\mathbf{x}}}\widehat{\mathbf{f}} + \lambda V \tag{5c}$$

$$\beta(\widehat{\Theta},\widehat{\mathbf{x}}) \doteq L_{\widehat{\mathbf{G}}}V(\widehat{\mathbf{x}})\widehat{\Theta},\ L_{\widehat{\mathbf{G}}}V = [\frac{\partial V}{\partial \widehat{\mathbf{x}}}\widehat{\mathbf{g}}^1 \cdots \frac{\partial V}{\partial \widehat{\mathbf{x}}}\widehat{\mathbf{g}}^m] \tag{5d}$$

where $L_{\widehat{\mathbf{f}}}V$ and $L_{\widehat{\mathbf{G}}}V$ are the Lie derivatives of V with respect to, $\widehat{\mathbf{f}}$ and $\widehat{\mathbf{G}}$, respectively; V is a control Lyapunov function that satisfies the following inequalities:

$$\alpha_1(\|\widehat{\mathbf{x}}\|) \leq V(\widehat{\mathbf{x}}) \leq \alpha_2(\|\widehat{\mathbf{x}}\|) \tag{6a}$$

$$\left\|\frac{\partial V(\widehat{\mathbf{x}})}{\partial \widehat{\mathbf{x}}}\right\| \leq \alpha_3(\|\widehat{\mathbf{x}}\|) \tag{6b}$$

$$\left\|\frac{\partial V(\widehat{\mathbf{x}})}{\partial \widehat{\mathbf{x}}}\widehat{\mathbf{G}}(\widehat{\mathbf{x}})\right\| \leq \alpha_4(\|\widehat{\mathbf{x}}\|) \tag{6c}$$

for some class $\mathcal{K}$ functions (A function $\alpha(\cdot)$ is said to be of class $\mathcal{K}$ if it is strictly increasing and $\alpha(0) = 0$) $\alpha_i(\cdot),\ i \in \{1,2,3,4\}$ and λ is a controller design parameter. The controller of Equation (5) belongs to the general class of constructive nonlinear controllers referred to in the literature as Sontag-type controllers. Similar to earlier bounded controller designs (see, for example, [16]), it is obtained by scaling Sontag's original universal formula to ensure that the control constraints are met within a certain well-defined region of the state space. The controller in Equation (5), however, differs from earlier designs in that it incorporates the fault explicitly into the controller synthesis formula.

It can be shown (see [16] for a similar proof) that the controller of Equation (5) satisfies the control constraints within a well-defined region in the state space, i.e.,:

$$\|\mathbf{k}(\widehat{\mathbf{x}},\widehat{\Theta})\| \leq u^{\max} \quad \forall\, \widehat{\mathbf{x}} \in \Psi(u^{\max},\widehat{\Theta}) \tag{7}$$

where

$$\Psi \doteq \{\widehat{\mathbf{x}} \in \mathbb{R}^{n_x} : L_{\widehat{\mathbf{f}}}V + \lambda V \leq u^{\max}\|\widehat{\Theta}^T L_{\widehat{\mathbf{G}}}V(\widehat{\mathbf{x}})^T\|\} \tag{8}$$

and that starting from any initial condition, $\widehat{\mathbf{x}}(0)$, within the compact set:

$$\Omega \doteq \{\widehat{\mathbf{x}} \in \Psi(u^{\max}, \widehat{\Theta}) : V(\widehat{\mathbf{x}}) \leq c\} \tag{9}$$

where $c > 0$ is the largest number for which $\Omega(u^{\max}, \widehat{\Theta}) \subset \Psi(u^{\max}, \widehat{\Theta})$, the time-derivative of the Lyapunov function, V, along the trajectories of the closed-loop model satisfies:

$$\dot{V}(\widehat{\mathbf{x}}) \leq -\lambda V(\widehat{\mathbf{x}}) \tag{10}$$

which implies that the origin of the closed-loop model under the auxiliary control law of Equation (5) is asymptotically stable in the presence of faults, with $\Omega(u^{\max}, \widehat{\Theta})$ as an estimate of the domain of attraction.

Remark 1. *The invariant set $\Omega(u^{\max}, \widehat{\Theta})$ defined in Equations (8) and (9) is an estimate of the state space region starting from where the origin of the closed-loop model is guaranteed to be asymptotically stable in the presence of control constraints and control actuator faults. As such, it represents an estimate of the fault-tolerant stabilization region. The expressions in Equations (8) and (9) capture the dependence of this region on both the magnitude of the control constraints and the magnitude of the fault estimate. Specifically, as the control constraints become tighter (i.e., $u^{\max}$ decreases), the fault-tolerant stability region is expected to shrink in size. In addition, as the severity of the fault increases (i.e., as $\widehat{\theta}_i$ tends to zero), the fault-tolerant stability region is expected to shrink in size. In the limit as $\widehat{\theta}_i \to 0$ for all i (i.e., total failure of all actuators), controllability is lost and asymptotic stabilization becomes impossible unless the system is open-loop stable (i.e., $L_{\widehat{\mathbf{f}}}V < 0$). Notice that the controller tuning parameter λ captures the classical tradeoff between stability and robustness. Specifically, as λ increases, Equation (10) predicts a higher dissipation rate of the Lyapunov function and thus a larger stability margin against small errors and perturbations. According to Equation (8), however, a larger value for λ leads to a smaller stability region in general.*

Remark 2. *The controller of Equation (5) is designed to account explicitly for faults, and enforce closed-loop stability by essentially canceling out the effect of the faults on the closed-loop dynamics. Notice, however, that, while the control action is an explicit function of the fault estimate, the upper bound on the dissipation rate of the Lyapunov function in Equation (10) is independent of the fault estimate.*

3.2. Characterization of Closed-Loop Stability under Discretely Sampled State Measurements

In this section, we analyze the stability properties of the closed-loop model when the auxiliary controller of Equation (5) is implemented using discretely-sampled measurements. This analysis is of interest given that MPC (to which the stability properties of the auxiliary controller will be transferred) is implemented in a discrete fashion. To this end, we consider the following sample-and-hold controller implementation:

$$\dot{\widehat{\mathbf{x}}}(t) = \widehat{\mathbf{f}}(\widehat{\mathbf{x}}(t)) + \widehat{\mathbf{G}}(\widehat{\mathbf{x}}(t))\widehat{\Theta}\mathbf{u}(t) \tag{11a}$$

$$\mathbf{u}(t) = \mathbf{k}(\widehat{\mathbf{x}}(t_k), \widehat{\Theta}), \; t \in [t_k, t_{k+1}), \; k \in \mathbb{N} \tag{11b}$$

where $t_{k+1} - t_k \triangleq \Delta$ is the sampling period. Owing to the non-vanishing errors introduced by the sample and hold implementation mechanism, only practical stability of the origin of the closed-loop model can be achieved in this case. Theorem 1 establishes that, provided a sufficiently small sampling period is used, the trajectory of the closed-loop model state can be made to converge in finite-time to an arbitrarily small terminal neighborhood of the origin, and that the size of this neighborhood depends on the magnitude of the fault as well as on the sampling period. To simplify the statement of the theorem, we first introduce some notation. Specifically, we use the symbols $\Phi_{\widehat{\mathbf{f}}}$ and $\Phi_{\widehat{\mathbf{G}}}$ to denote the Lipschitz constants of the functions $L_{\hat{\mathbf{f}}}V(\hat{\mathbf{x}})$ and $L_{\hat{\mathbf{G}}}V(\hat{\mathbf{x}})$, respectively, over the domain of interest, Ω, where:

$$\|L_{\hat{\mathbf{f}}}V(\hat{\mathbf{x}}(t)) - L_{\hat{\mathbf{f}}}V(\hat{\mathbf{x}}_0)\| \leq \Phi_{\hat{\mathbf{f}}}\|\hat{\mathbf{x}}(t) - \hat{\mathbf{x}}_0\| \tag{12a}$$
$$\|L_{\hat{\mathbf{G}}}V(\hat{\mathbf{x}}(t)) - L_{\hat{\mathbf{G}}}V(\hat{\mathbf{x}}_0)\| \leq \Phi_{\hat{\mathbf{G}}}\|\hat{\mathbf{x}}(t) - \hat{\mathbf{x}}_0\| \tag{12b}$$

for $\hat{\mathbf{x}}(t), \hat{\mathbf{x}}_0 \in \Omega$. We also define the following positive constants:

$$\gamma = K_{\hat{\mathbf{f}}} + K_{\hat{\mathbf{G}}}\|\widehat{\Theta}\|u^{\max} \tag{13a}$$
$$K_{\hat{\mathbf{f}}} = \max_{\hat{\mathbf{x}}\in\Omega}\|\hat{\mathbf{f}}(\hat{\mathbf{x}})\|, \; K_{\hat{\mathbf{G}}} = \max_{\hat{\mathbf{x}}\in\Omega}\|\widehat{\mathbf{G}}(\hat{\mathbf{x}})\| \tag{13b}$$

where $K_{\hat{\mathbf{f}}}$ and $K_{\hat{\mathbf{G}}}$ are guaranteed to exist due to the compactness of Ω.

Theorem 1. *Consider the closed-loop model of Equations (2)–(5), with a sample-and-hold implementation as described in Equation (11). Given any real positive number $\delta' \in (0, c)$, where c is defined in Equations (8) and (9), there exists a positive real number Δ^* such that if $\hat{\mathbf{x}}(t_0) \doteq \hat{\mathbf{x}}_0 \in \Omega(u^{\max}, \widehat{\Theta})$ and Δ is chosen such that $\Delta \in (0, \Delta^*]$, then the closed-loop model state trajectories are ultimately bounded and satisfy:*

$$\limsup_{t\to\infty} V(\hat{\mathbf{x}}(t)) \leq \delta' \tag{14}$$

where $\Delta^* = \min\{\bar{\Delta}, \Delta'\}$, $\bar{\Delta}$ and Δ' satisfy:

$$-\lambda\delta^f + (\Phi_{\hat{\mathbf{f}}} + \Phi_{\hat{\mathbf{G}}}\|\widehat{\Theta}\|u^{\max})\gamma\bar{\Delta} < -\epsilon \tag{15a}$$
$$\delta^f + (\Phi_{\hat{\mathbf{f}}} + \Phi_{\hat{\mathbf{G}}}\|\widehat{\Theta}\|u^{\max})\gamma(\Delta')^2 \leq \delta' \tag{15b}$$

for some $\epsilon > 0$ and $0 < \delta^f < \delta'$, where γ, $\Phi_{\hat{\mathbf{f}}}$ and $\Phi_{\hat{\mathbf{G}}}$ are defined in Equations (12) and (13). Furthermore, when $\hat{\mathbf{x}}(t_k) \in \Omega\backslash\Omega^f$ where $\Omega^f \triangleq \{\hat{\mathbf{x}} \in \mathbb{R}^{n_x} : V(\hat{\mathbf{x}}) \leq \delta^f\}$, $\dot{V}(\hat{\mathbf{x}}(t)) \leq -\epsilon$, $\forall\, t \in [t_k, t_{k+1})$.

Proof. Consider the following compact set:

$$\mathcal{M} \triangleq \{\hat{\mathbf{x}} \in \mathbb{R}^{n_x} : \delta^f \leq V(\hat{\mathbf{x}}) \leq c\} \tag{16}$$

for some $0 < \delta^f < c$. Let the control action be computed for some $\hat{\mathbf{x}}(t_k) := \hat{\mathbf{x}}_k \in \mathcal{M}$, and held constant until a time $\bar{\Delta}$, where $\bar{\Delta}$ is a positive real number, i.e.,

$$\mathbf{u}(t) = \mathbf{u}(\hat{\mathbf{x}}_k) \doteq \mathbf{u}_k, \; \forall\, t \in [t_k, t_k + \bar{\Delta}] \tag{17}$$

Then, for all $t \in [t_k, t_k + \bar{\Delta}]$, we have:

$$\dot{V}(\hat{\mathbf{x}}(t)) = L_{\hat{\mathbf{f}}}V(\hat{\mathbf{x}}_k) + L_{\hat{\mathbf{G}}}V(\hat{\mathbf{x}}_k)\widehat{\Theta}\mathbf{u}_k + [L_{\hat{\mathbf{f}}}V(\hat{\mathbf{x}}(t)) - L_{\hat{\mathbf{f}}}V(\hat{\mathbf{x}}_k)] + [L_{\hat{\mathbf{G}}}V(\hat{\mathbf{x}}(t))\widehat{\Theta}\mathbf{u}_k - L_{\hat{\mathbf{G}}}V(\hat{\mathbf{x}}_k)\widehat{\Theta}\mathbf{u}_k] \tag{18}$$

Since the control action is computed based on the model states in $\mathcal{M} \subset \Omega$, we have from Equation (10):

$$L_{\hat{\mathbf{f}}}V(\hat{\mathbf{x}}_k) + L_{\hat{\mathbf{G}}}V(\hat{\mathbf{x}}_k)\widehat{\Theta}\mathbf{u}_k = \dot{V}(\hat{\mathbf{x}}_k) \leq -\lambda V(\hat{\mathbf{x}}_k) \tag{19}$$

By definition, for all $\hat{\mathbf{x}}_k \in \mathcal{M}$, $V(\hat{\mathbf{x}}_k) \geq \delta^f$, and therefore:

$$L_{\hat{\mathbf{f}}}V(\hat{\mathbf{x}}_k) + L_{\hat{\mathbf{G}}}V(\hat{\mathbf{x}}_k)\widehat{\Theta}\mathbf{u}_k \leq -\lambda\delta^f \tag{20}$$

Given that $\widehat{\mathbf{f}}(\cdot)$ and the elements of $\widehat{\mathbf{G}}(\cdot)$ are smooth functions, and given that $\|\mathbf{u}\| \leq u^{\max}$ within Ω, and that $\mathcal{M}$ is bounded, one can find, for all $\widehat{\mathbf{x}}_k \in \mathcal{M}$ and a fixed $\bar{\Delta}$, a positive real number γ, such that:

$$\|\widehat{\mathbf{x}}(t) - \widehat{\mathbf{x}}_k\| \leq \gamma\bar{\Delta}, \ \forall\, t \in [t_k, t_k + \bar{\Delta}) \tag{21}$$

where γ is defined in Equation (13). Based on this and Equation (18), the following bound can be obtained:

$$\dot{V}(\widehat{\mathbf{x}}(t)) \leq -\lambda\delta^f + (\Phi_{\widehat{\mathbf{f}}} + \Phi_{\widehat{\mathbf{G}}}\|\widehat{\Theta}\|u^{\max})\|\widehat{\mathbf{x}}(t) - \widehat{\mathbf{x}}_k\| \leq -\lambda\delta^f + (\Phi_{\widehat{\mathbf{f}}} + \Phi_{\widehat{\mathbf{G}}}\|\widehat{\Theta}\|u^{\max})\gamma\bar{\Delta} \tag{22}$$

If we choose $\bar{\Delta} < (\lambda\delta^f - \epsilon)/(\Phi_{\widehat{\mathbf{f}}} + \Phi_{\widehat{\mathbf{G}}}\|\widehat{\Theta}\|u^{\max})\gamma$, we get:

$$\dot{V}(\widehat{\mathbf{x}}(t)) \leq -\epsilon < 0, \ \ \forall\, t \in [t_k, t_k + \bar{\Delta}) \tag{23}$$

This implies that, given any $0 < \delta' < c$, if δ^f is chosen such that $0 < \delta^f < \delta'$ and a corresponding value for $\bar{\Delta}$ is found, then if the control action is computed for any $\widehat{\mathbf{x}} \in \mathcal{M}$, and the hold time is less than $\bar{\Delta}$, $\dot{V}$ is guaranteed to remain negative over this time period and, therefore, $\widehat{\mathbf{x}}$ cannot escape Ω (since Ω is a level set of V).

Now, let us consider the case when, at the sampling time t_k, the model state is within $\Omega^f \triangleq \{\widehat{\mathbf{x}} \in \mathbb{R}^{n_x} : V(\widehat{\mathbf{x}}) \leq \delta^f\}$, i.e., $V(\widehat{\mathbf{x}}(t_k)) \leq \delta^f$. We have already shown that:

$$\dot{V}(\widehat{\mathbf{x}}(t)) \leq -\lambda V(\widehat{\mathbf{x}}_k) + (\Phi_{\widehat{\mathbf{f}}} + \Phi_{\widehat{\mathbf{G}}}\|\widehat{\Theta}\|u^{\max})\gamma\bar{\Delta} \tag{24}$$

which implies that:

$$\dot{V}(\widehat{\mathbf{x}}(t)) \leq (\Phi_{\widehat{\mathbf{f}}} + \Phi_{\widehat{\mathbf{G}}}\|\widehat{\Theta}\|u^{\max})\gamma\bar{\Delta} \tag{25}$$

Integrating both sides of the differential inequality above yields:

$$\begin{aligned} V(\widehat{\mathbf{x}}(t)) &= V(\widehat{\mathbf{x}}(t_k)) + \int_{t_k}^{t} \dot{V}(\widehat{\mathbf{x}}(\tau))d\tau \\ V(\widehat{\mathbf{x}}(t_k + \bar{\Delta})) &\leq \delta^f + (\Phi_{\widehat{\mathbf{f}}} + \Phi_{\widehat{\mathbf{G}}}\|\widehat{\Theta}\|u^{\max})\gamma(\bar{\Delta})^2 \end{aligned} \tag{26}$$

Based on the last bound above, given any positive real number δ', one can find a sampling period Δ' small enough such that the trajectory is trapped in Ω', i.e.,

$$V(\widehat{\mathbf{x}}(t_k + \Delta')) \leq \delta^f + (\Phi_{\widehat{\mathbf{f}}} + \Phi_{\widehat{\mathbf{G}}}\|\widehat{\Theta}\|u^{\max})\gamma(\Delta')^2 \leq \delta' \tag{27}$$

To summarize, if the sampling period Δ is chosen such that $\Delta \in (0, \Delta^*]$, where $\Delta^* \triangleq \min\{\bar{\Delta}, \Delta'\}$, then the closed-loop model state is ultimately bounded within the terminal set Ω' in finite time. This completes the proof of the theorem. □

Remark 3. *The result of Theorem 1 establishes the robustness of the controller of Equation (5) to bounded measurement errors introduced through the sample-and-hold implementation scheme. The controller is robust in the sense that the closed-loop model trajectory remains bounded and converges in finite-time to a terminal neighborhood centered at the origin, the size of which can be made arbitrarily small by choosing the sampling period to be sufficiently small. It should be noted that the bound on the dissipation rate of the Lyapunov function,* ϵ, *and the ultimate bound on the model state,* δ', *are both dependent on the size of the sampling period,* Δ, *and on the size of the fault estimate,* $\widehat{\Theta}$. *This dependence is captured by Equation (15). As expected, a sampling period that is too large could lead to instability.*

Remark 4. *By inspection of the inequality in Equation (24), it can be seen that as the norm of the fault matrix decreases the bound on the dissipation rate becomes tighter (more negative), potentially implying a faster decay of the Lyapunov function. To the extent that the norm of the fault matrix can be taken as a measure of fault* *severity (with a smaller norm indicating a more severe fault), this seems to suggest that increased fault severity actually helps speed up (rather than retard) the dissipation rate, which at first glance may seem counter-intuitive. To get some insight into this apparent discrepancy, it should first be noted that in obtaining the inequality in Equation (24) the control action term is essentially regarded as a disturbance that perturbs the nominal (uncontrolled) part of the plant, and is majorized using a convenient upper bound which includes the norm of the fault matrix as well as the magnitude of the control constraints. Based on this representation, a decrease in the norm of the fault matrix (due to a more severe fault) implies a reduction in the controller authority and, therefore, a decrease in the size of the disturbance which helps tighten the upper bound and potentially speed up the dissipation rate of the Lyapunov function. A similar reasoning can be applied when analyzing the dependence of the ultimate bound in Equation (27) on the fault size. An important caveat in making these observations is that what is impacted by the norm of the fault matrix is only the upper bound (either on the time-derivative of the Lyapunov function as in Equation (24) or on the Lyapunov function itself as in Equation (27)). A larger upper bound does not necessarily translate into slower decay.*

4. Design and Analysis of Lyapunov-Based Fault-Tolerant MPC

This section introduces a Lyapunov-based MPC formulation that retains the stability and fault-tolerance characteristics of the auxiliary bounded controller presented in the previous section. The main idea is to embed the conditions that characterize the fault-tolerant closed-loop stability properties of the auxiliary bounded controller as constraints within the finite-horizon optimal control problem in MPC. This idea of linking the auxiliary controller and MPC designs—and thus transferring the stability properties from one to the other—has it roots in the original Lyapunov-based MPC formulation presented in [17]. In the present work, we go beyond the original formulation to analyze its robustness with respect to implementation on the plant and derive explicit conditions that account explicitly for plant-model mismatch and control actuator faults.

To this end, we consider the following Lyapunov-based MPC formulation, where the control action is obtained by repeatedly solving the following finite-horizon optimal control problem:

$$\min_{\mathbf{u}\in\mathcal{U}} \int_{t_k}^{t_{k+N}} [\|\widehat{\mathbf{x}}(\tau)\|_{\mathbf{Q}}^2 + \|\mathbf{u}(\tau)\|_{\mathbf{R}}^2]\, d\tau \tag{28a}$$

$$\text{Subject to :}$$

$$\|\mathbf{u}(t)\| \leq u^{\max},\ \forall\, t \in [t_k, t_{k+N}) \tag{28b}$$

$$\dot{\widehat{\mathbf{x}}}(t) = \widehat{\mathbf{f}}(\widehat{\mathbf{x}}(t)) + \widehat{\mathbf{G}}(\widehat{\mathbf{x}}(t))\widehat{\Theta}\mathbf{u}(t) \tag{28c}$$

$$\widehat{\mathbf{x}}(t_k) = \mathbf{x}(t_k) \tag{28d}$$

$$\dot{V}(\widehat{\mathbf{x}}(t)) \leq -\epsilon,\ \forall\, t \in [t_k, t_{k+1}),\ \text{if } V(\widehat{\mathbf{x}}(t_k)) > \delta^f \tag{28e}$$

$$V(\widehat{\mathbf{x}}(t)) \leq \delta',\ \forall\, t \in [t_k, t_{k+1}),\ \text{if } V(\widehat{\mathbf{x}}(t_k)) \leq \delta^f \tag{28f}$$

where N represents the length of the prediction and control horizons; $\mathbf{Q}$ and $\mathbf{R}$ are positive-definite matrices that represent weights on the state and control penalties, respectively; and V is the control Lyapunov function used in the design of the bounded controller in Equations (5) and (6). The constraints in Equations (28e) and (28f) are imposed to ensure that this MPC enforces the same stability properties that the bounded controller enforces in the closed-loop model, and retains the same stability region estimate, $\Omega(u^{\max}, \widehat{\Theta})$. Theorem 2 provides a characterization of the closed-loop stability properties when the above MPC is applied to the plant of Equation (1) in the presence of plant-model mismatch and control actuator faults.

Theorem 2. *Consider the closed-loop system of Equation (1) subject to the MPC law of Equation (28) with a sampling period $\tilde{\Delta} < \Delta^*$, where Δ^* is defined in Theorem 1, that satisfies:*

$$-\epsilon + \rho_1 \mu(\bar{\delta}_1, \bar{\delta}_2, \widehat{\Theta}, u^{\max}, L_1, L_2, \tilde{\Delta}) + \rho_2 \|\Theta - \widehat{\Theta}\| u^{\max} \leq -\omega \tag{29}$$

for some $\omega > 0$, where ϵ satisfies Equation (15a) and

$$\begin{aligned}
\rho_1 &= \alpha_3(\alpha_1^{-1}(c)),\ \rho_2 = \alpha_4(\alpha_1^{-1}(c)) && \text{(30a)}\\
\mu &= \bar{\delta}_1 + \bar{\delta}_2\|\widehat{\Theta}\|u^{\max} + (L_1 + L_2\|\Theta\|u^{\max})\zeta(\tilde{\Delta}) && \text{(30b)}\\
\zeta(\tilde{\Delta}) &\triangleq c_1\tilde{\Delta}e^{c_2\tilde{\Delta}} && \text{(30c)}\\
c_1 &= \bar{\delta}_1 + \bar{\delta}_2\|\widehat{\Theta}\|u^{\max} + L_2\|\Theta - \widehat{\Theta}\|u^{\max}\|\widehat{\mathbf{x}}_0\| && \text{(30d)}\\
c_2 &= L_1 + L_2 u^{\max} && \text{(30e)}\\
\bar{\delta}_1 &= \max_{\mathbf{x}\in\Omega}\|\delta_1(\mathbf{x})\|,\ \bar{\delta}_2 = \max_{\mathbf{x}\in\Omega}\|\delta_2(\mathbf{x})\| && \text{(30f)}
\end{aligned}$$

where α_i, $i \in \{1,2,3,4\}$, are defined in Equation (6), L_1 and L_2 are the Lipschitz constants of $\mathbf{f}(\cdot)$ and $\mathbf{G}(\cdot)$ on Ω, respectively. Then, given any positive real number $\delta'' < c$, there exists a positive real number Δ^{**} such that, if $\widehat{\mathbf{x}}(t_0) = \mathbf{x}(t_0) \in \Omega$, $\tilde{\Delta} \in (0, \Delta^{**}]$, the closed-loop trajectories are ultimately bounded and:

$$\limsup_{t\to\infty} V(\mathbf{x}(t)) \le \delta' + \rho_1\zeta(\tilde{\Delta}) + \xi\zeta^2(\tilde{\Delta}) \le \delta'' < c \tag{31}$$

for some $\xi > 0$, where δ' satisfies Equation (15b). Furthermore, when $\mathbf{x}(t_k) \in \Omega\backslash\Omega'$ where $\Omega' \triangleq \{\mathbf{x} \in \mathbb{R}^{n_x} : V(\mathbf{x}) \le \delta'\}$, $\dot{V}(\mathbf{x}(t)) \le -\omega$, $\forall t \in [t_k, t_{k+1})$.

Proof. Defining the model estimation error as $\mathbf{e}(t) \doteq \widehat{\mathbf{x}}(t) - \mathbf{x}(t)$, the dynamics of the model estimation error are governed by:

$$\dot{\mathbf{e}} = [\widehat{\mathbf{f}}(\widehat{\mathbf{x}}) - \mathbf{f}(\widehat{\mathbf{x}})] + [\mathbf{f}(\widehat{\mathbf{x}}) - \mathbf{f}(\mathbf{x})] + [\widehat{\mathbf{G}}(\widehat{\mathbf{x}})\widehat{\Theta} - \mathbf{G}(\widehat{\mathbf{x}})\widehat{\Theta}]\mathbf{u} + [\mathbf{G}(\widehat{\mathbf{x}})\widehat{\Theta} - \mathbf{G}(\mathbf{x})\widehat{\Theta}]\mathbf{u} + [\mathbf{G}(\mathbf{x})\widehat{\Theta} - \mathbf{G}(\mathbf{x})\Theta]\mathbf{u} \tag{32}$$

Given $\widehat{\mathbf{x}}(t_0) = \mathbf{x}(t_0) \in \Omega$, $\widehat{\mathbf{x}}(t)$ will remain within Ω for all $t \in [t_0, t_0 + \Delta)$ because of the enforced stability constraints (which ensure boundedness of $\widehat{\mathbf{x}}$). If $\mathbf{x}(t)$ also remains within Ω during this interval, then the following bound on $\mathbf{e}(t)$, for $t \in [t_0, t + \Delta)$, can be derived:

$$\|\mathbf{e}(t)\| \le \|\mathbf{e}(t_0)\| + (\bar{\delta}_1 + \bar{\delta}_2\widehat{\Theta}\mathbf{u})(t - t_0) + \int_{t_0}^{t}[L_1 + L_2\Theta\mathbf{u}]\|\mathbf{e}(\tau)\|]d\tau + \int_{t_0}^{t}[L_2\|\widehat{\mathbf{x}}(t)\|(\Theta - \widehat{\Theta})\mathbf{u}]d\tau$$

where we have used Equation (3) and the Lipschitz properties of the various functions involved. In view of the model update policy in Equation (28d), we have $\mathbf{e}(t_0) = 0$, and, together with the fact that $t - t_0 \le \tilde{\Delta}$, the above bound simplifies to:

$$\|\mathbf{e}(t)\| \le c_1\tilde{\Delta} + c_2\int_{t_0}^{t+\tilde{\Delta}}\|\mathbf{e}(\tau)\|d\tau \tag{33}$$

Applying the Gronwall–Bellman inequality yields:

$$\|\mathbf{e}(t)\| \le c_1\tilde{\Delta}e^{c_2\tilde{\Delta}} = \zeta(\Delta), \text{ for } t \in [t_0, t_0 + \Delta) \tag{34}$$

Evaluating the time-derivative of the Lyapunov function along the trajectories of the closed-loop system yields:

$$\begin{aligned}
\dot{V}(\mathbf{x}) &= \frac{\partial V}{\partial \widehat{\mathbf{x}}}\widehat{\mathbf{f}}(\widehat{\mathbf{x}}) + \frac{\partial V}{\partial \widehat{\mathbf{x}}}\widehat{\mathbf{G}}(\widehat{\mathbf{x}})\widehat{\Theta}\mathbf{u} + \frac{\partial V}{\partial \mathbf{x}}\mathbf{f}(\mathbf{x}) - \frac{\partial V}{\partial \widehat{\mathbf{x}}}\widehat{\mathbf{f}}(\widehat{\mathbf{x}}) + \frac{\partial V}{\partial \mathbf{x}}\mathbf{G}(\mathbf{x})\Theta\mathbf{u} - \frac{\partial V}{\partial \widehat{\mathbf{x}}}\widehat{\mathbf{G}}(\widehat{\mathbf{x}})\widehat{\Theta}\mathbf{u}\\
&\le \dot{V}(\widehat{\mathbf{x}}(t)) + \frac{\partial V}{\partial \mathbf{x}}\mathbf{f}(\mathbf{x}) - \frac{\partial V}{\partial \mathbf{x}}\widehat{\mathbf{f}}(\mathbf{x}) + \frac{\partial V}{\partial \mathbf{x}}\widehat{\mathbf{f}}(\mathbf{x}) - \frac{\partial V}{\partial \widehat{\mathbf{x}}}\widehat{\mathbf{f}}(\widehat{\mathbf{x}}) + \frac{\partial V}{\partial \mathbf{x}}\mathbf{G}(\mathbf{x})\Theta\mathbf{u} - \frac{\partial V}{\partial \widehat{\mathbf{x}}}\widehat{\mathbf{G}}(\widehat{\mathbf{x}})\widehat{\Theta}\mathbf{u}
\end{aligned} \tag{35}$$

For $\hat{\mathbf{x}}(t_0) = \mathbf{x}(t_0) \in \Omega \setminus \Omega'$ and $\tilde{\Delta} < \Delta^*$, it can be shown upon substituting Equations (3), (4), (6) and (7) into Equation (35), and using the notation in Equation (30), that:

$$\dot{V}(\mathbf{x}) \leq -\epsilon + \rho_1\mu + \rho_2\|\Theta - \hat{\Theta}\|u^{\max} \tag{36}$$

Therefore, if Equation (29) holds, we have:

$$\dot{V}(\mathbf{x}(t)) \leq -\omega, \ \forall t \in [t_0, t_0 + \tilde{\Delta}) \tag{37}$$

For the case when $\hat{\mathbf{x}}(t_0) = \mathbf{x}(t_0) \in \Omega'$, we use the following inequality derived from a Taylor series expansion of $V(\mathbf{x})$:

$$V(\mathbf{x}) \leq V(\hat{\mathbf{x}}) + \frac{\partial V}{\partial \hat{\mathbf{x}}}\|\mathbf{x} - \hat{\mathbf{x}}\| + \xi\|\mathbf{x} - \hat{\mathbf{x}}\|^2 \tag{38}$$

where $\xi > 0$, and the term $\xi\|\mathbf{x} - \hat{\mathbf{x}}\|^2$ bounds the second and higher-order terms of the expansion. Together with Equations (6), (28f), and (34), it can be shown that:

$$V(\mathbf{x}) \leq \delta' + \alpha_3(\alpha_1^{-1}(c))\zeta(\tilde{\Delta}) + \xi\zeta^2(\tilde{\Delta}) \tag{39}$$

which implies that given any positive real number $\delta'' < c$, one can find a small enough $\tilde{\Delta}$ such that $V(\mathbf{x}(t)) \leq \delta''$ for all $t \in [t_0, t_0 + \tilde{\Delta})$.

The above analysis for the initial interval can be performed recursively for all subsequent intervals to show that the closed-loop state $\mathbf{x}(t)$ remains bounded within Ω, for all $t \geq t_0$, thus validating the initial assumption made on the boundedness of $\mathbf{x}$. Therefore, if Equation (29) is satisfied, we conclude that given any $\hat{\mathbf{x}}(t_0) = \mathbf{x}(t_0) \in \Omega$, we have for sufficiently small $\tilde{\Delta}$ that $\mathbf{x}(t) \in \Omega$ for all $t \in [t_0, \infty)$, and that the ultimate bound in Equation (31) holds. Furthermore, when $\hat{\mathbf{x}}(t_k) = \mathbf{x}(t_k) \in \Omega \setminus \Omega'$, we have $\dot{V}(\mathbf{x}(t)) \leq -\omega$, for all $t \in [t_k, t_{k+1})$. This completes the proof of the theorem. □

Remark 5. *The conditions in Equations (29)–(31) provide a characterization of the stability and performance properties of the closed-loop system under the MPC law of Equation (28). Specifically, the condition in Equations (29) and (30) characterize the upper bound on the dissipation rate of $\dot{V}$ along the trajectories of the closed-loop outside the terminal set. A comparison between this bound, ω, and the one enforced by the nominal MPC in the closed-loop model in Equation (28e), ϵ, shows that the actual rate is slower than the nominal one due to the combined influences of the plant-model mismatch, the faults and the discrepancy between the actual and estimated values of the faults. While some tuning of the discrepancy between the two dissipation rates can be exercised by adjusting the sampling period (note from Equations (29) and (30) that reducing $\tilde{\Delta}$ reduces μ), the difference between the two rates is ultimately dictated by the size of the plant-model mismatch and the magnitudes of the faults. Similarly, it can be seen that compared to the nominal ultimate bound enforced by MPC in the closed-loop model in Equation (28f), δ', the actual ultimate bound for the closed-loop system, δ'', is larger due to the effects of the model uncertainty and the faults. Again, while the discrepancy between the two bounds (i.e., between the two terminal sets) can be made smaller if $\tilde{\Delta}$ is chosen small enough, it is not possible in general to make that discrepancy arbitrarily small owing to the fact that the uncertainty and fault magnitudes are not adjustable parameters. The comparison between the nominal and actual bounds points to the fundamental limitations that model uncertainty and faults impose on the achievable closed-loop performance.*

Remark 6. *Note that if a fault, Θ, that satisfies the conditions in Equations (29)–(31) takes place, the closed-loop system will be inherently stable in the presence of such fault, and the MPC is said to be passively fault-tolerant. The conditions in Equations (29)–(31) suggest that, while mitigation of the fault effects is not necessary in this case given that stability is not jeopardized, it may still be desirable to actively accommodate the fault by adjusting the model parameter $\hat{\Theta}$ to enhance closed-loop performance. In particular, note from Equations (29)–(31) (see also Equation (36)) that when the actual fault size can be determined, adjusting the fault estimate used in the model to match the actual fault (i.e., setting $\Theta - \hat{\Theta} = 0$) helps tighten the dissipation rate bound on $\dot{V}$ and* *reduce the size of the ultimate bound, which helps improve closed-loop performance. The implementation of this fault accommodation measure requires knowledge of the magnitude of the fault, which in general can be obtained using fault estimation and identification techniques (see, for example, [18]). While exact knowledge of the fault size is not required, errors in estimating the fault magnitude (which lead to a nonzero mismatch between* Θ *and* $\widehat{\Theta}$*) can limit the extent to which the dissipation rate bound on* $\dot{V}$ *can tightened and the ultimate bound reduced, and therefore can limit the achievable performance benefits of fault accommodation.*

Remark 7. *The dependence of the fault-tolerant stabilization region associated with the proposed MPC formulation on the size of the control constraints points to an interesting link between the fault-tolerant MPC formulation and process design considerations. This connection stems from the fact that control constraints, which are typically the result of limitations on the capacity of control actuators, are dictated in part by equipment design considerations. As a result, an a priori process design choice that fixes the capacity of the control equipment automatically imposes limitations on the fault-tolerance capabilities of the MPC system. This connection can be used the other way around in order to aid the selection of a suitable process design that can enhance the fault-tolerance capabilities of the control system. Specifically, given a desired region of fault-tolerant operation for the MPC, one can use the characterization in Equations (8) and (9) to determine the corresponding size of the control constraints, and hence the capacity of the control equipment. It is worth noting that the integration of process design and control in the context of MPC has been the subject of several previous works (see, for example, [19–21]). However, the problem of integrating process design and fault-tolerant MPC under uncertainty has not been addressed in these prior works. The results in this paper shed some light on this gap and provide a general framework for examining the interactions between process design and control in the context of fault-tolerant MPC.*

5. Fault-Tolerant MPC Implementation Using Forecast-Triggered Communication

To implement the MPC law of Equation (28), the state measurement must be transmitted to the controller at every sampling time in order to update the model state. To reduce the frequency of sensor–controller information transfer, we proceed in this section to present a forecast-triggered sensor–controller communication strategy that optimizes network resource utilization without compromising closed-loop stability. The basic idea is to forecast at each sampling time the expected evolution (or rate of evolution) of the Lyapunov function over the following sampling interval based on the available state data and the worst-case uncertainty, and to trigger an update of the model state only in the event that the forecast indicates a potential increase in the Lyapunov function or a potential deterioration in the dissipation rate.

To explain how this communication strategy works, we assume that a copy of the MPC law is embedded within the sensors side to provide the control input trajectory and aid the forecasting process. At the same time, the state measurement, **x**, which is available from the sensors is monitored at the sampling times, and then the model estimation error **e** can be computed at each sampling instance. To perform the forecast for $t \in [t_k, t_{k+1})$, the bounds in Equations (36) and (39) are modified as follows:

$$\dot{V}(\mathbf{x}(t)) \leq -\epsilon + \rho_1[\bar{\delta}_1 + \bar{\delta}_2\|\widehat{\Theta}\|u^{\max} + \rho_2\|\Theta - \widehat{\Theta}\|u^{\max} + (L_1 + L_2\|\Theta\|u^{\max})\zeta(t - t_p)] \tag{40a}$$

$$V(\mathbf{x}(t)) \leq \delta' + \rho_1\zeta(t - t_p) + \xi\zeta^2(t - t_p) \tag{40b}$$

where t_p denotes the time that the last update prior to t_k took place. By comparing the above bounds with the original ones developed in Equations (36) and (39) for the case of periodic model updates, it can be seen that the sampling interval, $\tilde{\Delta}$, in the original bounds has now been replaced by the more general interval $t - t_p$. This modification is introduced to allow assessment of the impact of sensor–controller communication suspension on the evolution of the Lyapunov function, and to determine if the suspension could be tolerated for longer than one sampling period.

Algorithm 1 and the flowchart in Figure 1 summarize the proposed forecast-triggered communication strategy. The notation $\bar{V}(\mathbf{x})$ is used to denote the upper bound on $V(\mathbf{x})$ resulting from the forecast strategy.

Algorithm 1. Forecast-triggered sensor–controller communication strategy

---

Initialize $\widehat{\mathbf{x}}(t_0) = \mathbf{x}(t_0) \in \Omega$ and **set** $k = 0$, $p = 0$
Solve Equation (28) for $[t_0, t_1)$ and **implement** the first step of the control sequence
if $\widehat{\mathbf{x}}(t_{k+1}) \in \Omega \backslash \Omega'$ **then**
  Calculate $\bar{V}(\mathbf{x}(t_{k+2}))$ (estimate of $V(\mathbf{x}(t_{k+2}))$) using Equation (40a) and $V(\mathbf{x}(t_{k+1}))$
else
  Calculate $\bar{V}(\mathbf{x}(t_{k+2}))$ (estimate of $V(\mathbf{x}(t_{k+2}))$) using Equation (40b) and $\mathbf{e}(t_{k+1})$
end if
if $\bar{V}(\mathbf{x}(t_{k+2})) < V(\mathbf{x}(t_{k+1}))$ **then**
  Solve Equation (28) without Equation (28d) for $[t_{k+1}, t_{k+2})$
else if $\bar{V}(\mathbf{x}(t_{k+2})) \geq V(\mathbf{x}(t_{k+1}))$ and $\bar{V}(\mathbf{x}(t_{k+2})) \leq \delta'$ **then**
  Solve Equation (28) without Equation (28d) for $[t_{k+1}, t_{k+2})$
else
  Solve Equation (28) for $[t_{k+1}, t_{k+2})$ and set $p = k + 1$
end if
Implement the first step of the control sequence on $[t_{k+1}, t_{k+2})$
Set $k = k + 1$ and **go to** step 3

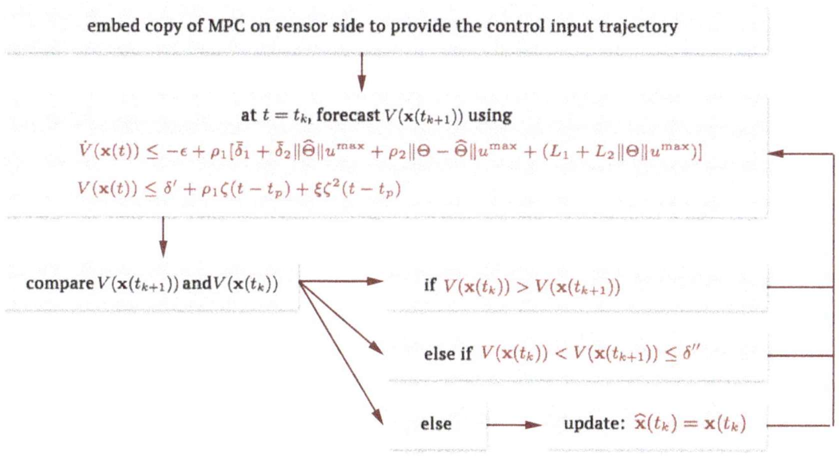

Figure 1. Flowchart of implementation of the forecast-triggered communication strategy. MPC: model predictive control.

Remark 8. *With regard to the implementation of Algorithm 1, the sensors need to obtain measurements of the state* $\mathbf{x}$ *at each sampling time,* t_k*, perform Steps 3–7 in the algorithm, and then determine whether or not to transmit the state to the controller to update the model state* $\widehat{\mathbf{x}}$ *based on the criteria described in Steps 8–14. Specifically, once the state arrives at* $t = t_{k+1}$*, the evolution of* $\mathbf{x}(t)$ *over the next sampling interval is forecasted using the actual value of* $V(\mathbf{x}(t_{k+1}))$ *and* $\mathbf{e}(t_{k+1})$*, as well as the constraint on the Lyapunov function that will become active over the next sampling interval,* $[t_{k+1}, t_{k+2})$*, which is dictated by the location of* $\widehat{\mathbf{x}}(t_{k+1})$ *within* Ω *relative to* Ω'*. If the projection resulting from the forecast indicates that* $\mathbf{x}$ *will enter a smaller level set of V or* *lie within Ω', no update of the model state needs to be performed at t_{k+1} since stability would still be guaranteed over the next sampling interval; otherwise, $\widehat{\mathbf{x}}(t_{k+1})$ must be reset to the actual state $\mathbf{x}(t_{k+1})$ to suppress the potential instability. Note that the decision to perform or skip a model state update at a given sampling instance is triggered by the prediction of a future event (potential breach of worst-case growth bounds on the Lyapunov function and its time-derivative) instead of a current event (i.e., a simple comparison of the situations at the current and previous sampling instants).*

Remark 9. *The condition that $\bar{V}(\mathbf{x}(t_{k+2})) < V(\mathbf{x}(t_{k+1}))$ in Step 8 of Algorithm 1 is used as a criterion for skipping an update at t_{k+1} since satisfying this requirement is sufficient to guarantee closed-loop stability and can also minimize the possibility of performing unnecessary model state updates that merely improve control system performance. When reducing sensor–controller communication is not that critical, or when improved control system performance is an equally important objective, a more stringent requirement on the decay rate of the Lyapunov function can be imposed to help avoid frequent skipping of model state updates and enhance closed-loop performance at the cost of increased sensor–controller communication.*

Remark 10. *Notice that the upper bounds used in performing the forecasts of Equation (40) depend explicitly on the magnitude of the fault, Θ, which implies that faults can influence the update rate of the model state and the sensor–controller communication frequency required to attain it. The impact of faults on the sensor–controller communication rate can be mitigated through the use of active fault accommodation and exploiting the dependence of the forecasting bounds on $\widehat{\Theta}$ which can be used as a fault accommodation parameter and adjusted to help reduce any potential increase in the sensor–controller communication rate caused by the faults. To see how this works, we first note that the term describing the mismatch between Θ and $\widehat{\Theta}$ in Equation (40a) (i.e., $\|\widehat{\Theta} - \Theta\|$) tends to increase the upper bounds on $\dot{V}$ and V, and therefore cause the projected values of $V(\mathbf{x})$ over the next sampling interval to be unnecessarily conservative and large which would trigger more frequent breaches of Step 8 or 10 in Algorithm 1, resulting in increased communication frequency. Actively accommodating the fault by setting $\widehat{\Theta} = \Theta$ helps reduce the forecasting bounds and decrease the projected values of V which, in turn, would increase the likelihood of satisfying Step 8 or 10 in Algorithm 1, resulting in the ability to skip more unnecessary update and communication instances. This analysis suggests that, in addition to enhancing closed-loop performance, fault accommodation is desirable in terms of optimizing sensor–controller communication needs (see the simulation example for an illustration of this point). As noted in Remark 6, however, possible errors in estimating the fault magnitude can impact the implementation of this fault accommodation strategy and potentially limit the achievable savings in sensor–controller communication costs.*

Remark 11. *The implementation of the forecast-triggered fault-tolerant MPC scheme developed in this work requires the availability of full-state measurements. When only incomplete state measurements are available, an appropriate state estimator with appropriate estimation error convergence properties needs to be designed and incorporated within the control system to provide estimates of the actual states based on the available measurements. The use of state estimates (in lieu of the actual states) in implementing the control and communication policies introduces errors that must be accounted for at the design stage to ensure robustness of the closed-loop system. This can generally be done by appropriately modifying the constraints in the MPC formulation and the communication-triggering thresholds based on the available characterization of the state estimation error. Extension of the proposed MPC framework to tackle the output feedback control problem is the subject of other research work.*

6. Simulation Case Study: Application to a Chemical Process

The objective of this section is to demonstrate the implementation of the forecast-triggered fault-tolerant MPC developed earlier using a chemical process example. To this end, we consider a non-isothermal continuous stirred tank reactor (CSTR) with an irreversible first-order exothermic reaction of the form $A \xrightarrow{k_0} B$, where A is the reactant and B is the product. The inlet stream feeds pure A at flow rate F, concentration C_{A0} and temperature T_{A0} into the reactor. The process dynamics are captured by the following set of ordinary differential equations resulting from standard mass and energy balances:

$$\dot{C}_A = \frac{F}{V}(C_{A0} - C_A) - k_0 \exp\left(\frac{-E}{RT}\right) C_A \tag{41a}$$

$$\dot{T} = \frac{F}{V}(T_{A0} - T) - \frac{\Delta H}{\rho c_p} k_0 \exp\left(\frac{-E}{RT}\right) C_A + \frac{Q}{\rho c_p V} \tag{41b}$$

where C_A is the concentration of A in the reactor; T is the reactor temperature; V is the reactor volume; k_0, E, and ΔH represent the pre-exponential factor, the activation energy, and the heat of reaction, respectively; R denotes the ideal gas constant; c_p and ρ are the heat capacity and density of the fluid in the reactor, respectively; and Q is the rate of heat transfer from the jacket to the reactor. The process parameter values are given in Table 1.

Table 1. Process and model parameter values for the continuous stirred tank reactor (CSTR) example in Equation (41).

Parameter	Process	Model
F (m^3/h)	3.34×10^{-3}	3.34×10^{-3}
V (m^3)	0.1	0.1
k_0 (h^{-1})	1.2×10^9	1.2×10^9
E (KJ/Kmol)	8.314×10^4	8.30×10^4
R (KJ/Kmol/K)	8.314	8.314
ρ (Kg/m^3)	1000	1010
C_p (KJ/Kg/K)	0.239	0.24
ΔH (KJ/Kmol)	-4.78×10^4	-4.8×10^4
C_{A0}^s (Kmol/m^3)	0.79	0.79
T_0^s (K)	352.6	352.6
Q^s (KJ/h)	0	0

The control objective is to stabilize the process state near the open-loop unstable steady-state ($C_A^s = 0.577$ Kmol/m^3, $T^s = 395.3$ K) in the presence of input constraints, control actuator faults and limited sensor–controller communication. The manipulated input is chosen as the inlet reactant concentration, i.e., $\mathbf{u} = C_{A0} - C_{A0}^s$, subject to the constraint $\|\mathbf{u}\| \leq 0.5\,\text{mol/m}^3$, where C_{A0}^s is the nominal steady state value of C_{A0}, and control actuator faults. We define the displacement variables $\mathbf{x} = [x_1\ x_2]^T = [C_A - C_A^s\ \ T - T^s]^T$, where the superscript s denotes the steady state value, which places the nominal equilibrium point of the system at the origin. A quadratic Lyapunov function candidate of the form $V(\mathbf{x}) = \mathbf{x}^T\mathbf{P}\mathbf{x}$, where:

$$\mathbf{P} = \begin{bmatrix} 37,400 & 1394.9 \\ 1394.9 & 63.5389 \end{bmatrix} \tag{42}$$

is a positive-definite matrix, is used for the synthesis of the controller in Equation (5) and the characterization of the closed-loop stability region in Equations (8) and (9). The value of the tuning parameter λ is fixed at 0.1 to ensure an adequate margin of robustness while providing an acceptable estimate of the stability region.

6.1. Characterization of the Fault-Tolerant Stabilization Region

Recall from Section 4 that the MPC formulation in Equation (28) inherits its closed-loop stability region from the auxiliary bounded controller, and that this region is explicitly dependent on the magnitude of the fault (see Equations (8) and (9)). Figure 2 depicts the dependence of the constrained stability region on fault severity. Specifically, the blue region refers to $\Psi(u^{\max}|\widehat{\theta} = 1)$, i.e., when the actuator is perfectly healthy; while the green and purple regions represent $\Psi(u^{\max})$ when $\widehat{\theta} = 0.8$, and $\widehat{\theta} = 0.5$, respectively. As all three regions are projected on a single plot, the purple region is completely contained within the green region which is fully contained within the blue region, i.e., $\Psi(u^{\max}|\hat{\theta} = 0.5) \subset \Psi(u^{\max}|\hat{\theta} = 0.8) \subset \Psi(u^{\max}|\hat{\theta} = 1)$. The largest level set Ω within each region $\Psi(u^{\max})$ is represented by the ellipse with the corresponding darker color. The three level sets form concentric ellipses and follow the same trend with $\Omega(u^{\max}|\hat{\theta} = 0.5) \subset \Omega(u^{\max}|\hat{\theta} = 0.8) \subset \Omega(u^{\max}|\hat{\theta} = 1)$. Figure 2 shows that the stability region shrinks in size as $\hat{\theta}$ decreases and the severity of the fault increases. Each level set in this figure provides an estimate of the set of initial states starting from which closed-loop stability is guaranteed in the presence of the corresponding fault.

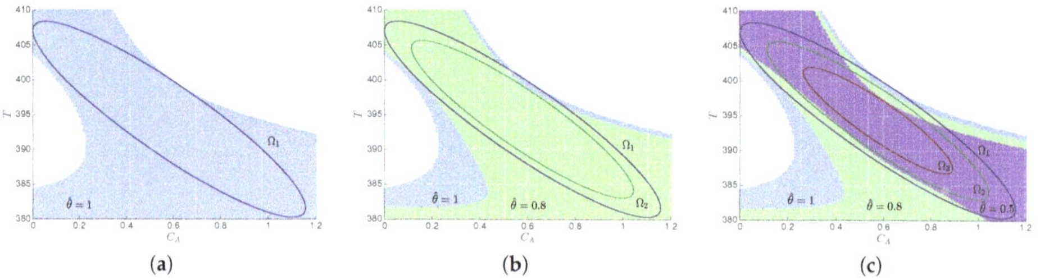

Figure 2. Estimates of the region of guaranteed fault-tolerant stabilization under MPC: (**a**) Ω_1 represents the estimate when $\hat{\theta} = 1$ (blue level set); (**b**) Ω_2 represents the estimate when $\hat{\theta} = 0.8$ (green level set); (**c**) Ω_3 represents the estimate when $\hat{\theta} = 0.5$ (purple level set).

6.2. Active Fault Accommodation in the Implementation of MPC

As discussed in Sections 3 and 4, the presence of control actuator faults generally reduces the stability region of the closed-loop system and enlarges the terminal set, which potentially compromises the stability and performance properties of the closed-loop system. The implementation of active fault accommodation measures such as adjusting the value of $\hat{\Theta}$ in the model, however, can help reduce the mismatch between the fault and its estimate used by the MPC, and therefore help reduce the size of the terminal set which can improves the closed-loop steady state performance. In this section, the MPC introduced in Equation (28) is implemented using the model parameters reported in Table 1 with an optimization horizon of 20 s and a sampling period of 2 s. The nonlinear optimization problem is solved using the standard "fmincon" algorithm in Matlab which generally yields locally optimal solutions. A step fault of $\theta = 0.8$ is introduced in the actuator at $t = 50$ s and persists thereafter.

Figure 3 compares the performance of the closed-loop system in the absence of faults (fault-free operation scenario in black) with the performance of the closed-loop system in the presence of faults (blue and red). The blue profiles depict the performance when the fault is accommodated, while the red profiles illustrate the performance when the fault is left unaccommodated. The dashed lines in Figure 3b,c represent the target steady-state values for the reactor temperature and reactant concentration, respectively. A steady-state offset resulting from the effect of discrete measurement sampling can be observed in Figure 3a–c, which indicates that with the uncertain model used, when the MPC is implemented in a sample-and-hold fashion, only ultimate boundedness can be achieved. The red profiles show that the fault pushes the closed-loop state trajectory away from the desired steady state and increases the size of the terminal set significantly. However, when accommodated by setting $\hat{\theta} = \theta = 0.8$ upon detection of the fault, a performance comparable to that obtained in the fault-free operation scenario can be achieved, as the blue profiles are very close to the black profiles in Figure 3a–c.

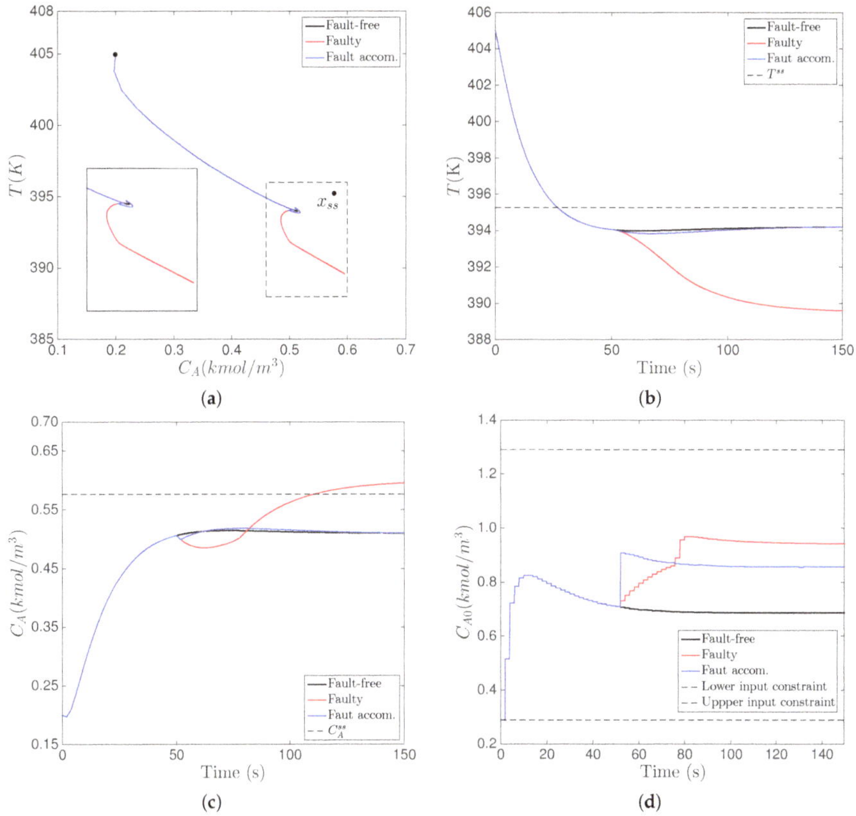

Figure 3. Comparison of the evolutions of: (**a**) the closed-loop state trajectory; (**b**) the closed-loop reactor temperature T; (**c**) the closed-loop reactant concentration C_A; and (**d**) the manipulated input, C_{A0}, for three different operating scenarios: one in the absence of any faults (black profiles); one in the presence of a fault but without implementing any fault accommodation (red profiles); and one in the presence of a fault and implementing fault accommodation (blue profiles).

6.3. Implementation of Fault-Tolerant MPC Using Forecast-Triggered Sensor–Controller Communication

In this section, we illustrate the forecast-triggered implementation strategy of MPC and highlight the resulting reduction in network resource utilization. To this end, we consider first the case of fault-free operation. Figure 4 illustrates the implementation of Algorithm 1. Each red square in the top plot represents the current value of $V(\mathbf{x})$ at the corresponding sampling instant t_k, and each blue circle represents the forecasted value of $V(\mathbf{x})$ calculated one sampling interval ahead. An update of the model state is triggered at a given sampling time t_k if either: (1) the forecasted $V(\mathbf{x}(t_{k+1}))$ is greater than the current $V(\mathbf{x}(t_k))$ (whenever $V(\mathbf{x}(t_k)) > \delta'$), or (2) $V(\mathbf{x}(t_k)) < \delta' < V(\mathbf{x}(t_{k+1}))$. The model state update events are depicted by the solid blue dots in the plot. The update profile shown in the bottom panel indicates the times when the model state updates take place. In this plot, a value of 1 denotes that an update event has occurred, while a value of zero indicates that an update has been skipped. Figure 4 captures only the case when $V(\mathbf{x}(t_k)) > \delta'$ (i.e., when the closed-loop trajectory lies outside the terminal set).

By examining Figure 4, it can be seen that at $t = 8$ s (the 4th sampling time), $V(\mathbf{x}(t_4)) = 586$ which is represented by the red square, and that $\bar{V}(\mathbf{x}(t_5))$ is forecasted to exceed the current value, which is represented by the blue dot at $t = 10$ s, which means that, without resetting the model state at $t = 8$ s, it is possible for $V(\mathbf{x}(t))$ to start to grow over the next sampling interval. To prevent the potential destabilizing tendency, a model state update is performed at $t = 8$ s as shown in the update profile at the bottom of the plot. Similarly, model state updates are triggered at $t = 20$ s and $t = 26$ s as a result of implementing the forecast-triggered communication strategy. At the other sampling instants, the condition of $V(\mathbf{x}(t_{k+1})) \leq V(\mathbf{x}(t_k))$ is satisfied and model state updates are not triggered.

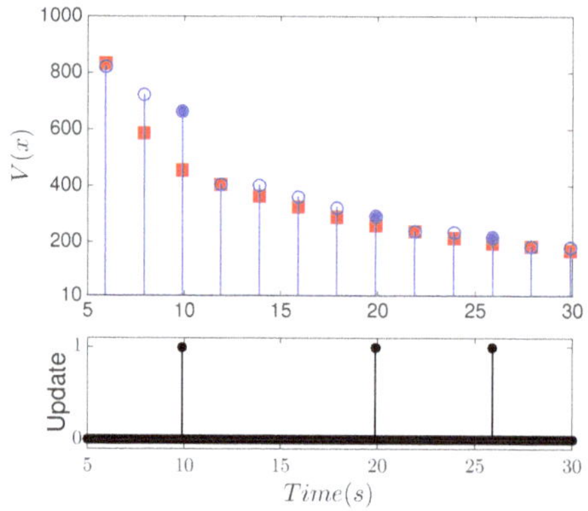

Figure 4. Illustration of how the forecast-triggered sensor–controller communication strategy is implemented. The top plot depicts current values of the Lyapunov function (red squares), projected values of the Lyapunov function (blue circles) and update events (solid blue dots) at different sampling times. The bottom plot depicts the time instances when the model state is updated.

The resulting closed-loop behavior is depicted by the red profile in Figure 5a which shows that the forecast-triggered communication strategy successfully stabilizes the reactor temperature near the desired steady state. Figure 5b compares the number of model state updates under a conventional MPC (where an update is performed at each sampling time) and the forecast-triggered MPC (where an update is performed only when triggered by a breach of the stability threshold). The comparison shows that stabilization using the forecast-triggered MPC requires only 14% of the model state updates over the same time interval, and is thus achieved with a significant reduction in the sensor–controller communication frequency.

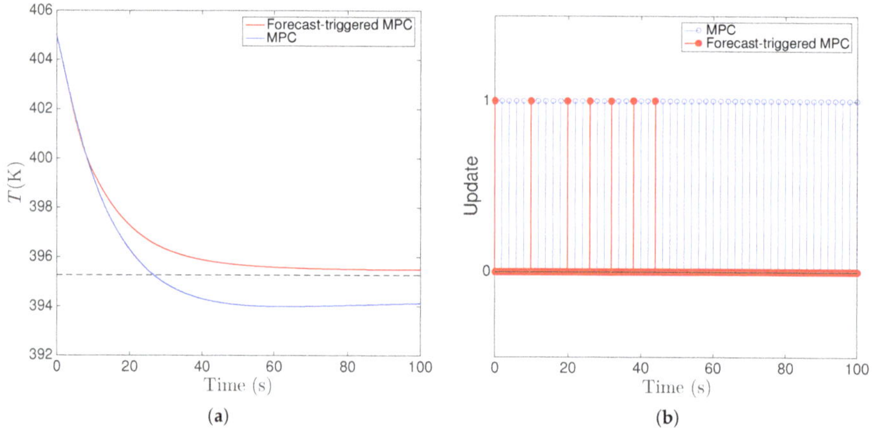

Figure 5. Closed-loop reactor temperature profiles (**a**); and model state update instances (**b**) under the conventional (blue) and forecast-triggered MPC schemes (red).

An examination of Figure 5a shows that the conventional MPC slightly outperforms the forecast-triggered MPC initially (i.e., during the transient stage) in the sense that it enforces a faster and more aggressive convergence of the closed-loop state. This is expected given the more frequent model state updates performed by the conventional MPC. It is interesting to note though that the forecast-triggered MPC exhibits a much smaller steady state offset (i.e., a smaller terminal set) despite the less frequent sensor–controller communication in this case. It should be noted, however, that the larger steady state offset achieved by the conventional MPC is not an indication of poorer performance, but is rather due to the different ways in which the two MPC schemes were implemented and the fact that for the event-triggered MPC a desired terminal set was specified a priori as part of the controller design and implementation logic, whereas for the conventional MPC a desired terminal set size was not specified. Specifically, for the forecast-triggered MPC, a small terminal set size was initially specified and then the sensor–controller communication logic was designed and implemented to keep the closed-loop state trajectory within that terminal set. For the conventional MPC, however, no specification of the desired terminal set was enforced. While it is possible, in principle, to specify the same terminal set for both MPC schemes, it was found that excessively fast sampling would be required to enforce the same tight convergence for the conventional MPC.

To demonstrate the benefits of active fault accommodation in the context of the forecast-triggered MPC scheme, we now consider the same fault scenario introduced earlier, where $\theta(t) = 0.8$ for $t \geq 50$ s. Figure 6 compares the performance of the closed-loop system in the fault-free scenario (shown in black) with those in the faulty operation cases, including the case when the actuator fault is accommodated (shown in blue) and the case when the actuator fault is left unaccommodated (shown in red). Similar to the result obtained in Figure 3, fault accommodation (realized by setting $\widehat{\theta} = \theta = 0.8$ at $t = 50$ s) reduces the steady-state offset and achieves closed-loop state profiles comparable to those obtained in the fault-free scenario. Figure 6d shows the corresponding model state update frequencies for the three cases. Recall that faults not only influence the closed-loop state performance, but can also negatively impact the projected bounds on the Lyapunov function or its derivative which are used in the forecasting strategy that triggers the model state updates. It can be seen from the middle plot in Figure 6d that when the fault is left unaccommodated an update of the model state is triggered at every sampling time after $t = 50$ s. In contrast, when the fault is appropriately accommodated, only three model state updates are needed after $t = 50$ s (see the bottom plot in Figure 6d) which yields an improved closed-loop performance. These results illustrate that the proposed fault accommodation strategy is beneficial for both performance improvement as well as network load reduction.

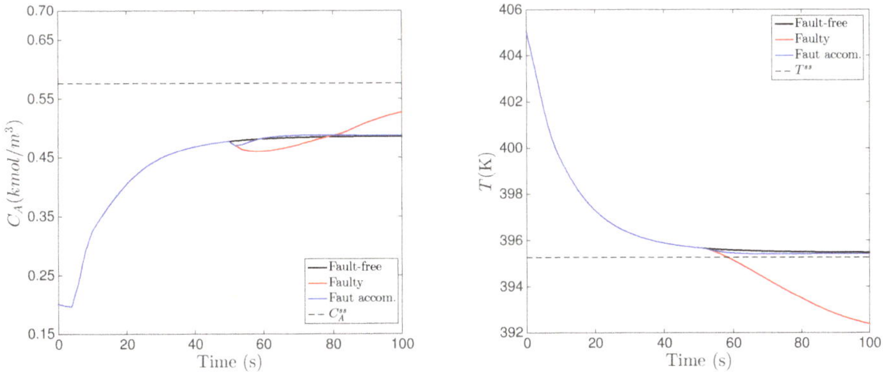

(a)

(b)

Figure 6. *Cont.*

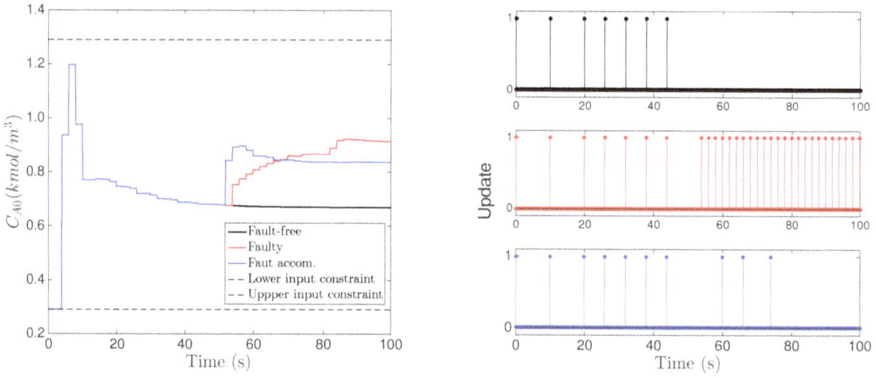

(c) (d)

Figure 6. Comparison of the performance of forecast-triggered MPC scheme under fault-free conditions (black), an accommodated fault scenario (blue) and an unaccommodated fault scenario (red): (**a**) closed-loop temperature profiles; (**b**) reactant concentration profiles; (**c**): manipulated input profile; and (**d**): model update frequency.

7. Conclusions

In this paper, a forecast-triggered fault-tolerant Lyapunonv-based MPC scheme is developed for constrained nonlinear systems with sensor–controller communication constraints. An auxiliary fault-tolerant bounded controller is initially designed to aid in the characterization of the region of fault-tolerant stabilization and subsequent design of the Lyapunov-based MPC. To handle sensor–controller communication constraints in the networked control system design, a forecast-triggered strategy for managing the sensor–controller information transfer is developed. In this strategy, model state updates using actual state measurements are triggered only when certain stability thresholds—derived based on a worst-case projection of the state trajectory—are breached. A simulation case study is presented to illustrate the implementation of the proposed MPC and its fault accommodation capabilities. The results show that the proposed design is effective in achieving closed-loop stability while simultaneously reducing communication network load.

Author Contributions: Conceptualization, N.H.E.-F. and D.X.; Methodology, N.H.E.-F. and D.X.; Validation, D.X.; Formal Analysis, D.X.; Writing—Original Draft Preparation, D.X.; Writing—Review & Editing, N.H.E.-F.; Supervision, N.H.E.-F.; Funding Acquisition, N.H.E.-F.

Funding: This research was funded by the US National Science Foundation, NSF, CBET-1438456.

Conflicts of Interest: The authors declare no conflict of interest.

References

1. Qin, S.J.; Badgwell, T.A. A survey of industrial model predictive control technology. *Control Eng. Pract.* **2003**, *11*, 733–764. [CrossRef]
2. Rawlings, J.B.; Mayne, D.Q. *Model Predictive Control: Theory and Design*; Nob Hill Publishing: Madison, WI, USA, 2009.
3. Christofides, P.D.; Liu, J.; de la Pena, D.M. *Networked and Distributed Predictive Control: Methods and Nonlinear Process Network Applications*; Springer: London, UK, 2011.
4. Ellis, M.; Liu, J.; Christofides, P.D. *Economic Model Predictive Control: Theory, Formulations and Chemical Process Applications*; Springer: London, UK, 2017.
5. Blanke, M.; Kinnaert, M.; Lunze, J.; Staroswiecki, M. *Diagnosis and Fault-Tolerant Control*; Springer: Berlin/Heidelberg, Germany, 2003.

6. Zhang, Y.; Jiang, J. Bibliographical Review on Reconfigurable Fault-Tolerant Control Systems. *Annu. Rev. Control* **2008**, *32*, 229–252. [CrossRef]
7. Mhaskar, P.; Liu, J.; Christofides, P.D. *Fault-Tolerant Process Control: Methods and Applications*; Springer: London, UK, 2013.
8. Christofides, P.D.; Davis, J.; El-Farra, N.H.; Clark, D.; Harris, K.; Gipson, J. Smart plant operations: Vision, progress and challenges. *AIChE J.* **2007**, *53*, 2734–2741. [CrossRef]
9. Mhaskar, P. Robust Model Predictive Control Design for Fault-Tolerant Control of Process Systems. *Ind. Eng. Chem. Res.* **2006**, *45*, 8565–8574. [CrossRef]
10. Dong, J.; Verhaegen, M.; Holweg, E. Closed-loop subspace predictive control for fault-tolerant MPC design. In Proceedings of the 17th IFAC World Congress, Seoul, Korea, 6–11 July 2008; pp. 3216–3221.
11. Camacho, E.F.; Alamo, T.; de la Pena, D.M. Fault-tolerant model predictive control. In Proceedings of the IEEE Conference on Emerging Technologies and Factory Automation, Bilbao, Spain, 13–16 September 2010; pp. 1–8.
12. Lao, L.; Ellis, M.; Christofides, P.D. Proactive Fault-Tolerant Model Predictive Control. *AIChE J.* **2013**, *59*, 2810–2820. [CrossRef]
13. Knudsen, B.R. Proactive Actuator Fault-Tolerance in Economic MPC for Nonlinear Process Plants. In Proceedings of the 11th IFAC Symposium on Dynamics and Control of Process Systems, Trondheim, Norway, 6–8 June 2016; pp. 1097–1102.
14. Hu, Y.; El-Farra, N.H. Quasi-decentralized output feedback model predictive control of networked process systems with forecast-triggered communication. In Proceedings of the American Control Conference, Washington, DC, USA, 17–19 June 2013; pp. 2612–2617.
15. Hu, Y.; El-Farra, N.H. Adaptive quasi-decentralized MPC of networked process systems. In *Distributed Model Predictive Control Made Easy*; Springer: Dordrecht, The Netherlands, 2014; Volume 69, pp. 209–223.
16. Christofides, P.D.; El-Farra, N.H. *Control of Nonlinear and Hybrid Process Systems: Designs for Uncertainty, Constraints and Time-Delays*; Springer: Berlin, Germany, 2005.
17. Mhaskar, P.; El-Farra, N.H.; Christofides, P.D. Stabilization of Nonlinear Systems with State and Control Constraints Using Lyapunov-Based Predictive Control. *Syst. Control Lett.* **2006**, *55*, 650–659. [CrossRef]
18. Allen, J.; El-Farra, N.H. A Model-based Framework for Fault Estimation and Accommodation Applied to Distributed Energy Resources. *Renew. Energy* **2017**, *100*, 35–43. [CrossRef]
19. Sanchez-Sanchez, K.B.; Ricardez-Sandoval, L.A. Simultaneous Design and Control under Uncertainty Using Model Predictive Control. *Ind. Eng. Chem. Res.* **2013**, *52*, 4815–4833. [CrossRef]
20. Bahakim, S.S.; Ricardez-Sandoval, L.A. Simultaneous design and MPC-based control for dynamic systems under uncertainty: A stochastic approach. *Comput. Chem. Eng.* **2014**, *63*, 66–81. [CrossRef]
21. Gutierrez, G.; Ricardez-Sandoval, L.A.; Budman, H.; Prada, C. An MPC-based control structure selection approach for simultaneous process and control design. *Comput. Chem. Eng.* **2014**, *70*, 11–21. [CrossRef]

mathematics

Article

Approximate Dynamic Programming Based Control of Proppant Concentration in Hydraulic Fracturing

Harwinder Singh Sidhu [1,2], **Prashanth Siddhamshetty** [1,2] **and Joseph S. Kwon** [1,2,*]

1 Artie McFerrin Department of Chemical Engineering, Texas A&M University, College Station, TX 77843, USA; harwindersingh289@gmail.com (H.S.S.); prashanth.s@tamu.edu (P.S.)
2 Texas A&M Energy Institute, Texas A&M University, College Station, TX 77843, USA
* Correspondence: kwonx075@tamu.edu; Tel.: +1-979-962-5930

Received: 16 June 2018; Accepted: 27 July 2018; Published: 1 August 2018

Abstract: Hydraulic fracturing has played a crucial role in enhancing the extraction of oil and gas from deep underground sources. The two main objectives of hydraulic fracturing are to produce fractures with a desired fracture geometry and to achieve the target proppant concentration inside the fracture. Recently, some efforts have been made to accomplish these objectives by the model predictive control (MPC) theory based on the assumption that the rock mechanical properties such as the Young's modulus are known and spatially homogenous. However, this approach may not be optimal if there is an uncertainty in the rock mechanical properties. Furthermore, the computational requirements associated with the MPC approach to calculate the control moves at each sampling time can be significantly high when the underlying process dynamics is described by a nonlinear large-scale system. To address these issues, the current work proposes an approximate dynamic programming (ADP) based approach for the closed-loop control of hydraulic fracturing to achieve the target proppant concentration at the end of pumping. ADP is a model-based control technique which combines a high-fidelity simulation and function approximator to alleviate the "curse-of-dimensionality" associated with the traditional dynamic programming (DP) approach. A series of simulations results is provided to demonstrate the performance of the ADP-based controller in achieving the target proppant concentration at the end of pumping at a fraction of the computational cost required by MPC while handling the uncertainty in the Young's modulus of the rock formation.

Keywords: approximate dynamic programming (ADP); model predictive control (MPC); hydraulic fracturing; model reduction; Kalman filter

1. Introduction

Petroleum and natural gas remain an important part of the global energy supply. Recently, the extraction of underground resources such as shale gas and oil, which are trapped in low porosity and ultra-low permeability formations, has become economically feasible due to the application of well-stimulation techniques such as hydraulic fracturing [1] and directional drilling [2]. Since its introduction in the 1940s, hydraulic fracturing has gradually developed as a standard practice and has been applied to various types of reservoir formations, and thus it has made significant contributions to the oil and gas industry [3].

In hydraulic fracturing, the ultimate goal is to enhance the productivity of a stimulated (i.e., fractured) well. The process begins with a step referred to as "perforation", in which small explosives are set off at spaced intervals at the wellbore to create initial fracture paths. Next, a fluid called pad is injected at a high pressure to initiate fractures of the rock at the perforated sites. Subsequently, a fracturing fluid called dirty volume consisting of water, additives, and proppant is pumped into the wellbore at sufficiently high pressure and flow rate to further propagate the fracture in the rock formation. Finally, the pumping is stopped, and the fractures are closed due to the natural stress of the rock formation. During the closure process, the remaining fluid seeps into the reservoir and the proppant is trapped inside the fracture. At the end of pumping, the concentration of the proppant should be uniform along the fracture to achieve a highly conductive channel which will result in effective extraction of oil and gas from the reservoir. The overall efficiency of the hydraulic fracturing process depends on (1) the uniformity of proppant concentration across the fracture at the end of pumping and (2) the final fracture geometry.

To produce a fracture with uniform proppant concentration across the fracture and the desired fracture geometry, it is important to generate an optimal pumping schedule. Several efforts in this direction were initially made by Nolte [4], Gu and Desroches [5] and Dontsov and Peirce [6]. Specifically, Nolte [4] developed a power-law type pumping schedule based on the conservation of fluid volume; Gu and Desroches [5] proposed a pumping schedule design technique using a detailed forward numerical simulator; and Dontsov and Peirce [6] designed a pumping schedule by taking into account proppant transport in their forward model.

The aforementioned techniques viewed hydraulic fracturing processes as an open-loop problem. Motivated by some advances in real-time measurement techniques such as downhole pressure analysis and microseismic monitoring, several attempts have recently been made to employ model predictive control (MPC) theory to regulate the fracture geometry and proppant concentration. Specifically, the limited availability of real-time measurements has been addressed by utilizing state estimators [7–9], and several model order-reduction (MOR) techniques [10–12] have been developed to handle the large computational requirements due to dynamic simulation of multiple highly-coupled partial differential equations (PDEs) defined over moving boundaries to describe the hydraulic fracturing process. However, there are two unresolved issues with the MPC approach. First, it is necessary to handle a potentially exorbitant online computational requirement due to the simulation of a nonlinear large-scale system at each sampling time (which is usually the case in hydraulic fracturing) and the use of a long prediction/control horizon to ensure satisfactory performance [13–15]. Second, the conventional MPC solves a deterministic open-loop optimal control problem at each sampling time. Therefore, it ignores the uncertainty and the feedback at future sampling times [16,17].

The above-mentioned limitations of MPC formulation can be handled by the approximate dynamic programming (ADP) approach [18], particularly circumventing the "curse-of-dimensionality" of the traditional dynamic programming (DP) approach. ADP is a model-based control technique and can be employed to derive an improved control policy, starting with some sub-optimal control policies (or, alternatively, closed-loop identification data). In recent years, ADP has been successfully applied to several applications such as a complex microbial cell reactor characterized by multiple steady states [19], Van de Vusse reaction in an isothermal CSTR [20], integrated plants with a reactor and a distillation column with a recycle [21], and systems described by hyperbolic PDEs [22–24]. Motivated by these earlier efforts, we present an ADP-based control framework for the closed-loop operation of a hydraulic fracturing process to achieve uniform proppant concentration across the fracture at the end of pumping.

The organization of this paper is as follows: first, a brief introduction of ADP is presented. Second, a high-fidelity model of hydraulic fracturing is constructed based on first-principles. Finally, we discuss the application of ADP to a hydraulic fracturing process and present a series of simulation results that demonstrates the superiority of the ADP-based controller over the standard MPC system in achieving uniform proppant concentration at the end of pumping, which is directly related to the overall productivity of a fractured well.

2. Approximate Dynamic Programming

Consider an optimal control problem with the set of all possible states and inputs represented by $\mathcal{X} \subset \mathbb{R}^{n_x}$ and $\mathcal{U} \subset \mathbb{R}^{n_u}$, respectively, where n_x and n_u are the number of state and manipulated input variables, respectively. For a deterministic state space system, an optimal state feedback policy can be determined by formulating an optimization problem as follows:

$$\min_{\mu \in \Pi} \quad \sum_{k=0}^{\infty} \Phi(x(t_k), u(t_k)) \tag{1a}$$

$$\text{s.t.} \quad u(t_k) = \mu(x(t_k)) \tag{1b}$$

$$g(x(t_k), u(t_k)) \geq 0 \quad k = 0, 1, \cdots \tag{1c}$$

$$x(t_{k+1}) = f(x(t_k), u(t_k)) \tag{1d}$$

where $x(t_k) \in \mathcal{X}$ is the system state vector and $u(t_k) \in \mathcal{U}$ is the manipulated input vector at $t = t_k$ (i.e., k^{th} sampling time), μ is the function mapping $x(t_k)$ to $u(t_k)$, Π is the set of all valid policies over which optimal μ is to be found, f is the model describing the time evolution of the system states, and $\Phi(x(t_k), u(t_k))$ is the single-stage cost incurred at state $x(t_k)$ at $t = t_k$ while implementing the control move $u(t_k)$.

DP is an alternative approach for solving multi-stage optimal control problems [25]. In DP, we define "cost-to-go" function, denoted by $J^{\mu}(x)$, of a starting state x as a sum of single-stage costs incurred from the state x under the control policy μ over the infinite horizon:

$$J^{\mu}(x) = \sum_{k=0}^{\infty} \Phi(x(t_k), u(t_k)), \quad x(t_0) = x \tag{2}$$

where it is assumed that $J^{\mu}(x)$ is well-defined over the entire $\mathcal{X}$. The objective of DP is to obtain the infinite horizon optimal cost-to-go function (J^{opt}) that satisfies the following Bellman equation:

$$J^{opt}(x(t_k)) = \min_{u(t_k) \in \mathcal{U}} [\Phi(x(t_k), u(t_k)) + J^{opt}(x(t_{k+1}))], \ \forall x \in \mathcal{X} \tag{3}$$

Once the optimal cost-to-go function J^{opt} is obtained, it can be subsequently employed to find the optimal input profile by solving the following point-wise single-stage optimization at every sampling time t_k:

$$\mu^{opt}(x(t_k)) = \arg \min_{u(t_k) \in \mathcal{U}} [\Phi(x(t_k), u(t_k)) + J^{opt}(x(t_{k+1}))] \tag{4}$$

There are very few problems such as linear quadratic (Gaussian) optimal control problem for which the Bellman equation can be solved analytically. Alternatively, numerical approaches such as "value iteration" or "policy iteration" can be employed. For systems with continuous state space, these numerical approaches can be employed either by discretizing the state space, or by using a finite dimensional parameterization. However, this can lead to potentially exorbitant computational requirements as the state dimension increases, which is referred to as "curse-of-dimensionality". Therefore, DP has been considered largely impractical for almost all the problems of practical interest.

The curse-of-dimensionality problem of DP can be handled by the ADP approach [18]. In contrast to the numerical approach to compute solutions for DP, which obtains the optimal cost-to-go function for the entire continuous state space, the ADP approach limits the control law calculations to only the relevant regions of the state space. These relevant regions are identified by performing closed-loop simulations of the system under some known heuristics or sub-optimal control policies. From the principle of optimality of DP, an improved control policy can be derived by solving the Bellman equation in a recursive manner within the sampled domain of the state space (which is a discrete set consisting of sampled states, and denoted by X_{sample}). For a system with continuous state space, it is not feasible to restrict the Bellman iterations to a set of discrete sampled states only. Therefore, a function approximator (denoted by $\tilde{J}(x)$) is used to interpolate cost-to-go values within the sampled states to approximate cost-to-go values in the original continuous state space. In this work, we used

K-nearest neighbors (KNN) as a function approximator. In KNN, the cost-to-go at any query point x^0 is computed as a distance-weighted average of its K-nearest neighbors:

$$\tilde{J}(x^0) = \sum_{j=1}^{K} w_j \tilde{J}(x_j) \tag{5}$$

where

$$w_j = \frac{1/d_j}{\sum_{i=1}^{K} 1/d_i} \tag{6}$$

and $d_1 \leq d_2 \leq \cdots$ are the Euclidean distances of $x_1, x_2, \cdots$ from the query point. x_1 is the closest point from x^0, followed by x_2, and so on.

3. Application of Approximate Dynamic Programming to a Hydraulic Fracturing Process

3.1. Dynamic Modeling of Hydraulic Fracturing

In this work, a large-scale process model of hydraulic fracturing is developed by adopting the following assumptions: (1) fracture propagation is described by Perkins, Kern, and Nordgren (PKN) model as shown in Figure 1 [26]; (2) the layers above and below have sufficiently large stresses such that the vertical fracture is confined within a single horizontal rock layer; (3) the fracture length is much greater than fracture width; (4) the fluid pressure is constant in the vertical direction; and (5) the fracture is surrounded by an isotropic homogenous elastic material.

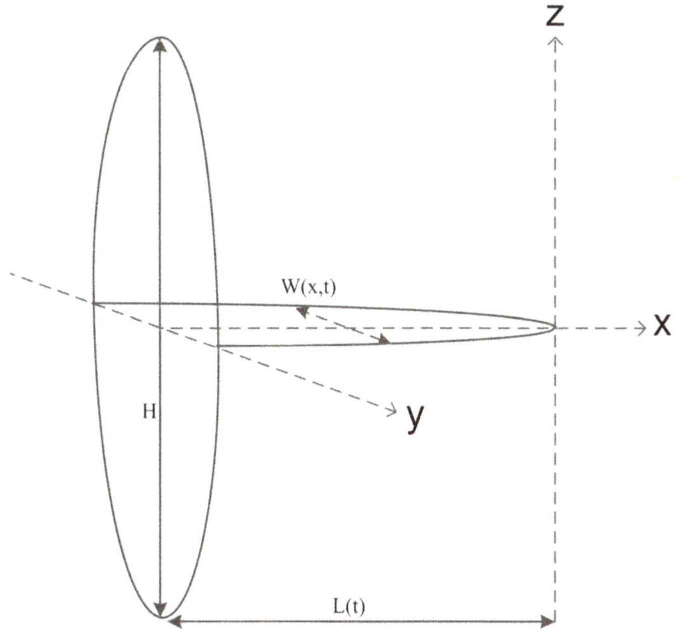

Figure 1. The PKN fracture model.

The dynamic modeling of hydraulic fracturing involves two sub-processes: fracture propagation and proppant transport. A brief description of the governing equations is presented below. The fluid flow rate in the horizontal direction is determined by the following equation for flow of a Newtonian fluid in an elliptical section [2,27]:

$$\frac{dP}{dx} = -\frac{64\mu Q}{\pi H W^3} \tag{7}$$

where P is the net pressure, $x \in [0, L(t)]$ is the time-dependent spatial coordinate in the horizontal direction, μ is the fluid viscosity, Q is the fluid flow rate in the horizontal direction, H is the predefined fracture height, and W is the fracture width.

For a crack under constant normal pressure, the fracture shape is elliptical as shown in Figure 1. The relationship between the maximum fracture width (i.e., the minor axis of the ellipse) and the net pressure is given as follows [28,29]:

$$W = \frac{2PH(1-\nu^2)}{E} \tag{8}$$

where ν is the Poisson ratio of the formation, and E is the Young's modulus of the formation.

By taking into account the fracture volume changes and the fluid leak-off into the reservoir, the continuity equation for an incompressible fluid flow inside the fracture is given by [27]:

$$\frac{\partial A}{\partial t} + \frac{\partial Q}{\partial x} + HU = 0 \tag{9}$$

where $A = \pi WH/4$ is the cross-sectional area of the elliptic fracture, and U is the fluid leak-off per unit height accounting for both fracture walls, which is determined by the following expression [2,30]:

$$U = \frac{2C_{leak}}{\sqrt{t - \tau(x)}} \tag{10}$$

where C_{leak} is the overall leak-off coefficient, t is the elapsed time since fracturing was initiated, and $\tau(x)$ is the time at which the fracture propagation has arrived at x for the first time.

Plugging Equations (7) and (8) into Equation (9) results in the following nonlinear parabolic PDE:

$$\frac{\pi H}{4}\frac{\partial W}{\partial t} - \frac{\pi E}{128\mu(1-\nu^2)}\left[3W^2\left(\frac{\partial W}{\partial x}\right)^2 + W^3\frac{\partial^2 W}{\partial x^2}\right] + HU = 0 \tag{11}$$

The initial condition for solving the above equations is that the fracture is closed, that is $W(x,0) = 0$. In addition, the two boundary conditions are considered as follows:

1. At the wellbore, the fluid flow rate is specified by $Q(x,t) = Q_0(t)$, where $Q_0(t)$ is the fluid injection rate (i.e., the manipulated variable).
2. At the fracture tip, $x = L(t)$, the fracture is always closed, that is $W(L(t),t) = 0$.

The modeling of the injected proppant transport is based on the following assumptions: (1) along the horizontal direction, the injected proppant will travel at the fracturing fluid's velocity; (2) the suspended proppant will settle towards the fracture bottom due to the gravitational force which will lead to the formation of a proppant bank; (3) proppant particles are sufficiently large so that the diffusive flux can be neglected when the convective flux is considered; and (4) because of low proppant concentration, the interactions between the individual particles are neglected, while the drag and gravitational forces acting on the proppant particles are still considered. Based on these assumptions, the advection of the suspended proppant can be expressed by the following equation:

$$\frac{\partial(WC)}{\partial t} + \nabla\cdot(WCV_p) = 0 \tag{12}$$

$$C(0,t) = C_0(t) \text{ and } C(x,0) = 0 \tag{13}$$

where C is the suspended proppant concentration, ∇ is the vector differential operator, V_p is the velocity with which the proppant particles are advected, and $C_0(t)$ is the inlet proppant concentration at the wellbore (i.e., the manipulated variable).

The net velocity of the proppant particles, V_p, is dependent on the suspended proppant concentration, C, the fluid velocity, V, and the gravitational settling velocity, V_s, which is given by the following expression [31]:

$$V_p = V - (1 - C)V_s \tag{14}$$

The gravity-induced proppant settling velocity in the fracturing fluid, V_s, is computed as follows [32]:

$$V_s = \frac{(1-C)^2}{10^{1.82C}} \frac{(\rho_{sd} - \rho_f)gd^2}{18\mu} \tag{15}$$

where ρ_{sd} is the proppant particle density, ρ_f is the pure fluid density, g is the gravitational acceleration constant, d is the proppant diameter, and μ is the fracture fluid viscosity where its relation with the proppant concentration can be described by the following empirical model [33]:

$$\mu(C) = \mu_0 \left(1 - \frac{C}{C_{max}}\right)^{-\beta} \tag{16}$$

where μ_0 is the pure fluid viscosity, β is an exponent in the range of 1.2 to 1.8, and C_{max} is the theoretical maximum proppant concentration determined by $C_{max} = (1-\phi)\rho_{sd}$ where ϕ is the proppant bank porosity. The particles settle out of the flow to the fracture bottom and form a proppant bank. The evolution of proppant bank height, δ, via the proppant settling is given as follows [34,35]:

$$(1-\phi)\frac{d(\delta W)}{dt} = CV_sW \tag{17}$$

where proppant bank is initially of vanishing thickness, so the initial condition is that $\delta(x,0) = 0$.

The above-mentioned PDE-ODE systems defined over time-dependent spatial domains will be solved using an in-house simulator developed by Siddhamshetty et al. [8].

3.2. *Obtaining Cost-to-Go Function Offline*

3.2.1. Simulation of Sub-Optimal Control Policies for ADP

In this section, a nonlinear MPC formulation is presented to obtain sub-optimal control policies for ADP. First, we employed unified fracture design (UFD) technique to obtain the optimal fracture length (L_{opt}) and width (W_{opt}) for a specific amount of proppant, M_{prop}, to be injected into a reservoir well [36]. Additionally, the target proppant concentration, C_{target}, at the end of hydraulic fracturing is calculated as follows:

$$C_{target} = \frac{M_{prop}}{HL_{opt}W_{opt}} \tag{18}$$

The high-fidelity process model described in Equations (7)–(17) demands high computational requirements, and thus, it cannot be directly employed for the controller design. While there are a variety of computationally efficient linear [11,37,38] and nonnlinear [10,39–42] MOR techniques available, in this work, we developed a reduced-order model (ROM) based on multivariate output error state space (MOESP) algorithm using the simulation results from the high-fidelity model as described in [8]. The developed ROM is presented as follows:

$$x(t_{k+1}) = Ax(t_k) + Bu(t_k) \tag{19a}$$

$$y(t_k) = Hx(t_k) \tag{19b}$$

where $x(t_k)$ represents the vector of states in the state space model at time instant t_k, A, B and H represent the system matrices, $y(t_k) = [W_0(t_k), L(t_k), C(x_1, t_k), C(x_2, t_k), C(x_3, t_k), C(x_4, t_k), C(x_5, t_k), C(x_6, t_k)]^T$ represents the vector of output variables, $W_0(t_k)$ is the fracture width at the wellbore, $L(t_k)$ is the fracture length, $C(x_1, t_k), \cdots, C(x_6, t_k)$ are the proppant concentrations at six different locations across the fracture, and $u(t_k) = [C_0(t_k), C_0(t_k - \zeta_{x_1}), C_0(t_k - \zeta_{x_2}), C_0(t_k - \zeta_{x_3}), C_0(t_k - \zeta_{x_4}), C_0(t_k - \zeta_{x_5}), C_0(t_k - \zeta_{x_6})]^T$ is the inlet proppant concentration at the wellbore (i.e., the manipulated input variable), ζ_{x_i} is the input time-delay due to the time required for the proppant to travel from the wellbore to a particular location x_i.

Remark 1. *Please note the linear discrete-time state space model is good for the purpose of this study. To develop the ROM, we varied the input profile so that we can cover the entire range of operating conditions that are being considered in the oil reservoir field. Alternatively, a nonlinear ROM can be used to improve the controller performance as the governing equation is indeed a nonlinear parabolic PDE with the moving boundary.*

In hydraulic fracturing, the readily available real-time measurements are limited to the fracture length and the fracture width at the wellbore, which are provided via the processed microseismic and downhole pressure data, respectively [43]. To estimate the remaining important state variables such as the proppant concentration across the fracture, we designed a Kalman filter by adding the process and measurement noise to the ROM presented in Equation (19) as described in [8]:

$$x(t_{k+1}) = Ax(t_k) + Bu(t_k - \zeta) + v(t_k) \tag{20a}$$

$$y(t_k) = Hx(t_k) + w(t_k) \tag{20b}$$

where v denotes the process noise, and w denotes the measurement noise. The process noise is assumed to be drawn from a zero mean multivariate normal distribution with covariance Q, and the measurement noise is assumed to be zero mean Gaussian white noise with covariance R.

The state estimator algorithm works in a two-step process: prediction and measurement update. Combining these two steps, the Kalman filter equations can be written as follows:

$$\hat{x}(t_{k+1}) = A\hat{x}(t_k) + Bu(t_k - \zeta) + M(t_k)(y_m(t_k) - \hat{y}(t_k)) \tag{21a}$$

$$M(t_k) = P(t_k)H^T(R(t_k) + HP(t_k)H^T)^{-1} \tag{21b}$$

$$P(t_{k+1}) = (I - M(t_k)H)P(t_k) \tag{21c}$$

where the operator $(\hat{\cdot})$ is used to denote the estimated variables, $M(t_k)$ is the Kalman filter gain, and $P(t_k)$ denotes the covariance of the state estimation error.

To determine sub-optimal control policies for ADP, we employed the following MPC scheme designed by Siddhamshetty et al. [8]:

$$\min_{C_{stage,k}} \quad (\mathbf{C}(t_f) - C_{target}\mathbb{1})^T Q_c (\mathbf{C}(t_f) - C_{target}\mathbb{1}) \tag{22a}$$

$$\text{s.t.} \quad \text{ROM, Equation (19)} \tag{22b}$$

$$\text{Kalman filter, Equation (21)} \tag{22c}$$

$$C_{min}\mathbb{1} \leq \mathbf{C}(t_k + j\Delta) \leq C_{max}\mathbb{1}, \ \forall j = 0, \cdots, 10 - k \tag{22d}$$

$$C_{stage,k-1+m} \leq C_{stage,k+m} \leq C_{stage,k-1+m} + 4 \text{ (ppga)}, \ m = 1, \cdots, 10 - k \tag{22e}$$

$$2Q_0\Delta\left(\sum_k C_{stage,k}\right) = M_{prop} \tag{22f}$$

$$L(t_f) = L_{opt}, \ W_0(t_f) \geq W_{opt} \tag{22g}$$

where t_f denotes the total treatment time, Q_c is a positive definite matrix used to compute the weighted norm, t_k is the current time, Δ is the time interval between sampling times,

$\mathbf{C}(t_k) = [C_1(t_k), C_2(t_k), C_3(t_k), C_4(t_k), C_5(t_k), C_6(t_k)]^T$ is the proppant concentration inside the fracture at six different locations at $t = t_k$, $\mathbb{1}$ is a 6×1 vector whose elements are all ones, $W_0(t_k)$ and $L(t_k)$ are the only readily available real-time measurements of the fracture width at the wellbore and the fracture length at $t = t_k$, respectively, and $C_{stage,k}$ is the inlet proppant concentration (i.e., the manipulated input) corresponding to the kth time interval i.e., $t \in [t_k, t_{k+1})$, which can be computed by solving Equation (22) with a shrinking prediction horizon $N_p = t_f - t_k$.

In the above optimization problem, the penalty function, Equation (22a), computes the squared deviation of the proppant concentration from the set-point at 6 different locations across the fracture at the end of pumping. At every sampling time t_k, the Kalman filter of Equation (21) is initialized to estimate the proppant concentration $\hat{\mathbf{C}}$ by using the real-time measurements of the fracture width at the wellbore and the fracture length. The constraint of Equation (22d) imposes limits on the concentration profiles to avoid premature termination of the hydraulic fracturing process. The constraint of Equation (22e) demands a monotonic increase in the input proppant concentration with an increment less than 4 ppga/stage, where ppga is a concentration unit used in petroleum engineering that refers to one pound of proppant added to a gallon of water. The constraint of Equation (22f) specifies the total amount of proppant to be injected. The terminal constraint of Equation (22g) employs the optimal fracture geometry, which is calculated by UFD scheme described in the preceding paragraph.

The dynamic model described in Section 3.1 was utilized to simulate the hydraulic fracturing process using the parameters listed in Table 1. In hydraulic fracturing, the characterization of rock mechanical properties is one of the key tasks that has to be performed prior to the model based controller design. This requires the availability of field data. Currently, in the field, a small-scale experiment called the mini-frac test is performed to collect preliminary data that can be used to characterize the geological properties. However, in this work, the model parameters utilized to simulate the hydraulic fracturing process are taken from literature [8,44]. Specifically, we considered 48,000 kg of proppant amount to be injected during the entire hydraulic fracturing process. For this fixed amount of proppant, we employed UFD scheme to obtain the corresponding optimal fracture length $L_{opt} = 135$ m and width $W_{opt} = 5.4$ mm, which were used as the optimal fracture geometry constraint (Equation (22g)) in the MPC formulation that has to be satisfied at the end of pumping. Then, the target proppant concentration at the end of pumping, $C_{target} = 9.765$ ppga, was calculated using Equation (18). The positive definite matrix, Q_c, was considered to be a diagonal matrix with the diagonal entries equal to 100. The pad time, t_p, was fixed to be 220 s and the constant flow rate of $Q_0 = 0.03$ m^3/s was used after the pad time. The Kalman filter and feedback control systems were initialized at the end of pad time (i.e., $t_k \geq t_p$). In the closed-loop simulation, t_f and Δ were chosen to be 1220 s and 100 s, respectively. The proppant pumping schedule was divided into 10 substages and the duration of each substage was identical to Δ. At each sampling time, the controller was called and the first input, $C_{stage,k}$, of the entire input profile ($C_{stage,k}, C_{stage,k+1}, \cdots$) obtained by solving the optimization problem over a prediction horizon length of N_p was applied to the dynamic model in a sample-and-hold fashion, and this procedure was repeated at each sampling time until the end of treatment. Please note that the controller performance can be improved by increasing the number of proppant pumping substages while maintaining the number of spatial locations across the fracture, at which we want to achieve the uniform proppant concentration [8]. However, it is not viable to have a large number of substages, and hence, we selected 10 substages because we did not observe significant improvement after that.

Remark 2. *Please note that Kalman filter is applied before the MPC optimization. Specifically, at every sampling time t_k, the real-time measurements of the fracture width at the wellbore, $W_0(t_k)$, and the fracture length, $L(t_k)$, were obtained from the high-fidelity model, which is a virtual experiment. These measurements were then utilized by Kalman filter of Equation (21) to estimate the (unmeasurable) proppant concentration inside the fracture at $t = t_k$. Finally, the inlet proppant concentration (i.e., the manipulated input), $C_{stage,k}$, was obtained by* *solving the optimization problem with the cost function of Equation (22a) under the constraints described by Equations (22b) and (22d)–(22g).*

Table 1. Model parameters used for the simulation.

Parameter	Symbol	Value
Leak-off coefficient	C_{leak}	6.3×10^{-5} m/s$^{1/2}$
Maximum concentration	C_{max}	0.64
Minimum concentration	C_{min}	0
Young's modulus	E	0.5×10^{10} Pa
Proppant permeability	k_f	60,000 mD
Formation permeability	k_r	1.5 mD
Vertical fracture height	H	20 m
Proppant particle density	ρ_{sd}	2648 kg/m^3
Pure fluid density	ρ_f	1000 kg/m^3
Fracture fluid viscosity	μ	0.56 Pa·s
Poisson ratio of formation	ν	0.2

The data covering the relevant regions in the continuous state space was obtained from the closed-loop simulations under the above-mentioned MPC scheme by initializing the Kalman filter and the feedback control systems with different fracture width at the wellbore and fracture length values at $t = t_p$, which were obtained by varying the flow rate Q_0 during the pad time—i.e., $t \in [0, t_p)$. In this work, the pad time t_p was fixed, and therefore, the fracture width at the wellbore and the fracture length at the end of pad time depended only on the flow rate Q_0. Every different Q_0 profile during the pad time would result in a unique combination of the fracture width at the wellbore and the fracture length at the end of pad time that allows the feedback system to compute a control input profile, which is dissimilar to other profiles, if their Q_0 during the pad time is different. In each closed-loop simulation, the real-time measurements of the fracture width at the wellbore and the fracture length were taken at 10 time instants (i.e., 10 substages). At each time instant, the real-time measurement was utilized by the Kalman filter to estimate the (unmeasurable) proppant concentration inside the fracture. Each measurement can be represented as $y(t_k) = [W_0(t_k), L(t_k)]^T$ and the corresponding estimated state can be represented as $\hat{\mathbf{C}}(t_k) = [\hat{C}_1(t_k), \hat{C}_2(t_k), \hat{C}_3(t_k), \hat{C}_4(t_k), \hat{C}_5(t_k), \hat{C}_6(t_k)]^T$ where $\hat{C}_i(t_k)$ is the proppant concentration inside the fracture at a specific location i at $t = t_k$. Twenty-four closed-loop simulations were performed by initializing the Kalman filter and the feedback control systems with 24 different initial conditions, resulting in 240 real-time measurements (denoted by $X_{measure}$) and the corresponding set of 240 estimated states (denoted by X_{est}).

Remark 3. *In hydraulic fracturing, the real-time measurement of the proppant concentration inside the fracture is not available due to the remote subterranean location where the fracture propagates. To overcome this challenge, we used the Kalman filter to estimate the proppant concentration based on the available real-time measurements of the fracture width at the wellbore and the fracture length. Therefore, we use the term "estimated states" in Section 3 instead of "sampled states" as described in Section 2.*

3.2.2. Initial Cost-to-Go Approximation

For all of the estimated states during the closed-loop simulations, the initial cost-to-go values (J^{μ^0}) were computed as follows:

$$J^{\mu^0}(\hat{\mathbf{C}}(t_k)) = \sum_{j=k}^{N_t} \Phi(\hat{\mathbf{C}}(t_j), u(t_j)) \tag{23}$$

where $N_t = 10$ is the number of time instants where the proppant concentrations are estimated by the Kalman filter during each closed-loop simulation, and $\Phi(\hat{\mathbf{C}}(t_j), u(t_j))$ is given by the following equation:

$$\Phi(\hat{\mathbf{C}}(t_j), u(t_j)) = (\hat{\mathbf{C}}(t_{j+1}) - C_{target}\mathbb{1})^T Q_c(\hat{\mathbf{C}}(t_{j+1}) - C_{target}\mathbb{1}) \tag{24}$$

A function approximator, denoted as $\tilde{J}^{\mu^0}(\hat{\mathbf{C}})$, was constructed to obtain the mapping between initial cost-to-go values and the estimated states (i.e., proppant concentration $\hat{\mathbf{C}}$) obtained during the closed-loop simulations. We used KNN (with $K = 5$) as a function approximator. In hydraulic fracturing, the real-time measurements readily available are limited to the fracture width at the wellbore and the fracture length. Therefore, we used only these two available measurements to determine the KNN. The function approximator works in the following way. Suppose there is a new measurement y. First, we determine the KNN of y from the set $X_{measure}$, which consists of 240 real-time measurements obtained during the closed-loop simulations. Let the KNN of y be $y_1, y_2, \cdots, y_K$ with the Euclidean distances $d_1, d_2, \cdots, d_K$, respectively, from y. Second, we select K proppant concentration vectors $\hat{\mathbf{C}}_1, \hat{\mathbf{C}}_2, \cdots, \hat{\mathbf{C}}_K$ from X_{est} which were estimated by the Kalman filter utilizing $y_1, y_2, \cdots, y_K$ during the closed-loop simulations, respectively. Then, the selected K proppant concentration vectors $\hat{\mathbf{C}}_1, \hat{\mathbf{C}}_2, \cdots, \hat{\mathbf{C}}_K$ are used in Equation (25) to determine the cost-to-go at a new state $\hat{\mathbf{C}}$ estimated by the Kalman filter utilizing the new measurement y.

$$\tilde{J}(\hat{\mathbf{C}}) = \sum_{j=1}^{K} w_j \tilde{J}(\hat{\mathbf{C}}_j) \tag{25}$$

where

$$w_j = \frac{1/d_j}{\sum_{i=1}^{K} 1/d_i} \tag{26}$$

The schematic diagram of the above procedure is shown in Figure 2.

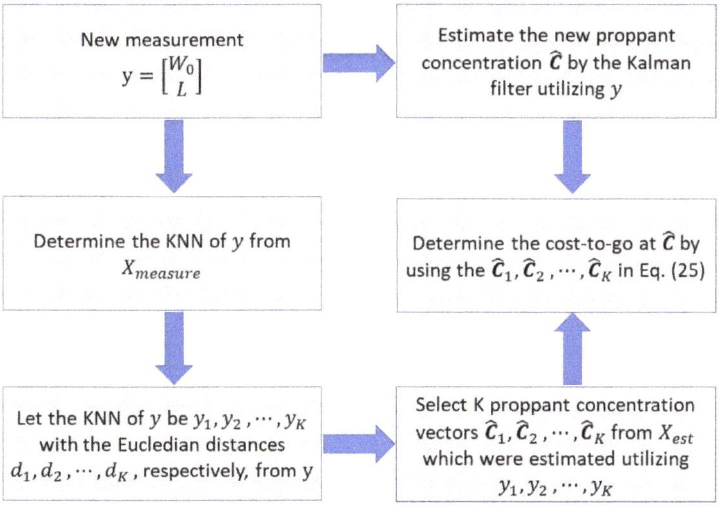

Figure 2. Schematic flow diagram to determine the cost-to-go at a new state $\hat{\mathbf{C}}$ estimated by the Kalman filter using a new measurement y.

3.2.3. Bellman Iteration

Value iteration is employed for offline cost-to-go improvement. In this approach, Bellman equation (Equation (3)) is solved iteratively for every estimated state in X_{est} until convergence. At each iteration step, we calculate J^{i+1} for every estimated state $\hat{\mathbf{C}}(t_k)$ by solving:

$$J^{i+1}(\hat{\mathbf{C}}(t_k)) = \min_{u(t_k)\in\mathcal{U}} [\Phi(\hat{\mathbf{C}}(t_k), u(t_k)) + \tilde{J}^i(\hat{\mathbf{C}}(t_{k+1}))] \tag{27}$$

where the superscript i denotes the iteration index, J^{i+1} is the updated cost-to-go value for $\hat{\mathbf{C}}(t_k)$, $\tilde{J}^i(\hat{\mathbf{C}}(t_{k+1}))$ is the estimate of cost-to-go value for the successor state $\hat{\mathbf{C}}(t_{k+1})$ and $\tilde{J}^0 = \tilde{J}^{\mu^0}$. After updating the cost-to-go values for all of the estimated states in X_{est}, we fit another function approximator to the resulting $\hat{\mathbf{C}}$ vs. $J^{i+1}(\hat{\mathbf{C}})$ data.

In this work, the Bellman iterations converged after six iterations with the following termination condition:

$$\frac{1}{N}\sum_{k=1}^{N} |J^{i+1}(x_k) - J^i(x_k)| < 0.35 \tag{28}$$

where $N = 240$ is the total number of estimated states during the closed-loop simulations. Figure 3 shows how the average absolute error, $\frac{1}{N}\sum_{k=1}^{N} |J^{i+1}(x_k) - J^i(x_k)|$, changes in subsequent iterations. In ADP, the time evolution of the system states is described by employing the ROM, developed for the MPC scheme, because of the high computational effort required to solve the high-fidelity model.

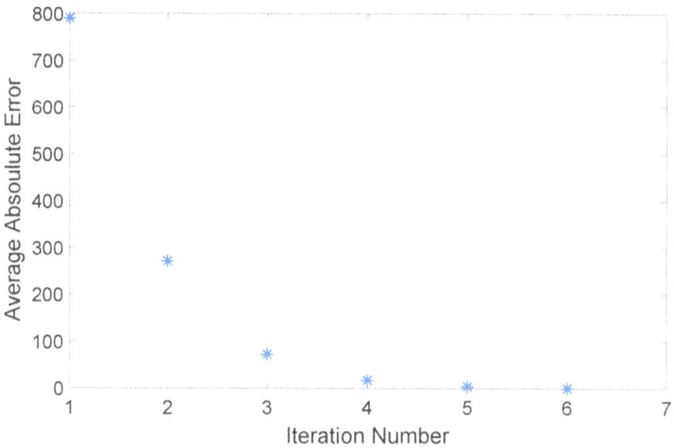

Figure 3. Profile of the average absolute error with iteration.

Remark 4. *In the ADP approach, the data covering the relevant regions in the continuous state space can be obtained by performing closed-loop simulations of the system under any known heuristics or sub-optimal control policies. The initial cost-to-go estimates do not affect the quality of the converged cost-to-go function [45,46]. Therefore, the performance of the ADP-based controller does not heavily rely on the MPC design. In this work, we employed the MPC scheme developed by Siddhamshetty et al. [8] because it provides a good input profile compared to other available input profiles such as Nolte's pumping schedule.*

3.3. Online Optimal Control

Once the cost-to-go iteration converges, we compute the control policy online by solving the following optimization problem:

$$u(t_k) = \arg \min_{u(t_k) \in \mathcal{U}} [\Phi(\hat{\mathbf{C}}(t_k), u(t_k)) + \tilde{J}^{N_c}(\hat{\mathbf{C}}(t_{k+1}))] \tag{29}$$

where $N_c = 6$ represents the number of iterations required for convergence, and $\tilde{J}^{N_c}$ denotes the converged cost-to-go function approximator.

The ADP-based controller is employed to achieve the uniform proppant concentration across the fracture at the end of pumping. To compare the performance of the ADP-based controller with the MPC system, we perform the closed-loop simulations under both of the control systems by initializing the Kalman filter with the fracture width at the wellbore and the fracture length values outside the "training set". We want to note that the "training set" refers to the set of values of the fracture width at the wellbore and the fracture length used to initialize the Kalman filter during the closed-loop simulations under the MPC system as described in Section 3.2.1. The profiles of the injected proppant concentration at the wellbore (i.e., the manipulated input) and proppant concentration across the fracture at the end of pumping using ADP-based controller and the MPC system are shown in Figures 4 and 5. The ADP-based controller shows an improvement over the MPC system in achieving the uniform proppant concentration across the fracture at the end of pumping. The performance of the closed-loop response can be understood based on the total cost-to-go values. Specifically, the cost-to-go values for the ADP-based controller and the MPC system are 179.66×10^3 and 181.58×10^3, respectively. Furthermore, the ADP-based controller takes less computational time than the MPC system to run a closed-loop simulation as shown in Figure 6. This is because ADP-based controller solves a single-stage optimization problem, Equation (29), whereas the MPC system solves a multi-stage optimization problem, Equation (22), at each sampling time. In MPC, the computational time required to solve the optimization problem keeps decreasing with time because the number of remaining stages to be considered keeps decreasing. However, in ADP, the computational time required at each sampling time is similar because ADP solves a single-stage optimization problem at every sampling time. Please note that the calculations were performed on a Dell workstation, powered by Intel(R) Core(TM) i74770 CPU@3.40 GHz, running the Windows 8 operating system. We would like to highlight that the reduction in the computational time can be very beneficial to enhance the productivity of the produced wells. Specifically, despite advances in measurement techniques, it still requires interruption by experienced engineers to distinguish useful information from noise. For example, it takes about 1–3 min to post-process the microseismic data. Therefore, having a computationally efficient controller such as ADP-based controllers would compensate for the time delay due to the interruption by experienced engineers.

Remark 5. *In the ADP-based controller, we did not include the constraints directly but used the data from 24 closed-loop simulations which satisfied the constraints. Specifically, for every sampling time t_k, we stored 24 values of the manipulated input obtained from the 24 closed-loop simulations. Let C_k^{LB} and C_k^{UB} be the minimum and the maximum value of the manipulated input at time instant t_k among the 24 stored values, respectively. Finally, at every sampling time t_k, Equations (27) and (29) were solved with C_k^{LB} and C_k^{UB} as the lower and the upper bound, respectively.*

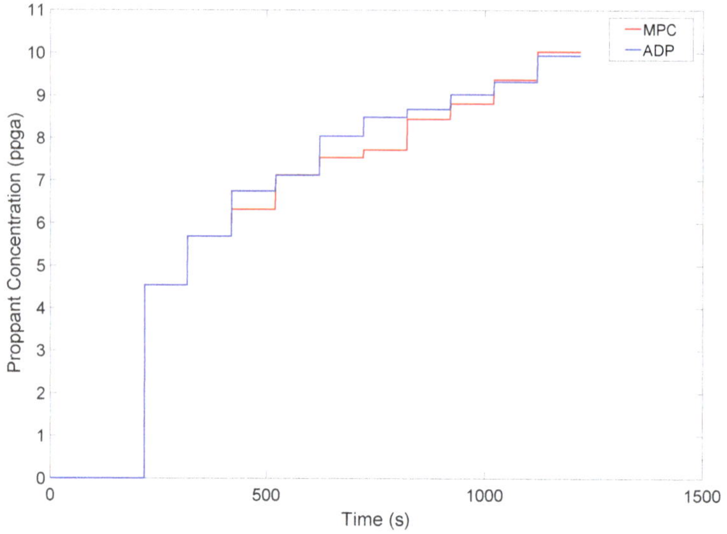

Figure 4. Comparison of the pumping schedule generated using the ADP-based controller and the MPC system.

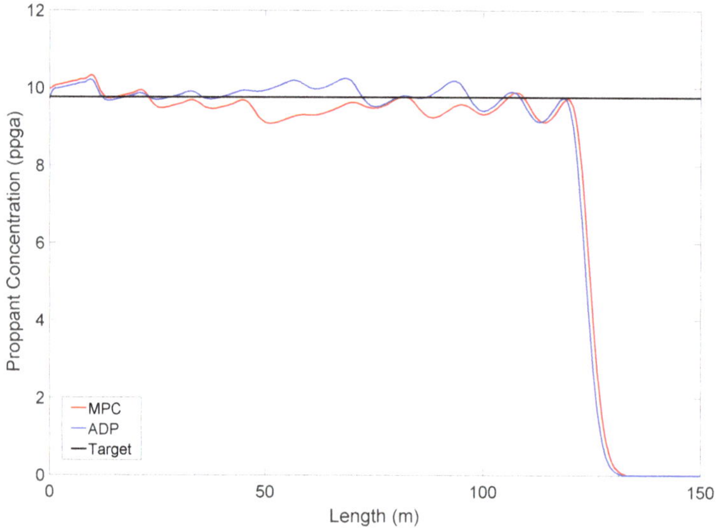

Figure 5. Comparison of spatial proppant concentration profiles obtained at the end of pumping using the ADP-based controller and the MPC system.

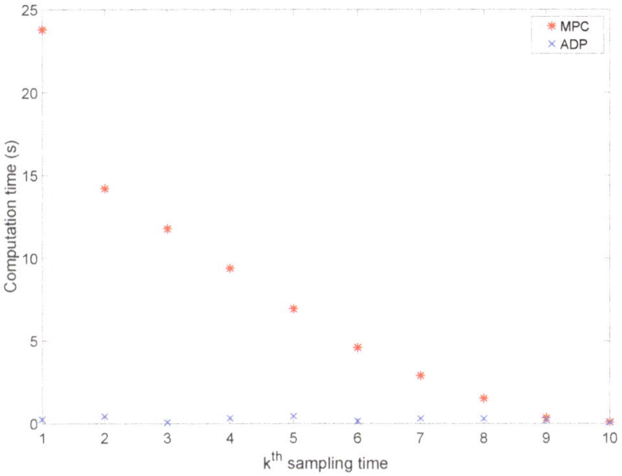

Figure 6. Comparison of the computation time to solve the optimization problem at each sampling time using the ADP-based controller and the MPC system.

3.4. ADP-Based Control with Plant–Model Mismatch

We also studied the performance of the ADP-based controller when the rock mechanical properties are not available *a priori*. Specifically, we considered a plant–model mismatch in Young's modulus, E, which significantly affects the controller performance with respect to achieving uniform proppant concentration across the fracture at the end of pumping [44,47].

In this case, the data covering the relevant regions in the state space was obtained from the closed-loop simulations under the MPC system described in Section 3.2.1 by considering the following five different scenarios: (1) no variation in E; (2) 15% increase in E; (3) 5% increase in E; (4) 5% decrease in E; and (5) 15% decrease in E. For each value of E, a total of five closed-loop simulations under the MPC system of Equation (22) were performed by initializing the Kalman filter with different values of the fracture width at the wellbore and the fracture length. As a result, a total of 250 estimated states were obtained from the five scenarios. For all the estimated states, the initial cost-to-go values (J^{μ^0}) were computed using Equation (23). Then, we employed KNN (with $K = 5$) as a function approximator to obtain the mapping between initial cost-to-go values and the estimated states as described in Section 3.2.2. This was followed by the value iteration step in which the Bellman iterations converged after five iterations with the following termination condition:

$$\frac{1}{N}\sum_{k=1}^{N} |J^{i+1}(x_k) - J^{i}(x_k)| < 0.35 \tag{30}$$

where $N = 250$ is the total number of estimated states obtained from the five scenarios. Figure 7 shows how the average absolute error, $\frac{1}{N}\sum_{k=1}^{N} |J^{i+1}(x_k) - J^{i}(x_k)|$, changes in subsequent iterations. Please note that the ROM used to run the closed-loop simulations under the MPC system for five different scenarios is developed with a fixed E, which is 0.5×10^{10} Pa (i.e., no variation in E). Therefore, it does not capture the plant–model mismatch due to uncertainty in E. In this work, the same ROM with a fixed E is used in the ADP-based controller to demonstrate its capability of handling a plant–model mismatch in E.

To test the performance of the ADP-based controller, we performed the closed-loop simulation with the converged cost-to-go function by considering 10% decrease in E and by initializing the Kalman filter with the fracture width at the wellbore and the fracture length values outside the "training set". Note that the "training set" refers to the set of values of the fracture width at the wellbore and the fracture length used to initialize the Kalman filter during the closed-loop simulations under the MPC system for five scenarios. The profiles of the injected proppant concentration at the wellbore (i.e., the manipulated input) and proppant concentration across the fracture at the end of pumping using the ADP-based controller are shown in Figures 8 and 9. It can be observed in Figure 9 that the ADP-based controller is able to achieve uniform proppant concentration across the fracture at the end of pumping. In other words, it effectively handled the plant–model mismatch in E. Please note that, even with the same rock formation, the performance of hydraulic fracturing can be significantly different [48,49]. This variability can, in part, be attributed to the spatially varying rock mechanical properties such as the Young's modulus, E. Therefore, having a controller such as the ADP-based control framework that can handle the plant–model mismatch in E can play a crucial role in enhancing the productivity of the produced well.

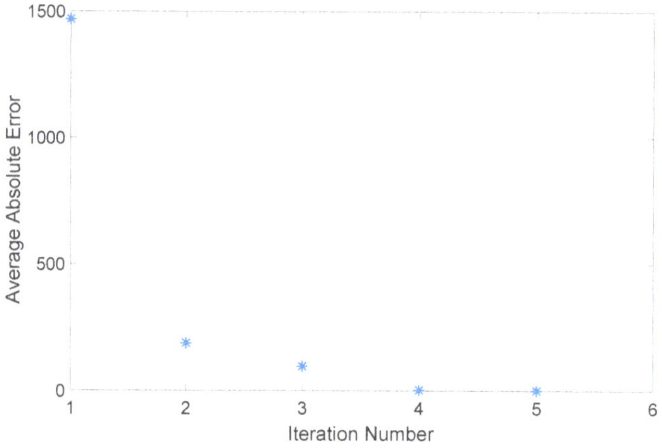

Figure 7. Profile of the average absolute error with iteration.

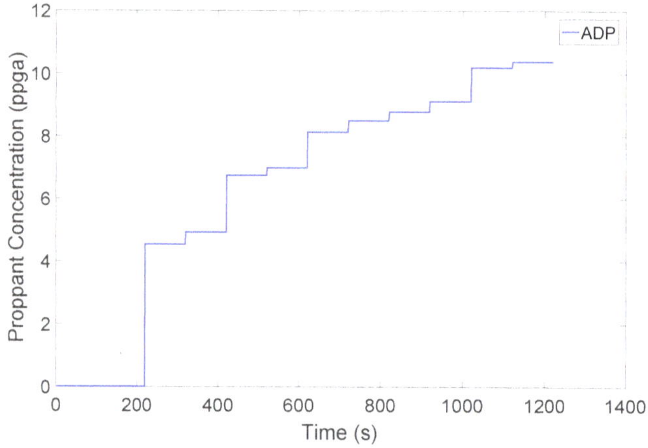

Figure 8. Pumping schedule generated using the ADP-based controller with plant–model mismatch.

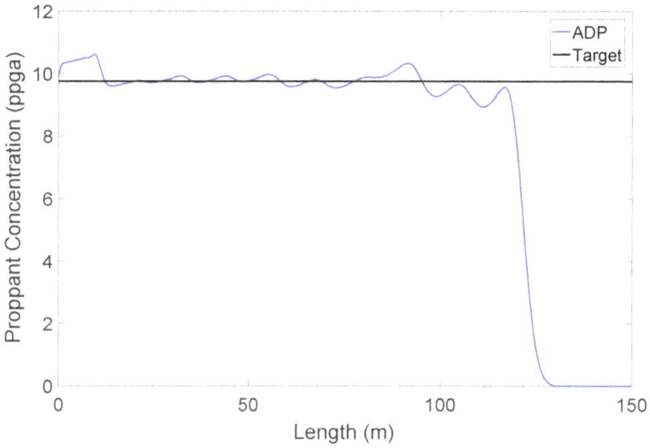

Figure 9. Spatial proppant concentration profile obtained at the end of pumping using the ADP-based controller with plant–model mismatch.

Remark 6. *Please note that in robust MPC-based methods [50–53], model parameter uncertainty has been directly considered in designing MPC. However, in the proposed ADP-based controller, the uncertainty in the model parameter (i.e., Young's modulus, E) is handled by developing an accurate cost-to-go function during the offline stage.*

4. Conclusions

In this work, we developed an ADP-based strategy to regulate the proppant concentration across the fracture at the end of pumping in hydraulic fracturing. First, we performed the closed-loop simulations under the standard MPC to obtain measurable outputs, which were then used to estimate unmeasurable states by the Kalman filter, and to initialize the cost-to-go function approximation in ADP. Second, we employed the value iteration for the cost-to-go improvement by performing iterations of the Bellman equation for all the estimated states during the closed-loop simulations under the standard MPC. Lastly, the ADP-based controller with the converged cost-to-go function approximator was employed for the design of a feedback control system to achieve uniform proppant concentration across the fracture at the end of pumping. The generated pumping schedule using the proposed ADP-based control framework was able to produce a uniform proppant concentration that was closer to the target concentration than the pumping schedule generated by the standard MPC. The ADP-based controller was able to generate an online pumping schedule at a fraction of the time required for the standard MPC as we only had to solve the single-stage optimization problem at each sampling time. Furthermore, the ADP-based controller was able to handle the plant–model mismatch in the Young's modulus of a rock formation. Therefore, this method holds promise to control the hydraulic fracturing process by handling uncertainties in the important geological properties.

Author Contributions: H.S.S. and J.S.K. conceived and designed the study; P.S. developed the mathematical model of hydraulic fracturing process and developed the MPC system; and H.S.S. and J.S.K. developed the ADP-based controller and wrote the paper.

Funding: This research received no external funding.

Acknowledgments: Financial support from the Artie McFerrin Department of Chemical Engineering and the Texas A&M Energy Institute is gratefully acknowledged.

Conflicts of Interest: The authors declare no conflict of interest. The funding sponsors had no role in the design of study; in the collection, analysis, or interpretation of data; in the writing of the manuscript, or in the decision to publish the results.

References

1. Economides, M.J.; Watters, L.T.; Dunn-Normall, S. *Petroleum Well Construction*; Wiley: New York, NY, USA, 1998.
2. Economides, M.J.; Nolte, K.G. *Reservoir Stimulation*; John Wiley & Sons: New York, NY, USA, 2000.
3. Economides, M.J.; Martin, T. *Modern Fracturing: Enhancing Natural Gas Production*; ET Publishing: Houston, TX, USA, 2007.
4. Nolte, K.G. Determination of proppant and fluid schedules from fracturing-pressure decline. *SPE Prod. Eng.* **1986**, *1*, 255–265. [CrossRef]
5. Gu, H.; Desroches, J. New pump schedule generator for hydraulic fracturing treatment design. In Proceedings of the SPE Latin American and Caribbean Petroleum Engineering Conference, Port-of-Spain, Trinidad and Tobago, 27–30 April 2003.
6. Dontsov, E.V.; Peirce, A.P. A new technique for proppant schedule design. *Hydraul. Fract. J.* **2014**, *1*, 1–8.
7. Gu, Q.; Hoo, K.A. Model-based closed-loop control of the hydraulic fracturing Process. *Ind. Eng. Chem. Res.* **2015**, *54*, 1585–1594. [CrossRef]
8. Siddhamshetty, P.; Yang, S.; Kwon, J.S. Modeling of hydraulic fracturing and designing of online pumping schedules to achieve uniform proppant concentration in conventional oil reservoirs. *Comput. Chem. Eng.* **2018**, *114*, 306–317. [CrossRef]
9. Siddhamshetty, P.; Kwon, J.S.; Liu, S.; Valkó, P.P. Feedback control of proppant bank heights during hydraulic fracturing for enhanced productivity in shale formations. *AIChE J.* **2018**, *64*, 1638–1650. [CrossRef]
10. Narasingam, A.; Siddhamshetty, P.; Kwon, J.S. Temporal clustering for order reduction of nonlinear parabolic PDE systems with time-dependent spatial domains: Application to a hydraulic fracturing process. *AIChE J.* **2017**, *63*, 3818–3831. [CrossRef]
11. Narasingam, A.; Kwon, J.S. Development of local dynamic mode decomposition with control: Application to model predictive control of hydraulic fracturing. *Comput. Chem. Eng.* **2017**, *106*, 501–511. [CrossRef]
12. Sidhu, H.S.; Narasingam, A.; Siddhamshetty, P.; Kwon, J.S. Model order reduction of nonlinear parabolic PDE systems with moving boundaries using sparse proper orthogonal decomposition: Application to hydraulic fracturing. *Comput. Chem. Eng.* **2018**, *112*, 92–100. [CrossRef]
13. Morari, M.; Lee, J.H. Model predictive control: Past, present and future. *Comput. Chem. Eng.* **1999**, *23*, 667–682. [CrossRef]
14. Mayne, D.Q.; Rawlings, J.B.; Rao, C.V.; Scokaert, P.O. Constrained model predictive control: Stability and optimality. *Automatica* **2000**, *36*, 789–814. [CrossRef]
15. Bemporad, A.; Morari, M. Control of systems integrating logic, dynamics and constraints. *Automatica* **1999**, *35*, 407–427. [CrossRef]
16. Lee, J.H.; Cooley, B. Recent advances in model predictive control and other related areas. In *AIChE Symposium Series*; 1971-c2002; American Institute of Chemical Engineers: New York, NY, USA, 1997; Volume 93, pp. 201–216.
17. Chikkula, Y.; Lee, J.H. Robust adaptive predictive control of nonlinear processes using nonlinear moving average system models. *Ind. Eng. Chem. Res.* **2000**, *39*, 2010–2023. [CrossRef]
18. Lee, J.H.; Lee, J.M. Approximate dynamic programming based approach to process control and scheduling. *Comput. Chem. Eng.* **2006**, *30*, 1603–1618. [CrossRef]
19. Kaisare, N.S.; Lee, J.M.; Lee, J.H. Simulation based strategy for nonlinear optimal control: Application to a microbial cell reactor. *Int. J. Robust Nonlinear Control* **2003**, *13*, 347–363. [CrossRef]
20. Lee, J.M.; Kaisare, N.S.; Lee, J.H. Choice of approximator and design of penalty function for an approximate dynamic programming based control approach. *J. Process Control* **2006**, *16*, 135–156. [CrossRef]
21. Tosukhowong, T.; Lee, J.H. Approximate dynamic programming based optimal control applied to an integrated plant with a reactor and a distillation column with recycle. *AIChE J.* **2009**, *55*, 919–930. [CrossRef]

22. Padhi, R.; Balakrishnan, S.N. Proper orthogonal decomposition based optimal neurocontrol synthesis of a chemical reactor process using approximate dynamic programming. *Neural Netw.* **2003**, *16*, 719–728. [CrossRef]
23. Joy, M.; Kaisare, N.S. Approximate dynamic programming-based control of distributed parameter systems. *Asia-Pac. J. Chem. Eng.* **2011**, *6*, 452–459. [CrossRef]
24. Munusamy, S.; Narasimhan, S.; Kaisare, N.S. Approximate dynamic programming based control of hyperbolic PDE systems using reduced-order models from method of characteristics. *Comput. Chem. Eng.* **2013**, *57*, 122–132. [CrossRef]
25. Bellman, R.E. *Dynamic Programming*; Princeton University Press: Princeton, NJ, USA, 1957.
26. Perkins, T.K.; Kern, L.R. Widths of Hydraulic Fractures. *J. Pet. Technol.* **1961**, *13*, 937–949. [CrossRef]
27. Nordgren, R. Propagation of a vertical hydraulic fracture. *Soc. Pet. Eng. J.* **1972**, *12*, 306–314. [CrossRef]
28. Sneddon, L.; Elliot, H. The opening of a Griffith crack under internal pressure. *Q. Appl. Math.* **1946**, *4*, 262–267. [CrossRef]
29. Gudmundsson, A. Stress estimate from the length/width ratios of fractures. *J. Struct. Geol.* **1983**, *5*, 623–626. [CrossRef]
30. Howard, G.C.; Fast, C.R. Optimum fluid characteristics for fracture extension. *Dril. Product. Pract.* **1957**, *24*, 261–270.
31. Adachi, J.; Siebrits, E.; Peirce, A.; Desroches, J. Computer simulation of hydraulic fractures. *Int. J. Rock Mech. Min. Sci.* **2007**, *44*, 739–757. [CrossRef]
32. Daneshy, A. Numerical solution of sand transport in hydraulic fracturing. *J. Pet. Technol.* **1978**, *30*, 132–140. [CrossRef]
33. Barree, R.; Conway, M. Experimental and numerical modeling of convective proppant transport. *J. Pet. Technol.* **1995**, *47*, 216–222. [CrossRef]
34. Gu, Q.; Hoo, K.A. Evaluating the performance of a fracturing treatment design. *Ind. Eng. Chem. Res.* **2014**, *53*, 10491–10503. [CrossRef]
35. Novotny, E.J. Proppant transport. In Proceedings of the SPE Annual Fall Technical Conference and Exhibition (SPE 6813), Denver, CO, USA, 9–12 October 1977.
36. Daal, J.A.; Economides, M.J. Optimization of hydraulic fracture well in irregularly shape drainage areas. In Proceedings of the SPE 98047 SPE International Symposium and Exhibition of Formation Flamage Control, Lafayette, LA, USA, 15–17 February 2006; pp. 15–17.
37. Corbett, B.; Mhaskar, P. Subspace identification for data-driven modeling and quality control of batch processes. *AIChE J.* **2016**, *62*, 1581–1601. [CrossRef]
38. Meidanshahi, V.; Corbett, B.; Adams, T.A., II; Mhaskar, P. Subspace model identification and model predictive control based cost analysis of a semicontinuous distillation process. *Comput. Chem. Eng.* **2017**, *103*, 39–57. [CrossRef]
39. Pourkargar, D.B.; Armaou, A. Modification to adaptive model reduction for regulation of distributed parameter systems with fast transients. *AIChE J.* **2013**, *59*, 4595–4611. [CrossRef]
40. Pourkargar, D.B.; Armaou, A. APOD-based control of linear distributed parameter systems under sensor/controller communication bandwidth limitations. *AIChE J.* **2015**, *61*, 434–447. [CrossRef]
41. Sahraei, M.H.; Duchesne, M.A.; Yandon, R.; Majeski, A.; Hughes, R.W.; Ricardez-Sandoval, L.A. Reduced order modeling of a short-residence time gasifier. *Fuel* **2015**, *161*, 222–232. [CrossRef]
42. Sahraei, M.H.; Duchesne, M.A.; Hughes, R.W.; Ricardez-Sandoval, L.A. Dynamic reduced order modeling of an entrained-flow slagging gasifier using a new recirculation ratio correlation. *Fuel* **2017**, *196*, 520–531. [CrossRef]
43. Quirein, J.A.; Grable, J.; Cornish, B.; Stamm, R.; Perkins, T. Microseismic fracture monitoring. In Proceedings of the SPWLA 47th Annual Logging Symposium, Veracruz, Mexico, 4–7 June 2006.
44. Narasingam, A.; Siddhamshetty, P.; Kwon, J.S. Handling Spatial Heterogeneity in Reservoir Parameters Using Proper Orthogonal Decomposition Based Ensemble Kalman Filter for Model-Based Feedback Control of Hydraulic Fracturing. *Ind. Eng. Chem. Res.* **2018**, *57*, 3977–3989. [CrossRef]
45. Bertsekas, D. *Dynamic Programming and Optimal Control*; Athena Scientific: Belmont, MA, USA, 2005; Volume 1.
46. Lee, J.M.; Lee, J.H. An approximate dynamic programming based approach to dual adaptive control. *J. Process Control* **2009**, *19*, 859–864. [CrossRef]

47. Jafarpour, B. Sparsity-promoting solution of subsurface flow model calibration inverse problems. *Adv. Hydrogeol.* **2013**, 73–94. [CrossRef]
48. Daniels, J.L.; Waters, G.A.; Le Calvez, J.H.; Bentley, D.; Lassek, J.T. Contacting more of the barnett shale through an integration of real-time microseismic monitoring, petrophysics, and hydraulic fracture design. In Proceedings of the SPE Annual Technical Conference and Exhibition, Anaheim, CA, USA, 11–14 November 2007.
49. King, G.E. Thirty years of gas shale fracturing: What have we learned? In Proceedings of the SPE Annual Technical Conference and Exhibition, Florence, Italy, 19–22 September 2010.
50. Lucia, S.; Finkler, T.; Engell, S. Multi-stage nonlinear model predictive control applied to a semi-batch polymerization reactor under uncertainty. *J. Process Control* **2013**, *23*, 1306–1319. [CrossRef]
51. Gutierrez, G.; Ricardez-Sandoval, L.A.; Budman, H.; Prada, C. An MPC-based control structure selection approach for simultaneous process and control design. *Comput. Chem. Eng.* **2014**, *70*, 11–21. [CrossRef]
52. Rodriguez-Perez, B.E.; Flores-Tlacuahuac, A.; Ricardez Sandoval, L.; Lozano, F.J. Optimal Water Quality Control of Sequencing Batch Reactors Under Uncertainty. *Ind. Eng. Chem. Res.* **2018**, *57*, 9571–9590. [CrossRef]
53. Lucia, S.; Tătulea-Codrean, A.; Schoppmeyer, C.; Engell, S. Rapid development of modular and sustainable nonlinear model predictive control solutions. *Control Eng. Pract.* **2017**, *60*, 51–62. [CrossRef]

Σ mathematics

MDPI

Article

A Nonlinear Systems Framework for Cyberattack Prevention for Chemical Process Control Systems †

Helen Durand ‡

Department of Chemical Engineering and Materials Science, Wayne State University, Detroit, MI 48202, USA; helen.durand@wayne.edu; Tel.: +1-313-577-3475

† This paper is an extended version of our paper published in the Proceedings of the 6th IFAC Conference on Nonlinear Model Predictive Control.

‡ Current address: 5050 Anthony Wayne Drive, Detroit, MI 48202, USA.

Received: 13 August 2018; Accepted: 12 September 2018; Published: 14 September 2018

Abstract: Recent cyberattacks against industrial control systems highlight the criticality of preventing future attacks from disrupting plants economically or, more critically, from impacting plant safety. This work develops a nonlinear systems framework for understanding cyberattack-resilience of process and control designs and indicates through an analysis of three control designs how control laws can be inspected for this property. A chemical process example illustrates that control approaches intended for cyberattack prevention which seem intuitive are not cyberattack-resilient unless they meet the requirements of a nonlinear systems description of this property.

Keywords: cybersecurity; process control; model predictive control (MPC); nonlinear systems theory; Lyapunov stability

1. Introduction

Accident prevention for chemical processes has been receiving increased attention in the process control literature as calls for a systems approach to chemical process safety [1–3] are being mathematically formalized and incorporated within control design [4–6]. Controllers have been formulated which compute control actions in a fashion that coordinates their actions with the actions of the safety systems [7], and several works have explored methods for keeping the closed-loop state of a nonlinear system away from unsafe conditions in state-space using controllers designed to avoid such regions [8–11]. In addition, several works have explored fault diagnosis and detection [12–14] or fault-tolerant control designs (e.g., [15–18]). Despite these advances in the integration of safety and control for handling safety issues which arise from faults or disturbances and are therefore not intended, the work which has explored the safety issues associated with cybersecurity [19] breaches of process control systems performed with the intent of bringing the plant to an unsafe, unprofitable, or under-producing condition to seek to hurt others has remained, for the most part, unexplored (with exploration of the topic in works such as [20]). This gap in the literature is notable given the increasing threat that cybersecurity breaches pose for safe process operation. For example, cyberattacks have been successful at creating power outages in the Ukraine [21], causing sewage to enter nearby land and water from a wastewater treatment plant [22] and damaging equipment at a uranium enrichment plant [23]. They have also recently targeted systems at a petrochemical plant [24,25] with the apparent goal of creating an explosion (though this attack thankfully failed). Unlike the most commonly discussed cyberattacks in the media and in the literature, which are primarily concerned with stealing information for the purpose of using that information to compromise companies or individuals economically or socially (e.g., [26]), cyberattacks against process control systems have the potential to seek to create physical damage, injury, or death or a lack of supply of products that are necessary for daily life and therefore are a critical problem to address.

A common technique for handling cybersecurity for control systems has been to rely on computer science/information technology, computer hardware, or networking solutions [27]. Example solutions in these categories include code randomization [28], limiting privileges in access or operation with respect to control systems [29], preventing types of information flow with unidirectional gateways [30], using redundant sensors [31], firewalls [32,33], and encryption [34]. Other approaches include changing library load locations [35] or network settings [36], or randomly selecting encrypted data from sensors to compare with unencrypted information [37]. However, the success of the recent attacks mentioned above on control systems, and the surprising methods by which some of them have been carried out (e.g., transmission via USB sticks and local area networks of the Stuxnet virus followed by its subsequent ability to evade detection with rootkits and zero-day vulnerabilities [20,23]) indicate that the traditional techniques for cyberattack prevention may not be enough. Furthermore, the use of wireless sensors in chemical process control networks can introduce cybersecurity vulnerabilities [38,39]. Given the efficiency gains and lower costs expected to be associated with developing technologies such as improved sensors, the Internet of Things [40], and Cloud computing [41], where increased connectivity and computing usage in the chemical process industries has the potential to pose new cybersecurity risks, the need for alternative techniques to the traditional approaches is growing. The topic of resilience of control designs against cyberattacks [42,43] has been explored in several works [44–47]. For example, in [48–50], resiliency of controllers to cyberattacks in the sense that they continue to function acceptably during and after cyberattacks has been explored in a game-theoretic context. Reliable state estimation also plays a part in resilience [51,52]. Approaches based on process models have been important in suggested attack detection policies [31,53,54] and in policies for preventing attacks that assume that the allowable (i.e., safe) state transitions can be enumerated and therefore that it can be checked whether a control action creates an allowable transition before applying it [55]). The ability of a controller to know the process condition/state has been considered to be an important part of cyberattack resilience of control systems as well [56].

Motivated by the above considerations, this work mathematically defines cyberattacks in a nonlinear systems framework and demonstrates how this framework should guide the development of process designs and controllers to prevent the success of cyberattacks of different types. We highlight the criticality of the nonlinear systems perspective, as opposed to seemingly intuitive approaches that follow more along the lines of traditional computing/networking cybersecurity concepts related to hiding or randomizing information, in preventing the success of cyberattacks, with a focus on those which impact sensor measurements. To demonstrate that intuitive approaches are insufficient for achieving cyberattack-resilience unless they cause specific mathematical properties to hold for the closed-loop system, we explore the pitfalls of two intuitive approaches that do not come with such guarantees and investigate a third approach for which the guarantees can be made for certain classes of nonlinear systems under sufficient conditions, showing that it may be possible to develop methods of operating a plant that meet these properties. This exploration of the properties of control designs that are and are not cyberattack-resilient elucidates key principles that are intended to guide the development of cyberattack-resilient controllers in the future: (a) cyberattack policies for simulation case studies have a potential to be determined computationally; (b) randomization in controller implementation can be introduced within frameworks such as model predictive control (MPC) [57,58] that are common in the process industries without compromising closed-loop stability; and (c) creative implementation strategies which trade off between control policies of different types may help with the development of cyberattack-resilient control designs. A chemical process example is used to demonstrate that controllers which do not meet the nonlinear systems definition of cyberattack resiliency may not be sufficient for preventing the closed-loop state from being brought to an unsafe operating condition. This paper extends the work in [59].

2. Preliminaries

2.1. Notation

The notation $|\cdot|$ denotes the Euclidean norm of a vector. A function $\alpha : [0, a) \rightarrow [0, \infty)$ is of class $\mathcal{K}$ if $\alpha(0) = 0$ and α is strictly increasing. The notation x^T represents the transpose of a vector x. The symbol " / " denotes set subtraction (i.e., $x \in A/B = \{x \in R^n : x \in A, x \notin B\}$). $\lceil\cdot\rceil$ signifies the ceiling function (i.e., the function that returns the nearest integer greater than its argument); $\lfloor\cdot\rfloor$ signifies the floor function (i.e., the function that returns the nearest integer less than its argument).

2.2. Class of Systems

The class of nonlinear systems under consideration in this work is:

$$\dot{x}(t) = f(x(t), u(t), w(t)) \tag{1}$$

where f is a locally Lipschitz nonlinear vector function of the state vector $x \in R^n$, input vector $u \in U \subset R^m$, and disturbance vector $w \in W \subset R^l$, where $W := \{w \in R^l : |w| \leq \theta\}$. We consider that X is a set of states considered to be safe to operate at in the sense that no safety incidents will occur if $x \in X$; therefore, we desire to maintain x within the set X. For the purposes of the developments below, we will assume that outside of X, the closed-loop state is in an unsafe region of state-space. We consider that the origin is an equilibrium of the system of Equation (1) (i.e., $f(0,0,0) = 0$). Furthermore, we make the following stabilizability assumption:

Assumption 1. *There exist n_p explicit stabilizing control laws $h_i(x)$, $i = 1, \ldots, n_p$, for the system of Equation (1), where $n_p \geq 1$, with corresponding sufficiently smooth positive definite Lyapunov functions $V_i : R^n \rightarrow R_+$ and functions $\alpha_{j,i}(\cdot)$, $j = 1, \ldots, 4$, of class $\mathcal{K}$ such that the following inequalities hold for all $x \in D_i \subset R^n$:*

$$\alpha_{1,i}(|x|) \leq V_i(x) \leq \alpha_{2,i}(|x|) \tag{2}$$

$$\frac{\partial V_i(x)}{\partial x} f(x, h_i(x), 0) \leq -\alpha_{3,i}(|x|) \tag{3}$$

$$\left|\frac{\partial V_i(x)}{\partial x}\right| \leq \alpha_{4,i}(|x|) \tag{4}$$

$$h_i(x) \in U \tag{5}$$

for $i = 1, \ldots, n_p$, where D_i is an open neighborhood of the origin.

We define a level set of V_i contained within D_i where $x \in X$ as a stability region Ω_{ρ_i} of the nominal ($w(t) \equiv 0$) system of Equation (1) under the controller $h_i(x)$ ($\Omega_{\rho_i} := \{x \in X \cap D_i : V_i(x) \leq \rho_i\}$).

By the smoothness of each V_i, the Lipschitz property of f, and the boundedness of x, u, and w, we obtain the following inequalities:

$$|f(x_1, u, w) - f(x_2, u, 0)| \leq L_x|x_1 - x_2| + L_w|w| \tag{6}$$

$$\left|\frac{\partial V_i(x_1)}{\partial x} f(x_1, u, w) - \frac{\partial V_i(x_2)}{\partial x} f(x_2, u, 0)\right| \leq L'_{x,i}|x_1 - x_2| + L'_{w,i}|w| \tag{7}$$

$$|f(x, u, w)| \leq M \tag{8}$$

for all $x, x_1, x_2 \in \Omega_{\rho_i}$, $i = 1, \ldots, n_p$, $u \in U$, and $w \in W$, where $L_x > 0$, $L_w > 0$, and $M > 0$ are selected such that the bounds in Equations (6) and (8) hold regardless of the value of i, and $L'_{x,i}$ and $L'_{w,i}$ are positive constants for $i = 1, \ldots, n_p$.

The instantaneous cost of the process of Equation (1) is assumed to be represented by a continuous function $L_e(x, u)$ (we do not require that L_e have its minimum at the origin/steady-state). We also assume that the instantaneous production rate of the desired product for the process is given by the continuous function $P_d(x, u)$ (which may be the same as L_e but is not required to be).

2.3. Model Predictive Control

MPC is an optimization-based control design formulated as:

$$\min_{u(t)\in S(\Delta)} \int_{t_k}^{t_{k+N}} L_e(\tilde{x}(\tau), u(\tau))\, d\tau \tag{9}$$

$$\text{s.t.} \quad \dot{\tilde{x}}(t) = f(\tilde{x}(t), u(t), 0) \tag{10}$$

$$\tilde{x}(t_k) = x(t_k) \tag{11}$$

$$\tilde{x}(t) \in X,\ \forall\, t \in [t_k, t_{k+N}) \tag{12}$$

$$u(t) \in U,\ \forall\, t \in [t_k, t_{k+N}) \tag{13}$$

where $u(t) \in S(\Delta)$ signifies that the input trajectories are members of the class of piecewise-constant vector functions with period Δ. The nominal (i.e., $w(t) \equiv 0$) model of Equation (1) (Equation (10)) is used by the MPC of Equations (9)–(13) to develop predictions $\tilde{x}$ of the process state, starting at a measurement of the process state at t_k (Equation (11); in this work, full state feedback is assumed to be available to an MPC), which are then used in computing the value of the stage cost L_e over the prediction horizon of N sampling periods (Equation (9)) and evaluating the state constraints (Equation (12)). The inputs computed by the MPC are required to meet the input constraint (Equation (13)). The inputs are applied in a receding horizon fashion.

2.4. Lyapunov-Based Model Predictive Control

Lyapunov-based model predictive control (LMPC) [60,61] is a variation on the MPC design of the prior section and is formulated as follows:

$$\min_{u(t)\in S(\Delta)} \int_{t_k}^{t_{k+N}} L_e(\tilde{x}(\tau), u(\tau))\, d\tau \tag{14}$$

$$\text{s.t.} \quad \dot{\tilde{x}}(t) = f(\tilde{x}(t), u(t), 0) \tag{15}$$

$$\tilde{x}(t_k) = x(t_k) \tag{16}$$

$$\tilde{x}(t) \in X,\ \forall\, t \in [t_k, t_{k+N}) \tag{17}$$

$$u(t) \in U,\ \forall\, t \in [t_k, t_{k+N}) \tag{18}$$

$$\begin{aligned} &V_1(\tilde{x}(t)) \le \rho_{e,1},\ \ \forall\, t \in [t_k, t_{k+N}), \\ &\text{if } x(t_k) \in \Omega_{\rho_{e,1}} \end{aligned} \tag{19}$$

$$\begin{aligned} &\frac{\partial V_1(x(t_k))}{\partial x} f(x(t_k), u(t_k), 0) \\ &\quad \le \frac{\partial V_1(x(t_k))}{\partial x} f(x(t_k), h_1(x(t_k)), 0) \\ &\text{if } x(t_k) \in \Omega_{\rho_1}/\Omega_{\rho_{e,1}} \text{ or } t_k \ge t' \end{aligned} \tag{20}$$

where the notation follows that of Equations (9)–(13). In LMPC, the predicted state is required to meet the Lyapunov-based stability constraint of Equation (19) when $x(t_k) \in \Omega_{\rho_{e,1}} \subset \Omega_{\rho_1}$ by maintaining the predicted state within the set $\Omega_{\rho_{e,1}}$ throughout the prediction horizon, and the input is required to meet the Lyapunov-based stability constraint of Equation (20) when $x(t_k) \notin \Omega_{\rho_{e,1}}$ to cause the closed-loop state to move toward a neighborhood of the origin throughout a sampling period. $\Omega_{\rho_{e,1}}$ is chosen to make Ω_{ρ_1} forward invariant under the LMPC of Equations (14)–(20) in the presence of sufficiently small disturbances and a sufficiently small Δ. t' is a time after which it is desired to enforce the constraint of Equation (20) for all times regardless of the position of $x(t_k)$ in state-space. Due to the closed-loop stability and robustness properties of $h_1(x)$ [62], $h_1(\tilde{x}(t_q))$, $q = k, \ldots, k+N-1$, $t \in [t_q, t_{q+1})$, is a feasible solution to the optimization problem of Equations (14)–(20) at every sampling time if $x(t_0) \in \Omega_{\rho_1}$ because it is guaranteed to cause $V_1(x)$ to decrease along the closed-loop state trajectories of the nonlinear process throughout each sampling period in the prediction horizon when Δ and θ are sufficiently small until the closed-loop state enters a neighborhood $\Omega_{\rho_{\min,1}}$ of the origin. Furthermore, the LMPC of Equations (14)–(20) is guaranteed to maintain the closed-loop state within Ω_{ρ_1} throughout all sampling periods of the prediction horizon when parameters such as $\rho_{e,1}$, Δ, and θ are sufficiently small through the design of the Lyapunov-based stability constraints of Equations (19) and (20) which take advantage of the stability properties of $h_1(x)$ [60]. It is furthermore guaranteed under sufficient conditions that V_1 decreases along the closed-loop state trajectory throughout a sampling period when the constraint of Equation (20) is activated at a sampling time.

3. A Nonlinear Dynamic Systems Perspective on Cyberattacks

Cybersecurity of chemical process control systems is fundamentally a chemical engineering problem - cyberattackers can find value in attacking plants because they can affect the economics of large companies, the supply of important chemicals, and the health and lives of plant workers and civilians if they are able to gain control over the process inputs, due to the nature of chemical processes and how chemical processes behave. The implication of this is that chemical engineers should be able to take steps during process and control design that can make cyberattacks more difficult or, ideally, make it impossible for them to be successful at affecting economics, production, or safety.

Cyberattacks against process control systems seek to use information flows in control loops to impact physical systems; the ultimate goal of a cyberattacker of a process control system, therefore, can be assumed to be changing the inputs to the process [20] from what they would otherwise be if the attack was not occurring. In this work, we assume that the plant controllers are feedback controllers. There are various means by which a cyberattacker may attempt to affect such a control loop which include providing false state measurements to a feedback controller, providing incorrect signals to the actuators (i.e., bypassing the controller) [31], falsifying stored process data, preventing information from flowing to some part of a control loop [63], manipulating the controller code itself [20], or directly falsifying the signals received by an operator [37,64] (so that he or she does not notice that the process inputs are abnormal). In summary, the electromagnetic signals in the control loop can be falsified. These signals cause physical elements like actuators to move, impacting the condition of the actual process. Contrary to the typical assumptions in feedback control, the association between the input physically implemented on the process and the process state is removed during a cyberattack. A mathematical definition for cyberattacks on feedback control systems is therefore as follows:

Definition 1. *A cyberattack on a feedback control system is a disruption of information flow in the loop such that any $u \in U$ can potentially be applied at any state x that is accessed by the plant over time.*

A process design that is resilient to cyberattacks attempting to influence process safety has many conceptual similarities to a process that is inherently safe [65–69]; the dynamic expression of this resilience property is as follows, where $\bar{X} \subseteq X$ represents a set of allowable initial conditions:

Definition 2. *A process design that is resilient to cyberattacks intended to affect process safety is one for which there exists no input policy $u(t) \in U$, $t \in [0, \infty)$, such that $x(t) \notin X$, for any $x(t_0) \in \bar{X}$ and $w(t) \in W$, $t \in [0, \infty)$.*

The resilience of the process design here depends on which variables are selected as manipulated inputs; a different input selection may lead to a different assessment of whether the process design is resilient to cyberattacks. Similarly, different designs will give a different dynamic model in Equation (1), which means that the inputs will impact the states differently over time (and whether $x \in X$); therefore, the design itself also plays a role in whether Definition 2 holds as well. Furthermore, the definition of resiliency is independent of the control laws used to control the process. This is because cyberattacks manipulate the process inputs such that they do not necessarily cause process constraints to be met (though the inputs are still physically constrained by the input bounds) and do not necessarily have any relationship to the actual state measurement (Definition 1). Therefore, resiliency of a process design to cyberattacks must be developed assuming that any input policy within the input bounds can be applied to the process.

We can also define cyberattack resilience of a process design against attacks on the plant economics. However, because of the minimal assumptions placed on L_e, it is not possible to require that resilience of a plant to cyberattacks on profitability means that the profit is not at all affected by a cyberattack. For example, consider the case that L_e has a global minimum (e.g., it may be a quadratic function of the states and inputs). In this case, if u is not equal to its value at the global minimum of L_e due to a cyberattack (which affects x), then it would not be expected that the long-term profit will be the same as it would be if the state always remained at its global minimum value. However, we would expect that if profit is minimally affected by a cyberattack, there are relatively small consequences to the attack occurring if it was to occur, and furthermore because of the minimal consequences, a cyberattacker may not find it worthwhile to attempt the attack. Therefore, we define lower and upper bounds on the asymptotic average value of L_e ($L_{e,lb}$ and $L_{e,ub}$, respectively) such that if the cost is within these bounds, the process is still considered highly profitable and the company suffers minimal consequences from an attack. This leads to the definition of a process design that is resilient to cyberattacks against plant profitability as follows (where it is still required that $x(t) \in X$ since safety during operation would be a prerequisite to production):

Definition 3. *A process design that is resilient to cyberattacks intended to affect process profit is one for which $x(t) \in X$ for $t \in [0, \infty)$ for any $x(t_0) \in \bar{X}$ and the following inequality holds:*

$$L_{e,lb} \leq \limsup_{T \to \infty} \frac{1}{T} \int_0^T L_e(x(t), u(t))dt \leq L_{e,ub} \tag{21}$$

for all $u(t) \in U$ and $w(t) \in W$, for $t \in [0, \infty)$.

Cyberattack resilience of a process design against production losses would be defined as in Definition 3, except that Equation (21) would be replaced by

$$P_{d,lb} \leq \liminf_{T \to \infty} \frac{1}{T} \int_0^T P_d(x(t), u(t))dt \leq P_{d,ub} \tag{22}$$

where $P_{d,lb}$ and $P_{d,ub}$ represent the minimum and maximum values in the allowable production range (or if there are n_q products instead of one, each with instantaneous production rate $P_{d,i}$, $i = 1, \ldots, n_q$, upper and lower bounds can be set on the time integral of each instantaneous production rate).

For the same reasons as noted for Definition 2, Definition 3 (and its extension to the production attack case) depends on the design and input selection, but not the control law. In general, it may be difficult to assess whether Definitions 2 and 3 or the production extension hold for a process, though closed-loop simulations for a variety of different values of $x(t_0) \in \bar{X}$, $u \in U$ and $w \in W$, with different sampling periods for each, may provide some sense of how the process behaves and potentially could help demonstrate that the process is not cyberattack resilient if there is an input found in the bounds that causes a lack of satisfaction of the conditions. However, not finding any such input during simulations does not necessarily mean that the process is resilient to cyberattacks unless every situation posed in the definitions has been tested.

Despite the difficulty of verifying whether Definitions 2 and 3 or its production extension hold for a process, the definitions serve an important role in clarifying what cyberattack resilience of a system would look like from a nonlinear systems perspective. At first, the independence of these definitions from the control law implies that cybersecure process systems are only possible to achieve if the process design itself with the selected inputs and their ranges causes Definitions 2 and 3 or the production extension to be satisfied, which would not be expected to be typical. Therefore, at first this seems to imply that chemical processes will generally be susceptible to cyberattacks. However, it also must be understood that the definitions are meant to express resilience against *any* cyberattack of any kind geared toward affecting the inputs, as they express cyberattacks in the most general sense as being related to inputs and states; different types of cyberattacks would need to be analyzed individually to see whether it is possible to design a process or control system that prevents cyberattack success.

Remark 1. *Though Definitions 2 and 3 and the production extension are presented such that any input policy can be chosen (e.g., continuous or sample-and-hold with different sampling periods), a knowledge that the inputs are only applied in sample-and-hold could be used to require that the definitions only hold for sample-and-hold input policies in the bounds with the sampling periods noted (assuming that the cyberattack cannot also impact the sampling period).*

Remark 2. *Other works have mathematically defined cyberattack-resilience concepts as well. For example, ref. [70] explores event triggering within the context of resilient control defined for input-affine nonlinear systems with disturbances to be the capacity of a controller to return the state to a set of safe states when it exits these in finite time. Ref. [71] also defines resiliency, for linear systems, as being related to the capacity of a controller to drive the closed-loop state to certain sets and maintain it in safe states (similar to the definitions above).*

4. Defining Cyberattack Resilience Against Specific Attack Types: Sensor Measurement Falsification in Feedback Control Loops

In the remainder of this work, we focus on attacks that provide false state measurements within X to feedback controllers with the goal of impacting process safety and will seek a better understanding of the properties of controllers that are cyberattack-resilient in such a case. The difference between what is required for cyberattack resilience in this case and what is required in Definition 2 is that the controller and its implementation strategy always play a role in state measurement falsification attacks (i.e., the controller is not bypassed completely to get to the actuators, so that the control law itself always plays a role in dictating what inputs can be computed for given falsified state measurements). Therefore, we would ideally like to develop controllers and their implementation strategies that ensure that the inputs which would be computed by these controllers, regardless of the state measurements they are provided, would over time guarantee that $x \in X$, $\forall t \geq 0$, if $x(t_0) \in \bar{X}$. The definition of cyberattack resilience becomes:

Definition 4. *Consider the system of Equation (1) under feedback controllers and their implementation strategies for which the set of all possible input policies which may be computed for $t \in [0, \infty)$ for all $x(t_0) \in \bar{X}$ given the control laws and their implementation strategies is denoted by $U_{allow,i}(t)$, $i = 1, \ldots, n_u$, $t \geq 0$, where $n_u \geq 1$ represents the number of possible input trajectories, with each covering the time horizon $t \in [0, \infty)$. The system of Equation (1) is resilient to cyberattacks that falsify state measurements with the goal of affecting process safety under these feedback control policies if there exists no possible input policy $u(t) \in U_{allow,i}(t)$, $i = 1, \ldots, n_u$, $t \in [0, \infty)$, such that $x \notin X$, for any $x(t_0) \in \bar{X}$ and $w(t) \in W$, $t \in [0, \infty)$.*

In Definition 4, n_u maybe ∞. Furthermore, sampling period lengths are taken into account in the definition of $U_{allow,i}(t)$. Though Definition 4 may appear difficult to use, we will later provide an operating policy which, for certain subclasses of the system of Equation (1), guarantees cyberattack resilience of the closed-loop system according to Definition 4, indicating that provably cyberattack-resilient control designs for false state measurements in X intended to affect process safety may be possible to develop, particularly if assumptions or restrictions are imposed.

5. Control Design Concepts for Deterring Sensor Measurement Falsification Cyberattacks on Safety: Benefits, Limitations, and Perspectives

In this section, we initially use a chemical process example to motivate the need for cyberattack-resilient control designs according to Definition 4, despite the non-constructive nature of the definition, by demonstrating that cyberattack-resilient control is preferable compared to strategies that detect attacks when they occur and subsequently compensate for them [20,72–77]. Subsequently, we will investigate in more detail what it takes for a control design to be cyberattack-resilient. To do this, we will present two "intuitive" concepts for operating a process in a manner intended to deter cyberattacks; however, through a chemical process example, we will illustrate that due to the definition of cyberattacks in a nonlinear systems context (Definition 1), these intuitive methods are not cyberattack-resilient according to Definition 4. Despite this, the study of the reasons that these designs fail to guarantee cyberattack resilience will develop important insights that may guide future work on cyberattack-resilient controllers. We close with an example of a control design that is cyberattack resilient according to Definition 4 for a subset of the class of systems of Equation (1), demonstrating that despite the non-constructive nature of Definition 4, it may be possible to find operating strategies that can be proven to meet this definition.

5.1. Motivating Example: The Need for Cyberattack-Resilient Control Designs

Consider the simplified Tennessee Eastman process, developed in [78] and used in [20] to explore the results of several cyberattacks on sensors for this process performed one sensor at a time. The process consists of a single vessel that serves as both a reaction vessel and a separator, in which the reaction $A + C \rightarrow D$ occurs in the presence of an inert B. The reactor has two feed streams with molar flow rates F_1 and F_2, where the former contains A, C, and trace B, and the latter contains pure A (these will be denoted in the following by Stream 1 and 2 (S1 and S2), respectively). A, B, and C are assumed to be in the vapor phase at the conditions in the reactor, with D as a nonvolatile liquid in which none of the other species is appreciably soluble, such that the streams leaving the reaction vessel are a vapor at molar flow rate F_3 containing only A, C, and B, and a liquid product at molar flow rate F_4 containing only D (the vapor and liquid streams will be denoted by Stream 3 and 4 (S3 and S4), respectively, in the following). The dynamic model describing the changes in the number of mols of each species in the reactor (N_A, N_B, N_C and N_D for species A, B, C, and D, respectively, each in kmol) is given as follows:

$$\frac{dN_A}{dt} = y_{A1}F_1 + F_2 - y_{A3}F_3 - r_1 \tag{23}$$

$$\frac{dN_B}{dt} = y_{B1}F_1 - y_{B3}F_3 \tag{24}$$

$$\frac{dN_C}{dt} = y_{C1}F_1 - y_{C3}F_3 - r_1 \tag{25}$$

$$\frac{dN_D}{dt} = r_1 - F_4 \tag{26}$$

where $y_{A1} = 0.485$, $y_{B1} = 0.005$, and $y_{C1} = 0.51$ are the mol fractions of A, B, and C, in S1, and y_{A3}, y_{B3}, and y_{C3} are the mol fractions of A, B, and C in S3 (i.e., $y_{i3} = \frac{N_i}{(N_A+N_B+N_C)}$, $i = A,\ B,\ C$). The units of both sides of Equations (23)–(26) are kmol/h. r_1 is the rate at which the reaction in the vessel takes place, and it is given by the following:

$$r_1 = 0.00117 y_{A3}^{1.2} y_{C3}^{0.4} P^{1.6} \tag{27}$$

where r_1 is given in units of kmol/h and P (in kPa) represents the pressure in the vessel and is computed via the ideal gas law as follows:

$$P = \frac{(N_A + N_B + N_C)R_g T}{V_v} \tag{28}$$

where $R_g = 8.314$ kJ/kmol·K and $T = 373$ K (i.e., isothermal operation is assumed). V_v represents the volume of vapor in the vessel, where the vessel has a fixed volume of V = 122 m^3 but the liquid has a time-varying volume that depends on N_D and the liquid molar density of 8.3 kmol/m^3 such that V_v is given (in m^3) as follows:

$$V_v = 122 - \frac{N_D}{8.3} \tag{29}$$

with N_D in kmol. It is desired that the liquid level in the tank not exceed 30 m^3 (the steady-state value of the liquid level is 44.18% of its maximum value).

Three process inputs are assumed (u_1, u_2, and u_3), which represent set-points for the percent opening of three valves that determine the flow rates F_1, F_2, and F_3 as follows:

$$\frac{dX_1}{dt} = 360(u_1 - X_1) \tag{30}$$

$$\frac{dX_2}{dt} = 360(u_2 - X_2) \tag{31}$$

$$\frac{dX_3}{dt} = 360(u_3 - X_3) \tag{32}$$

$$F_1 = 330.46\frac{X_1}{100} \tag{33}$$

$$F_2 = 22.46\frac{X_2}{100} \tag{34}$$

$$F_3 = 0.00352X_3\sqrt{P - 100} \tag{35}$$

where the units of time in Equations (30)–(32) are h and the units of flow in Equations (33)–(35) are kmol/h, and X_1, X_2, and X_3 represent the percentage opening of each valve (with an allowable range between 0% and 100%, such that the valve output would saturate if it hits these bounds). A fourth valve is also available for S4 for which the set-point for the valve position is adjusted with a proportional controller based on the error between the percentage of the 30 m^3 of available liquid volume that is used in the tank ($V_{\%,used}$) and the desired (steady-state) value of the percentage of the available liquid volume ($V_{\%,sp}$) as follows:

$$\frac{dX_4}{dt} = 360([X_{4,s} + K_c(V_{\%,sp} - V_{\%,used})] - X_4) \tag{36}$$

where $X_{4,s}$ represents the steady-state value of the percentage opening of the valve for S4, X_4 represents the percentage opening of the valve for S4, $K_c = -1.4$ is the tuning parameter of the proportional controller used in setting the set-point value for X_4, and $V_{\%,used} = \frac{(100)(N_D)}{(8.3)(30)}$. The molar flow rate of S4 is given in terms of X_4 as follows:

$$F_4 = 0.0417X_4\sqrt{P - 100} \tag{37}$$

The steady-state values for the variables and associated inputs are presented in Table 1, with the subscript s denoting the steady-state value of each variable.

For this process, it is desired to maintain the value of the pressure in the reaction vessel below $P_{max} = 3000$ kPa. To regulate the process at its steady-state value, where $P_s < P_{max}$ as required as shown in Table 1, different control laws can be considered. We first consider the proportional-integral (PI) control laws developed in [78], which were applied in cyberattack scenarios involving attacks on sensors in [20]. In this case, the input u_1 is adjusted in a manner that seeks to modify the flow rate of the product D, u_2 is adjusted in a manner that seeks to modify the composition of A in S3 to avoid losing more reactant than necessary, and u_3 is adjusted in a manner that seeks to modify the pressure in the vessel since it can directly affect how much vapor flow can exit the vessel. To account for physical limitations on the maximum value of S3, an additional mechanism is also added to help with pressure control by allowing pressures greater than 2900 kPa to result in the set-point value for F_4 that u_1 uses in computing how large F_1 should be being lowered to avoid providing reactants to the reactor and thereby decreasing the outlet pressure. This is achieved through a fourth PI controller that computes a signal u_4 used in adjusting the set-point of F_4. The control laws, in sample-and-hold with a sampling period of $\Delta = 0.1$ h, are as follows:

$$u_1(t_k) = u_1(t_{k-1}) + K_{c,1}(e_1(t_k) - e_1(t_{k-1}) + \frac{\Delta}{\tau_{I,1}}e_1(t_k)) \quad (38)$$

$$e_1(t_k) = F_{4,sp,adj}(t_k) - F_4(t_k) \quad (39)$$

$$u_2(t_k) = u_2(t_{k-1}) + K_{c,2}(e_2(t_k) - e_2(t_{k-1}) + \frac{\Delta}{\tau_{I,2}}e_2(t_k)) \quad (40)$$

$$e_2(t_k) = 100(y_{A3,s} - y_{A3}(t_k)) \quad (41)$$

$$u_3(t_k) = u_3(t_{k-1}) + K_{c,3}(e_3(t_k) - e_3(t_{k-1}) + \frac{\Delta}{\tau_{I,3}}e_3(t_k)) \quad (42)$$

$$e_3(t_k) = P_s - P(t_k) \quad (43)$$

$$u_4(t_k) = u_4(t_{k-1}) + K_{c,4}(e_4(t_k) - e_4(t_{k-1}) + \frac{\Delta}{\tau_{I,4}}e_4(t_k)) \quad (44)$$

$$e_4(t_k) = P_{bound} - P(t_k) \quad (45)$$

where P_{bound} = 2900 kPa and the controller parameters are given in Table 1. $F_{4,sp,adj}$ represents the adjusted set-point for F_4 set to $F_{4,s}$ if $u_4 > 0$ but to $F_{4,sp,adj} = F_{4,s} + u_4$ otherwise. u_1, u_2, and u_3 would saturate at 0 or 100% if these limits were reached.

In [20], several cyberattacks are proposed on the sensors associated with the controllers described above (i.e., incorrect measurements are provided to the controllers, causing them to compute inputs for the process which they would not otherwise have computed), with one sensor being attacked at a time. The results in [20] indicate that some types of attacks are successful at driving the pressure above its maximum bound, whereas others are not. For example, the authors of [20] comment that it was difficult in the simulations to achieve problematic pressures in the vessel with the measured values of y_{A3} or F_4 being falsified for the controllers computing u_1 and u_2, whereas it is possible with a falsification of the measurement of P for the controllers computing u_3 and u_4 to achieve a pressure in the reactor above its limit. For example, Figure 1 shows the results of setting the measurement of y_{A3} received by the controller computing u_1 to its maximum value (i.e., a mol fraction of 1) between 10 h and 30 h of operation after initializing the process at the steady-state. In both this case and in simulations with the measurement of F_4 received by the controller computing u_2 set to its minimum value (i.e., 0 kmol/h) between 10 h and 30 h of operation after initializing the process at the steady-state, the pressure during the simulations did not exceed 3000 kPa. However, if we simulate the process with the P measurement set to its minimum value of 0 kPa to affect the controllers computing u_3 and u_4, the pressure does exceed 3000 kPa (i.e., the cyberattack succeeds in bringing the plant to an unsafe condition; in this case, the simulation was performed only for 30 h as the unsafe condition was already reached within this timeframe). The simulations were performed with an integration step size of 10^{-4} h for simulating the dynamic process model of Equations (23)–(45). The simulations were performed in MATLAB R2016a by MathWorks®.

Table 1. Steady-state values for the states of the Tennessee Eastman Process [78].

Parameter	Value	Unit
$N_{A,s}$	44.49999958429348	kmol
$N_{B,s}$	13.53296996509594	kmol
$N_{C,s}$	36.64788062995841	kmol
$N_{D,s}$	110.0	kmol
$X_{1,s}$	60.95327313484253	%
$X_{2,s}$	25.02232231706676	%
$X_{3,s}$	39.25777017606444	%
$X_{4,s}$	47.03024823457651	%
$u_{1,s}$	60.95327313484253	%
$u_{2,s}$	25.02232231706676	%
$u_{3,s}$	39.25777017606444	%
$V_{\%,sp}$	44.17670682730923	%
$F_{1,s}$	201.43	kmol/h
$F_{2,s}$	5.62	kmol/h
$F_{3,s}$	7.05	kmol/h
$F_{4,s}$	100	kmol/h
P_s	2700	kPa
$y_{A3,s}$	0.47	-
$y_{B3,s}$	0.1429	-
$y_{C3,s}$	0.3871	-
$K_{c,1}$	0.1	% h/kmol
$\tau_{I,1}$	1	h
$K_{c,2}$	2	%
$\tau_{I,2}$	3	h
$K_{c,3}$	–0.25	%/kPa
$\tau_{I,3}$	1.5	h
$K_{c,4}$	0.7	kmol/kPa·h
$\tau_{I,4}$	3	h

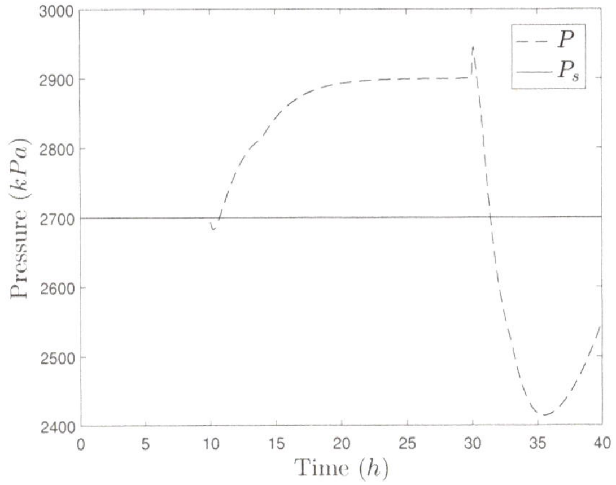

Figure 1. Pressure trajectory for the system of Equations (23)–(45) for a falsified y_{A3} measurement set at 1 between 10 and 30 h of operation under proportional-integral (PI) control.

The differences in the results based on the attack performed indicate the complexities of closed-loop nonlinear systems that can make it difficult to predict every possible attack at a plant to develop appropriate detection and compensation strategies for attacks. In each case, a nonlinear system evolves over time under different input policies, and its response is therefore difficult to predict *a priori*. In addition to the dynamics of the process itself, the dynamics of the other controllers that are not receiving falsified measurements and how they interact with the inputs computed by controllers that are receiving false measurements impact the success of the attack. For example, in Figure 1, the pressure measurement has not been compromised, and mechanisms are in place (through u_3 and u_4) for adjusting the pressure if it increases. Those come into play once the pressure increases significantly, and are able to maintain the pressure below the problematic value of 3000 kPa. A similar mechanism prevents the pressure from exceeding its threshold when the F_4 measurement is falsified; when the measurement of P is falsified, however, the controllers which provided the robustness against the attack success in the other two cases are compromised and the attacks succeed. The number of sensors and which sensors are compromised also play a role (i.e., as shown by the attack on P, if the right sensors are compromised, an unsafe situation can be set up in this process). Furthermore, Figure 1 demonstrates that attack scenarios can be non-obvious. In this figure, the highest value of the pressure occurs not when the value of y_{A3} received by the controller which manipulates u_2 is being falsified, but in the transient after it ceases to be falsified. If the maximum pressure bound had been lower, the pressure in this transient could have exceeded it by creating a rapid change in direction of the inputs once the actual state measurement of y_{A3} becomes available again. In such a case, an attack could focus on the falsification followed by the removal of the falsification as an attack, rather than only on the falsified measurement.

5.2. Deterring Sensor Measurement Falsification Cyberattacks on Safety: Creating Non-Intuitive Controller Outputs

The simplified Tennessee Eastman Process demonstrates that control designs with theoretical guarantees regarding cyberattack-resilience would be a valuable alternative to approaches which assume cyberattacks can be detected. In the next several sections, we seek to better understand how such controllers might be developed by examining two "intuitive" approaches which fail to meet the definition of cyberattack-resilience despite the logic behind their design, followed by an approach which meets the cyberattack-resilience definition. The first "intuitive" approach to be discussed is based on the concept that if the control law can be kept hidden from an attacker and the control law is sufficiently complex such that it is difficult for an attacker to postulate what input will be computed for a given state measurement without knowing the control law, the attacker may have difficulty in performing an attack. The control design that we will explore in this regard is an MPC with a sufficient number of and/or types of constraints in the controller such that it may become difficult to predict, without solving the optimization problem, what input may be computed for a given state measurement. The LMPC of Equations (14)–(20) is an example of a controller which might be considered. In that controller, the constraints of Equations (19) and (20) may cause the inputs computed by the LMPC of Equations (14)–(20) to be different from those computed by the MPC of Equations (9)–(13); therefore, if the same falsified state measurement was provided to both, it is possible that one might compute a control action that could drive the closed-loop state to an unsafe condition, whereas the other may not. If the cyberattacker did not know the control law being used, the presence of additional constraints like the stability-based constraints may cause inputs to be computed which an attacker does not expect. Furthermore, due to the closed-loop stability guarantees which can be made for LMPC (i.e., the closed-loop state remains in Ω_{ρ_1} at all times under sufficient conditions) [60], a check at each sampling time on whether the measured state is in Ω_{ρ_1} may provide a type of detection mechanism for cyberattacks that may make it more difficult for them to succeed. Specifically, under normal operating conditions, the state measurement should never be outside Ω_{ρ_1}; if it is, it may be considered that there is a potential the state measurement has been falsified. If a cyberattacker is unaware of the value of ρ_1, he or she may provide a false state measurement to the controller which triggers detection; on the other hand, if he or she is only able to attack a limited number of sensors, unless the attacker knows or can predict the readings of the unattacked sensors at each sampling time, the attacker does not know how close the full state measurement being received by the controller (incorporating the attacked and unattacked measurements) is to being outside of Ω_{ρ_1}. Again, an attack may be detected or deterred in this case.

Difficulties with this approach include, however: (1) if the cyberattacker did not know the control law being used, it is questionable whether a high-impact attack would be attempted regardless of the control law being used (i.e., it may not matter whether it has Lyapunov-based stability constraints or not), because in any case the control law is not known and therefore attempting to randomly attack the controller may be considered overly risky and unlikely to avoid detection; (2) the attacker may gain access to all of the sensors and learn the value of ρ_1, and thereby be able to maintain the falsified state measurement always in Ω_{ρ_1} to avoid detection.

Remark 3. *We note that closed-loop stability of an approach like LMPC under normal operation (no cyberattacks) is proven elsewhere (e.g., [60]). The proof in [60] relies on the state measurement being accurate; therefore, this proof does not allow us to prove closed-loop stability in the presence of a cyberattack.*

5.2.1. Problems with Creating Non-Intuitive Controller Outputs

The pitfall of this approach from a nonlinear dynamic systems perspective is that it does not make any attempt to prevent policies from existing that could create unsafe operating conditions if the control law becomes known (i.e., Definition 4 is violated); it essentially assumes luckiness by hoping that the cyberattacker will never be able to figure out enough about the control design to be able to attack it. If the attacker does figure out the control law, it may not be overly difficult for them to develop an attack policy that could drive the closed-loop state to an unsafe condition while maintaining the falsified state measurement in Ω_{ρ_1}, despite the many constraints. For example, it may be possible to develop an optimization problem in some cases that can be used in helping develop attack policies, and then those can be assessed within closed-loop simulations to see whether they may be likely to produce a problematic state trajectory.

To see this, consider a continuous stirred tank reactor (CSTR) in which the reactant A is converted to the product B via an irreversible second-order reaction. The feed and outlet volumetric flow rates of the CSTR are F, with the feed concentration C_{A0} and feed temperature T_0. The CSTR is operated non-isothermally with a jacket used to remove or add heat to the reactor at heat rate Q. Constant liquid density ρ_L, heat capacity C_p, and liquid volume V are assumed, with the constants (from [79]) in Table 2. The dynamic process model is:

$$\dot{C}_A = \frac{F}{V}(C_{A0} - C_A) - k_0 e^{-\frac{E}{R_g T}} C_A^2 \tag{46}$$

$$\dot{T} = \frac{F}{V}(T_0 - T) - \frac{\Delta H k_0}{\rho_L C_p} e^{-\frac{E}{R_g T}} C_A^2 + \frac{Q}{\rho_L C_p V} \tag{47}$$

where C_A and T represent the concentration and temperature in the reactor, respectively, E is the activation energy of the reaction, k_0 is the pre-exponential constant, R_g is the ideal gas constant, and ΔH is the enthalpy of reaction. We develop the following vectors for the states and inputs in deviation form: $x = [x_1\ x_2]^T = [C_A - C_{As}\ T - T_s]^T$ and $u = [u_1\ u_2]^T = [C_{A0} - C_{A0s}\ Q - Q_s]^T$, where $C_{As} = 1.22$ kmol/m^3, $T_s = 438.2$ K, $C_{A0s} = 4$ kmol/m^3, and $Q_s = 0$ kJ/h are the steady-state values of C_A, T, C_{A0}, and Q at the operating steady-state.

The control objective is to maximize the following profit-based stage cost for the process of Equations (46) and (47) representing the production rate of the product B while computing control actions which meet the input constraints $0.5 \leq C_{A0} \leq 7.5$ kmol/m^3 and $-5 \times 10^5 \leq Q \leq 5 \times 10^5$ kJ/h and maintain closed-loop stability:

$$L_e = k_0 e^{-\frac{E}{R_g T(\tau)}} C_A(\tau)^2 \tag{48}$$

We will use an LMPC with the stage cost in Equation (48) to control this process. We choose a Lyapunov function $V_1 = x^T P x$, where $P = [1200\ 5;\ 5\ 0.1]$, $h_{1,1}(x) = 0$ kmol/m^3 for simplicity, and $h_{1,2}(x)$ is determined by Sontag's control law [80] as follows:

$$h_{1,2}(x) = \begin{cases} -\frac{L_{\tilde{f}}V_1 + \sqrt{L_{\tilde{f}}V_1^2 + L_{\tilde{g}_2}V_1^4}}{L_{\tilde{g}_2}V_1}, & \text{if } L_{\tilde{g}_2}V_1 \neq 0 \\ 0, & \text{if } L_{\tilde{g}_2}V_1 = 0 \end{cases} \tag{49}$$

where if $h_{1,2}$ fell below or exceeded the upper or lower bound on u_2, $h_{1,2}$ was saturated at the respective bound. In Equation (49), $\tilde{f}$ represents the vector containing the terms in Equations (46) and (47) (after the model has been rewritten in deviation variable form in terms of x_1 and x_2) that do not contain any inputs, and $\tilde{g}$ represents the matrix that multiplies the vector of inputs u_1 and u_2 in the equation. $L_{\tilde{f}}V_1$ and $L_{\tilde{g}_k}V_1$ represent the Lie derivatives of V_1 with respect to $\tilde{f}$ and $\tilde{g}_k$, $k = 1, 2$. The state-space was discretized and the locations where $\dot{V}_1 < 0$ under the controller $h_1(x)$ were examined and used to set $\rho_1 = 180$. $\rho_{e,1}$ was set to be less than ρ_1, and was (heuristically) chosen to be 144. The process is initialized at $x_{init} = [-0.4 \text{ kmol/m}^3\ \ 20 \text{ K}]^T$ and simulated with the integration step of 10^{-4} h, with N set to 10, and with Δ set to 0.01 h. The Lyapunov-based stability constraint activated when $x(t_k) \in \Omega_{\rho_{e,1}}$ was enforced at the end of every sampling period in the prediction horizon, and whenever the Lyapunov-based stability constraint involving the time-derivative of the Lyapunov function was enforced, the other Lyapunov-based constraint was implemented at the end of the sampling periods after the first. The simulations were implemented in MATLAB using fmincon. The initial guess provided to fmincon was the steady-state input vector. The maximum and minimum values of u_2 were multiplied by 10^{-5} within the optimization problem due to the large magnitudes of the upper and lower bounds allowed for this optimization variable.

Table 2. Parameters for the continuous stirred tank reactor (CSTR) process.

Parameter	Value	Unit
V	1	m^3
T_0	300	K
C_p	0.231	kJ/kg·K
k_0	8.46×10^6	m^3/h·kmol
F	5	m^3/h
ρ_L	1000	kg/m^3
E	5×10^4	kJ/kmol
R_g	8.314	kJ/kmol·K
ΔH	-1.15×10^4	kJ/kmol

To consider an attack on the safety of this process, we assume that we do not want the temperature in the reactor to go 55 K above T_s (because no temperature at any point in the stability region is this high, the controller should, under normal operation, have no trouble achieving this). However, if we assume that the cyberattacker knows the control law and can access the state measurements, he or she could exploit this to design an attack policy specific to the closed-loop system under consideration. To demonstrate that this can be possible, we will computationally develop an attack policy for this process through two optimization problems, the first of which tries to compute control actions within the input bounds which maximize the temperature reached within $N\Delta$ time units from the (actual) current state measurement, and the second of which finds a state measurement (to use as the false value in an attack) which can generate control actions that, ideally, are as close as possible to those developed in the first optimization problem and also ensure that there is a feasible solution to the constraints which will be employed in the LMPC. The first optimization problem is as follows:

$$\min_{u(t)\in S(\Delta)} \quad -(x_2(t_N)+T_s) \tag{50}$$

$$\text{s.t.} \quad \dot{\tilde{x}}(t) = \tilde{f}(\tilde{x}(t)) + \tilde{g}u(t) \tag{51}$$

$$\tilde{x}(t_0) = x_{init} \tag{52}$$

$$-3.5 \leq u_1(t) \leq 3.5, \ \forall\, t \in [t_0, t_N) \tag{53}$$

$$-10^5 \leq u_2(t) \leq 10^5, \ \forall\, t \in [t_0, t_N) \tag{54}$$

Equations (50)–(54) are designed such that the solution of this optimization problem is a piecewise-constant input trajectory that meets the process input constraints (Equations (53) and (54)) and drives the temperature in the reactor as high as possible in $N\Delta$ time units (Equation (50)) according to the dynamics of the process (Equation (51)) starting from the state measurement at the current time (Equation (52); the current time is denoted by t_0 in this optimization problem since this problem is solved only once instead of in a receding horizon fashion). The solution of this optimization problem for the process of Equations (46) and (47) is a piecewise-constant input trajectory with u_1 varying between 3.4975 and 3.4983 kmol/m^3 and u_2 varying between 499856.52 and 499908.01 kJ/h over the $N\Delta$ time units.

Because the inputs are approximately constant throughout the $N\Delta$ time units in the solution to Equations (50)–(54), this suggests that a single initial condition may be sufficient for causing the problematic input policy to be generated at each sampling time. Specifically, the only information that the LMPC of Equations (14)–(20) receives from an external source at each time that it is solved is the state measurement in Equation (16); because it uses a deterministic process model and deterministic constraints, the LMPC of Equations (14)–(20) has a single solution for a given state measurement. Therefore, if a cyberattacker determines that an attack policy which applies the same input at every sampling time is desirable, he or she can cause the controller to compute this input at every sampling time by determining a state measurement value for which the problematic input is the solution to Equations (14)–(20), and then providing that same state measurement to the LMPC at every sampling time to cause it to keep computing the same problematic input.

The following second optimization problem finds the initial condition to use at each of the next N sampling periods that may cause the values of u_1 and u_2 in the first sampling period of the prediction horizon to be close to the averages of the N values of u_1 ($u_{1,desired}$) and the N values of u_2 ($u_{2,desired}$), respectively, determined by Equations (50)–(54), while allowing the constraints of Equations (14)–(20) to be met:

$$\min_{u(t)\in S(\Delta), x_{meas}} \quad \int_{t_0}^{t_1} \left[(u_1(\tau) - u_{1,desired})^2 + 10^{-10}(u_2(\tau) - u_{2,desired})^2 \right] d\tau \tag{55}$$

$$\text{s.t.} \quad \dot{\tilde{x}}(t) = \tilde{f}(\tilde{x}(t)) + \tilde{g}u(t) \tag{56}$$

$$\tilde{x}(t_0) = x_{meas} \tag{57}$$

$$-3.5 \leq u_1(t) \leq 3.5, \ \forall\, t \in [t_0, t_N) \tag{58}$$

$$-10^5 \leq u_2(t) \leq 10^5, \ \forall\, t \in [t_0, t_N) \tag{59}$$

$$V_1(\tilde{x}(t_j)) \leq \rho_{e,1}, \ \ j = 0, \ldots, N \tag{60}$$

This optimization problem reverse engineers the LMPC of Equations (14)–(20) (except that it neglects the objective function of the controller) in the sense that it seeks to find an initial condition x_{meas} (Equation (57)) to provide to the LMPC of Equations (14)–(20) for which there exists a feasible input policy for the N sampling periods of the prediction horizon that meets the process input constraints (Equations (58) and (59)) as well as the Lyapunov-based stability constraint of Equation (19) (Equation (60)) while allowing this feasible trajectory to include u_1 and u_2 in the first sampling period of the prediction horizon taking values as close to the problematic values $u_{1,desired}$ and $u_{2,undesired}$

as possible. The reason for only requiring u_1 and u_2 in the first sampling period of the prediction horizon to be as close as possible to the attack values is that though the optimization problem of Equations (55)–(60) is being solved only once to obtain the sensor attack policy x_{meas} to provide to the LMPC at each subsequent sampling time, the LMPC will be solved at every sampling time and will only apply the input for the first sampling period of the prediction horizon in each case. The formulation of Equation (60) assumes that the attacker knows the exact manner in which this constraint is enforced in the LMPC, where, as noted above, it will be enforced at the end of every sampling period in the prediction horizon. The addition of the requirement in Equation (60) that $V_1(\tilde{x}(t_0)) \leq \rho_{e,1}$ is used to pre-select that x_{meas} should be within $\Omega_{\rho_{e,1}}$. This eliminates the need to try to solve a disjunctive or mixed integer nonlinear program [81] that allows the initial condition to be either in $\Omega_{\rho_{e,1}}$ or $\Omega_{\rho_1}/\Omega_{\rho_{e,1}}$ such that the constraint to be employed (i.e., Equation (19) or Equation (20)) depends on the optimization variables that are the components of x_{meas}. The components of x_{meas} were essentially unconstrained in Equations (55)–(60).

In solving Equations (50)–(60), the bounds on u_2 were multiplied by 10^{-5}. The false state measurement determined from Equations (55)–(60) was $x_1 = -0.05207$ kmol/m^3 and $x_2 = -8.3934$ K. Figure 2 demonstrates that when this state measurement is used at every sampling period for 10 sampling periods, the inputs computed are able to drive the temperature significantly above its threshold value $x_2 = 55$ K within a short time. When disturbances are added (specifically, simulations were performed with disturbances added to the right-hand sides of Equations (46) for w_1 and (47) for w_2) generated using the MATLAB functions rng(10) to generate a seed with normrnd to generate a pseudorandom number from a normal distribution with mean of zero and a standard deviation of 30 kmol/h (for w_1) and 3200 K/h (for w_2), with both inputs clipped when necessary to bound them such that $|w_1| \leq 90$ and $|w_2| \leq 9600$, an unsafe situation is again set up in 10 sampling periods in which x_2 approaches 300 K as in Figure 2. The LMPC only receives state measurements, regardless of whether there are disturbances or not; therefore, if the same state measurement is given every time, it computes the same solution to the optimization problem every time and when this solution is able to drive the closed-loop state to an unsafe condition if continuously applied, the cyberattacker succeeds. The attack-defining concept posed here could be attempted for other attack goals as well, such as minimizing a profit-based objective function in Equations (50)–(54) to seek to compute an attack policy that financially attacks the plant or minimizing a production-based objective function to seek to attack the chemical supply from the plant.

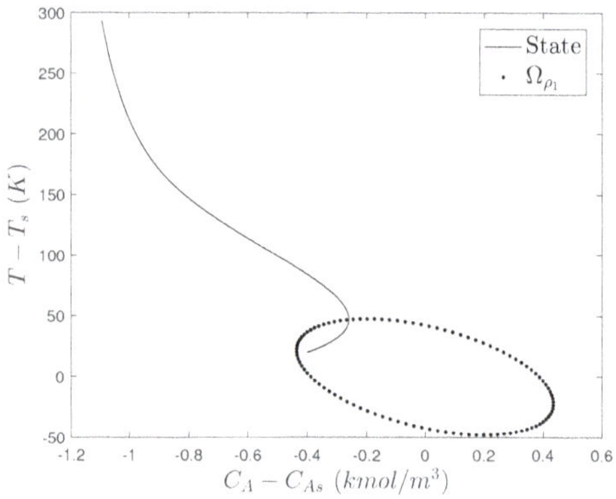

Figure 2. State-space trajectory showing the state trajectory in 10 sampling periods with the falsified state measurements determined through optimization applied at every sampling time, in the absence of disturbances.

Remark 4. *The CSTR example indicates an important difference between traditional safety thinking and thinking about cyberattacks. In traditional safety thinking, there will be unsafe operating conditions that might be considered very unlikely to be achieved; when considering cyberattacks, there can be deliberate attempts on the part of the attacker to set up unsafe operating conditions that might otherwise be very unlikely to be achieved. It is therefore important to seriously consider unlikely scenarios at the hazard analysis stage from the perspective of whether a cyberattack on the control system could lead them to occur.*

Remark 5. *Though the cyberattack design methodology presented in this section suggests that cyberattacks on specific control designs might be developed computationally, the framework used in Equations (50)–(60) may not always achieve expected effects. Specifically, the initial condition determined by Equations (55)–(60) may not actually result in the control actions of Equations (50)–(54) being computed at each sampling time by the controller because the only feature of Equations (55)–(60) that seeks to associate* x_{meas} *with* $u_{1,desired}$ *and* $u_{2,desired}$ *is a soft constraint rather than a hard constraint, and it is, therefore, not guaranteed to be met. Furthermore, Equations (55)–(60) do not account for the role of the objective function in affecting which inputs would actually be computed for a given state measurement. In this example, the false state measurement determined through Equations (50)–(60) was able to rapidly set up an unsafe scenario when used to cyberattack the LMPC; to develop attacks for other systems, it may be necessary to develop a more sophisticated method for determining the false state measurements or to use closed-loop simulations to determine if the false state measurements determined computationally provide an appropriate attack scenario with which to test research results. Finally, it should be noted that Equations (50)–(54) are not guaranteed to find an input that drives* x_2 *above its threshold in N sampling periods; whether or not this occurs may depend on the process dynamics, the input bounds, the initial condition, and also the number of sampling periods N over which the increase in* x_2 *is allowed to occur.*

5.3. Deterring Sensor Measurement Falsification Cyberattacks on Safety: Creating Unpredictable Controller Outputs

The second "intuitive" approach seeks to address a perceived deficiency in the first "intuitive" approach, namely that the success of the cyberattacks in Section 5.2.1 was related to the fact that the cyberattacker could figure out the mapping between $x(t_k)$ and u by learning the control law. One idea for addressing this would be to develop sets of stabilizing control laws for a process and choose only one, randomly, at each sampling time. Then, if the inputs which the various potential control laws would compute for the same state measurement are significantly different, it may be more difficult for an attacker to determine an attack policy that, regardless of the control law chosen at a sampling time, will drive the closed-loop state to an unsafe condition (even if the attacker knew every potential control law).

Before we can consider such an approach, it must be established that randomization in the controller selection process can be achieved without impacting closed-loop stability under normal operation (i.e., in the absence of a cyberattack). Theory-based control designs with stability guarantees from an explicitly characterizable region of attraction even in the presence of disturbances (e.g., LMPC) are therefore attractive options for use in randomization strategies for control laws. In the remainder of this section, we present an example of a control design and implementation strategy that uses LMPC to incorporate randomness in process operation (with the goal of deterring cyberattacks by obscuring the mapping between a state measurement at a given sampling time and the input to be computed) with closed-loop stability guarantees under normal operation even in the presence of the randomness. However, like the design in Section 5.2, this design and its implementation strategy do not fundamentally prevent the existence of an input policy which could create an unsafe condition for some $x(t_0) \in \bar{X}$ (when, for example, $\bar{X} = \Omega_{\rho_1}$), and therefore if this design succeeds in preventing or delaying the impacts of cyberattacks, it does so more on the basis of chance than rigor, which is demonstrated below using the CSTR example.

5.3.1. Creating Unpredictable Controller Outputs: Incorporating Randomness in LMPC Design

The randomized LMPC design involves the development of n_p controllers of the form of Equations (14)–(20) but where each can have a different Lyapunov function, Lyapunov function upper bound, and Lyapunov-based controller as follows:

$$\min_{u(t)\in S(\Delta)} \int_{t_k}^{t_{k+N}} L_e(\tilde{x}(\tau),u(\tau))\,d\tau \tag{61}$$

$$\text{s.t.} \quad \dot{\tilde{x}}(t) = f(\tilde{x}(t),u(t),0) \tag{62}$$

$$\tilde{x}(t_k) = x(t_k) \tag{63}$$

$$\tilde{x}(t) \in X,\ \forall\, t \in [t_k, t_{k+N}) \tag{64}$$

$$u(t) \in U,\ \forall\, t \in [t_k, t_{k+N}) \tag{65}$$

$$V_i(\tilde{x}(t)) \le \rho_{e,i},\ \ \forall\, t \in [t_k, t_{k+N}), \quad \text{if } x(t_k) \in \Omega_{\rho_{e,i}} \tag{66}$$

$$\frac{\partial V_i(x(t_k))}{\partial x} f(x(t_k),u(t_k),0) \le \frac{\partial V_i(x(t_k))}{\partial x} f(x(t_k),h_i(x(t_k)),0) \quad \text{if } x(t_k) \in \Omega_{\rho_i}/\Omega_{\rho_{e,i}} \text{ or } t_k \ge t' \text{ or } \delta = 1 \tag{67}$$

where V_i, $\rho_{e,i}$, ρ_i, and h_i, $i = 1,\ldots,n_p$, play the roles in Equations (61)–(67) of V_1, $\rho_{e,1}$, ρ_1, and h_1, respectively, from Equations (14)–(20). Each combination of V_i and h_i is assumed to satisfy Equations (2)–(5) $\forall x \in \Omega_{\rho_i}$ and $\Omega_{\rho_{e,i}} \subset \Omega_{\rho_i}$. For $j = 2,\ldots,n_p$, the Ω_{ρ_j} should be subsets of Ω_{ρ_1} for reasons that will be clarified in Section 5.3.1.1. To introduce an additional aspect of randomness at each sampling time, the parameter δ is introduced in Equation (67). It can take a value of either 0 or 1, and one of those two values is randomly selected for it at each sampling time. $\delta = 1$ corresponds to activation of the constraint of Equation (67) even when $t_k < t'$ or $x(t_k) \in \Omega_{\rho_{e,i}}$.

With the n_p controllers of the form of Equations (61)–(67) and the two possible values of δ in each of these LMPC's at every sampling time, Equations (61)–(67) represent $2n_p$ potential controllers which may be selected at every sampling time (though if $x(t_k) \in \Omega_{\rho_i}/\Omega_{\rho_{e,i}}$ for n_q of these controllers, Equations (61)–(67) with $\delta = 0$ and $\delta = 1$ are the same, such that the number of control laws is $2n_p - n_q$). One could consider other potential control options in addition, such as the Lyapunov-based controllers $h_i(x)$, $i = 1,\ldots,n_p$. However, though all of these controllers are designed and are available in principle, they could cause closed-loop stability issues that require that not all of them be available to be randomly selected between at each sampling time. The conditions which determine which controllers are possibilities at a given sampling time should rely on the position of $x(t_k)$ in state-space and specifically whether $x(t_k) \in \Omega_{\rho_i}$ for the i-th controller to be considered as a candidate.

To exemplify this, consider the two level sets Ω_{ρ_1} and Ω_{ρ_2} and their subsets $\Omega_{\rho_{e,1}}$ and $\Omega_{\rho_{e,2}}$ shown in Figure 3. Two potential values of $x(t_k)$ are presented (x_a and x_b) to exemplify the role that the state-space location of $x(t_k)$ should play in determining which of the n_p controllers of the form of Equations (61)–(67) or the Lyapunov-based controllers of the form $h_i(x(t_k))$ should be considered as candidates to randomly select between at a given sampling time. Consider first that $x(t_k) = x_a$. In this case, $x(t_k) \in \Omega_{\rho_1}/\Omega_{\rho_{e,1}}$, and therefore, as described in Section 2.4, the LMPC of Equations (61)–(67) with $i = 1$ would be able to maintain the closed-loop state in Ω_{ρ_1} throughout the subsequent sampling period. It is also true that $x(t_k) \notin \Omega_{\rho_{e,2}}$, so it may at first seem reasonable to consider that if the LMPC of Equations (61)–(67) is used with $i = 2$, the constraint of Equation (67) could be activated to decrease the value of the Lyapunov function between two sampling periods and thereby drive the closed-loop state toward the origin using the properties of the Lyapunov-based controller and the constraint of the form of Equation (67) described in Section 2.4. However, the closed-loop stability properties delivered by the constraint of Equation (67) are developed with the requirement that Equations (2)–(5) must hold within the stability region and that $x(t_k)$ must be in this stability region. When $x(t_k) \notin \Omega_{\rho_2}$, these properties are not guaranteed to hold. Therefore, when $x(t_k) = x_a$ in Figure 3, the LMPC of Equations (61)–(67) with $i = 2$ would not be a wise choice to randomly select at a given sampling time. Similarly, $h_2(x(t_k))$ is guaranteed to maintain closed-loop stability when $x(t_k) \in \Omega_{\rho_2}$, but if $h_2(x(t_k))$ is applied when $x(t_k) = x_a$, $x(t_k) \notin \Omega_{\rho_2}$ and therefore the stability properties are not guaranteed to hold.

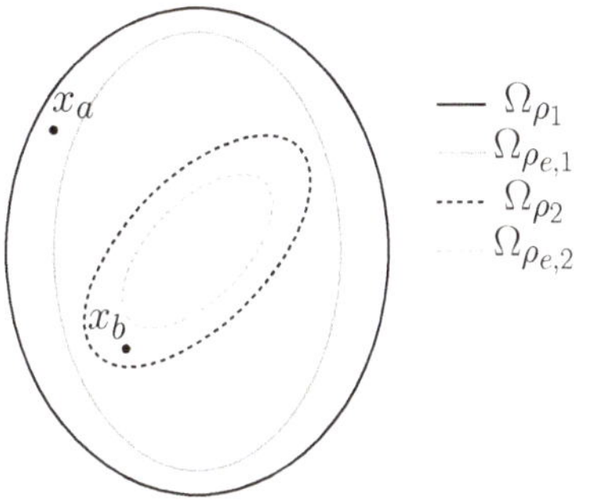

Figure 3. Intersecting stability regions with two different potential initial conditions $x(t_k) = x_a$ and $x(t_k) = x_b$.

In contrast, consider the potential initial condition $x(t_k) = x_b$. In this case, $x(t_k) \in \Omega_{\rho_1}$ and Ω_{ρ_2}. Consequently, Equations (61)–(67) with $i = 1$ or $i = 2$ (for $\delta = 1$ or $\delta = 0$), $h_1(x(t_k))$, and $h_2(x(t_k))$ can all maintain closed-loop stability of the process of Equation (1), and therefore all could be considered as potential control designs between which to randomly select at t_k. This indicates that the location of $x(t_k)$ in state-space should be checked with respect to Ω_{ρ_i}, $i = 1, \ldots, n_p$, before developing a candidate set of controllers to randomly select between at t_k. It should be noted, however, that if Ω_{ρ_i}, $i = 2, \ldots, n_p$, are subsets of Ω_{ρ_1}, then at each sampling time, Equations (61)–(67) with $i = 1$ and $\delta = 0$, Equations (61)–(67) with $i = 1$ and $\delta = 1$, and $h_1(x(t_k))$ are all candidate controllers that can maintain closed-loop stability. If $x(t_k)$ is in the intersection of additional level sets, there are additional candidate controllers which could be randomly selected between. Therefore, the minimum number of candidate controllers is 3 (or 2 if $x(t_k) \in \Omega_{\rho_1} / \Omega_{\rho_{e,1}}$ such that Equations (61)–(67) with $\delta = 0$ and $\delta = 1$ are equivalent), with more potentially being possible, especially as more stability regions with more intersections are developed.

Taking the above considerations into account, the implementation strategy for the LMPC design of Equations (61)–(67) is proposed as follows:

Step 1. At t_k, a random integer j between 1 and $2n_p$ is selected, and δ is randomly selected to be zero or one.

Step 2. If $j \in \{2, \ldots, n_p\}$, set $i = j$. If $j \in \{n_p + 2, \ldots, 2n_p\}$, set $i = j - n_p$. Verify that $V_i(x(t_k)) \in \Omega_{\rho_i}$. If yes, move to Step 3. If not, return to Step 1.

Step 3. If j is a number between 1 and n_p, use the LMPC of Equations (61)–(67) with $i = j$ and the selected value of δ. If $j = n_p + d$, $d = 1, \ldots, n_p$, set $u = h_d(x(t_k))$.

Step 4. Apply the control action computed for t_k to the process of Equation (1).

Step 5. $t_k \leftarrow t_{k+1}$. Return to Step 1.

Remark 6. *To prevent the possibility that the same index that is found to not meet the conditions in Step 2 at t_k will be selected multiple times as Steps 1 and 2 are repeated until a value of j is found for which $V_i(x(t_k)) \in \Omega_{\rho_i}$,* *indexes that cause $V_i(x(t_k)) \notin \Omega_{\rho_i}$ can be removed in the random integer selection procedure in Step 1 at t_k as they are identified before they force the algorithm to return to Step 1.*

5.3.1.1. Stability Analysis of Randomized LMPC

In this section, we develop sufficient conditions required for the randomized LMPC implementation strategy to provide closed-loop stability of the nonlinear process of Equation (1) under this implementation strategy and feasibility of the LMPC of Equations (61)–(67) when it is selected via the implementation strategy in the absence of a cyberattack in Section 5.3.1 to be used in determining a control action at a given sampling time. We first introduce two propositions that will then be used in proving the main results.

Proposition 1. *Refs. [60,82] Consider the systems*

$$\dot{x}_a(t) = f(x_a(t), u(t), w(t)) \tag{68}$$

$$\dot{x}_b(t) = f(x_b(t), u(t), 0) \tag{69}$$

with initial states $x_a(t_0) = x_b(t_0) \in \Omega_{\rho_1}$. There exists a function f_W of class $\mathcal{K}$ such that:

$$|x_a(t) - x_b(t)| \leq f_W(t - t_0) \tag{70}$$

for all $x_a(t), x_b(t) \in \Omega_{\rho_1}$ and all $w(t) \in W$ with:

$$f_W(\tau) = \frac{L_w \theta}{L_x}(e^{L_x \tau} - 1) \tag{71}$$

Proposition 2. *Refs. [60,82] Consider the Lyapunov function $V_i(\cdot)$ of the system of Equation (1). There exists a quadratic function $f_{V,i}(\cdot)$ such that:*

$$V_i(x) \leq V_i(\hat{x}) + f_{V,i}(|x - \hat{x}|) \tag{72}$$

for all $x, \hat{x} \in \Omega_{\rho_i}$ with

$$f_{V,i}(s) = \alpha_{4,i}(\alpha_{1,i}^{-1}(\rho_i))s + M_{v,i}s^2 \tag{73}$$

where $M_{v,i} > 0$ is a constant.

Proposition 3. *Ref. [62] Consider the Lyapunov-based controller $h_i(x)$ that meets Equations (2)–(5) with Lyapunov function $V_i(\cdot)$, applied in sample-and-hold to the system of Equation (1). If $\rho_i > \rho_{e,i} > \rho_{\min,i} > \rho_{s,i}$, and $\theta > 0$, $\Delta > 0$, and $\epsilon_{w,i} > 0$ satisfy:*

$$-\alpha_{3,i}(\alpha_{2,i}^{-1}(\rho_{s,i})) + L'_{x,i}M\Delta + L'_{w,i}\theta \leq -\epsilon_{w,i}/\Delta \tag{74}$$

then $\forall x(t_k) \in \Omega_{\rho_i}/\Omega_{\rho_{s,i}}$,

$$V_i(x(t)) \leq V_i(x(t_k)) \tag{75}$$

and $x(t) \in \Omega_{\rho_i}$ for $t \in [t_k, t_{k+1})$. Furthermore, if $\rho_{\min,i}$ is defined as follows:

$$\rho_{\min,i} = \max\{V_i(x(t+\Delta)) : V_i(x(t)) \leq \rho_{s,i}\} \tag{76}$$

then the closed-loop state is ultimately bounded in $\Omega_{\rho_{\min,i}}$ in the sense that:

$$\limsup_{t \to \infty} |x(t)| \in \Omega_{\rho_{\min,i}} \tag{77}$$

Theorem 1. *Consider the system of Equation (1) in closed-loop under the implementation strategy of Section 5.3.1 based on controllers $h_i(x)$ that satisfy Equations (2)–(5), and consider that the conditions in Proposition 3 hold. Let $\epsilon_{w,i} > 0$, $\Delta > 0$, $\rho_i > \rho_{e,i} > \rho_{\min,i} > \rho_{s,i}$ satisfy:*

$$\rho_{e,i} \leq \rho_i - f_{V,i}(f_W(\Delta)) \tag{78}$$

and Equations (74) and (76), for $i = 1, \ldots, n_p$, and $\Omega_{\rho_{e,j}} \subset \Omega_{\rho_{e,1}}$, $j = 2, \ldots, n_p$. If $x(t_0) \in \Omega_{\rho_1}$ and $N \geq 1$, then the state $x(t)$ of the closed-loop system is always bounded in Ω_{ρ_1}.

Proof. The proof consists of two parts. In the first part, we demonstrate that despite the random selection of a control law in Step 1 of the implementation strategy in Section 5.3.1, a characterizable control action is applied at every sampling time, and the LMPC of Equations (61)–(67) is feasible at every sampling time at which it is used for determining the control action to apply to the process. In the second part, we prove the results of Theorem 1.

Part 1. To demonstrate that an input with characterizable properties is returned by the implementation strategy of Section 5.3.1 at every sampling time to be applied to the process, we note that one of two inputs is returned at every sampling time: a) a control action computed by the LMPC of Equations (61)–(67) with $i = j$ where $x(t_k) \in \Omega_{\rho_j}$ or b) a Lyapunov-based controller $h_j(x(t_k))$ where $x(t_k) \in \Omega_{\rho_j}$.

In case (a), a solution to the LMPC of Equations (61)–(67) must have the characterizable property that it met the constraints of the LMPC because the LMPC always has at least one feasible solution. Specifically, $h_i(\tilde{x}(t_q))$, $q = k, \ldots, k+N-1$, $t \in [t_q, t_{q+1})$, with $i = j$, is a feasible solution to the optimization problem of Equations (61)–(67) when $x(t_k) \in \Omega_{\rho_j}$. It causes the constraint of Equation (64) to be met because $h_i(\tilde{x}(t_q))$, $q = k, \ldots, k+N-1$, $t \in [t_q, t_{q+1})$, maintains the closed-loop state in $\Omega_{\rho_j} \subseteq \Omega_{\rho_1}$ by Proposition 3, and the state constraint of Equation (64) is met for all states in Ω_{ρ_1}. $h_i(x)$ in sample-and-hold also satisfies the input constraint of Equation (65) by Equation (5). From Proposition 3, it causes the constraint of Equation (66) to be met when $x(t_k) \in \Omega_{\rho_j}$, and it trivially satisfies the constraint of Equation (67). Notably, the feasibility of $h_i(x)$ in sample-and-hold is true regardless of whether $\delta = 1$ or $\delta = 0$ because this is a feasible solution to all constraints of the optimization problem.

In case (b), the control action applied to the process is also characterizable because it is a control action that meets Proposition 3. Therefore, regardless of the control action applied at t_k, the control action has characterizable properties which can be used in establishing closed-loop stability. Furthermore, whenever Equations (61)–(67) are used to determine an input at a given sampling time, a feasible solution to this optimization problem always exists because it is ensured that $x(t_k) \in \Omega_{\rho_i}$ before the solution is obtained, and the feasibility of $h_i(\tilde{x}(t_q))$, $q = k, \ldots, k+N-1$, $t \in [t_q, t_{q+1})$ was demonstrated to hold above as long as $x(t_k) \in \Omega_{\rho_i}$.

Part 2. In this part, we prove that even with a control law randomly selected at every sampling time according to the implementation strategy in Section 5.3.1, the closed-loop state is maintained within Ω_{ρ_1} for all times if $x(t_0) \in \Omega_{\rho_1}$.

To demonstrate this, we first consider the case that at a given sampling time, a control law of the form of Equations (61)–(67) with $i = j$ when $x(t_k) \in \Omega_{\rho_j}$ is selected. In this case, either the constraint of Equation (66) is activated (if $x(t_k) \in \Omega_{\rho_{e,i}}$), the constraint of Equation (67) is activated (if $x(t_k) \in \Omega_{\rho_i} / \Omega_{\rho_{e,i}}$, $t_k \geq t'$, or $\delta = 1$), or both are activated (as may occur, for example, if $t_k \geq t'$ or $\delta = 1$ but $x(t_k) \in \Omega_{\rho_{e,i}}$).

Consider first the case that Equation (66) is activated. In this case, application of Proposition 2 (assuming that $x(t) \in \Omega_{\rho_i}$ for $t \in [t_k, t_{k+1})$) gives:

$$V_i(x(t)) \leq V_i(\tilde{x}(t)) + f_{V,i}(|x(t) - \tilde{x}(t)|) \tag{79}$$

for $t \in [t_k, t_{k+1})$. Applying the constraint of Equation (66) and Proposition 1, we obtain that:

$$V_i(x(t)) \leq \rho_{e,i} + f_{V,i}(f_W(|t - t_k|)) \leq \rho_{e,i} + f_{V,i}(f_W(\Delta)) \tag{80}$$

for $t \in [t_k, t_{k+1})$. When Equation (78) holds, $V_i(x(t)) \leq \rho_i$, for $t \in [t_k, t_{k+1})$, which validates the assumption used in deriving this result and guarantees that $x(t) \in \Omega_{\rho_i}$ for $t \in [t_k, t_{k+1})$ when $x(t_k) \in \Omega_{\rho_{e,i}}$ and the LMPC of Equations (61)–(67) is used to determine the input to the process of Equation (1). Because $\Omega_{\rho_i} \subseteq \Omega_{\rho_1}$, $x(t) \in \Omega_{\rho_1}$ for $t \in [t_k, t_{k+1})$.

Consider now the case that the constraint of Equation (67) is activated. In this case, we have from this constraint and Equation (3) that

$$\begin{aligned} &\frac{\partial V_i(x(t_k))}{\partial x} f(x(t_k), u(t_k), 0) \\ &\leq \frac{\partial V_i(x(t_k))}{\partial x} f(x(t_k), h_i(x(t_k)), 0) \leq -\alpha_{3,i}(|x(t_k)|) \end{aligned} \tag{81}$$

from which we can obtain:

$$\begin{aligned} &\frac{\partial V_i(x(t))}{\partial x} f(x(t), u(t_k), w(t)) \\ &= \frac{\partial V_i(x(t))}{\partial x} f(x(t), u(t_k), w(t)) \\ &\quad - \frac{\partial V_i(x(t_k))}{\partial x} f(x(t_k), u(t_k), 0) \\ &\quad + \frac{\partial V_i(x(t_k))}{\partial x} f(x(t_k), u(t_k), 0) \\ &\leq \left| \frac{\partial V_i(x(t))}{\partial x} f(x(t), u(t_k), w(t)) \right. \\ &\quad \left. - \frac{\partial V_i(x(t_k))}{\partial x} f(x(t_k), u(t_k), 0) \right| - \alpha_{3,i}(|x(t_k)|) \\ &\leq L'_{x,i}|x(t) - x(t_k)| + L'_{w,i}|w| - \alpha_{3,i}(|x(t_k)|) \\ &\leq L'_{x,i} M\Delta + L'_{w,i}\theta - \alpha_{3,i}(|x(t_k)|) \end{aligned} \tag{82}$$

for $t \in [t_k, t_{k+1})$, where the last inequality follows from Equations (7) and (8). Furthermore, if $x(t_k) \in \Omega_{\rho_i} / \Omega_{\rho_{s,i}}$, we can obtain from Equation (82) that:

$$\begin{aligned} &\frac{\partial V_i(x(t))}{\partial x} f(x(t), u(t_k), w(t)) \\ &\qquad \leq L'_{x,i} M\Delta + L'_{w,i}\theta - \alpha_{3,i}(\alpha_{2,i}^{-1}(\rho_{s,i})) \end{aligned} \tag{83}$$

If Equation (74) holds, then

$$\frac{\partial V_i(x(t))}{\partial x} f(x(t), u(t_k), w(t)) \leq -\epsilon_{w,i}/\Delta \tag{84}$$

Integrating Equation (84) gives that $V_i(x(t)) \leq V_i(x(t_k))$, $\forall t \in [t_k, t_{k+1})$, such that if $x(t_k) \in \Omega_{\rho_i} / \Omega_{\rho_{s,i}}$, then $x(t) \in \Omega_{\rho_i}$, $\forall t \in [t_k, t_{k+1})$.

If instead $x(t_k) \in \Omega_{\rho_{s,i}} \subset \Omega_{\rho_i}$, then from Equation (76), $x(t) \in \Omega_{\rho_{\min,i}} \subset \Omega_{\rho_i}$ for $t \in [t_k, t_{k+1})$. Therefore, if Equations (61)–(67) are used to compute the input trajectory at t_k and $x(t_k) \in \Omega_{\rho_i}$ and Equation (67) is applied, $x(t) \in \Omega_{\rho_i}$ for $t \in [t_k, t_{k+1})$ (this holds regardless of whether Equation (66) is simultaneously applied since this proof relied only on whether Equation (67) is applied and not whether the other constraints were simultaneously applied). Because $\Omega_{\rho_i} \subseteq \Omega_{\rho_1}$, this indicates that when the LMPC of Equations (61)–(67) is used with the constraint of Equation (67) activated to determine the control action at t_k when $x(t_k) \in \Omega_{\rho_i}$, then $x(t) \in \Omega_{\rho_1}$ for $t \in [t_k, t_{k+1})$.

Finally, consider the case that $x(t_k) \in \Omega_{\rho_i}$ and $h_i(x(t_k))$ is used to control the process of Equation (1) from t_k to t_{k+1}. In this case, the following holds:

$$\frac{\partial V_i(x(t_k))}{\partial x} f(x(t_k), h(x(t_k)), 0) \leq -\alpha_{3,i}(|x(t_k)|) \tag{85}$$

as follows from Equation (3). Using a similar series of steps as in Equation (82), we obtain:

$$\begin{aligned}&\frac{\partial V_i(x(t))}{\partial x} f(x(t), h(x(t_k)), w(t)) \\ &\leq L'_{x,i} M\Delta + L'_{w,i}\theta - \alpha_{3,i}(|x(t_k)|)\end{aligned} \tag{86}$$

If $x(t_k) \in \Omega_{\rho_i} / \Omega_{\rho_{s,i}}$, then as for Equation (83), we obtain:

$$\begin{aligned}&\frac{\partial V_i(x(t))}{\partial x} f(x(t), h(x(t_k)), w(t)) \\ &\leq L'_{x,i} M\Delta + L'_{w,i}\theta - \alpha_{3,i}(\alpha_{2,i}^{-1}(\rho_{s,i}))\end{aligned} \tag{87}$$

If Equation (74) holds, then we can use a similar series of steps as for Equation (84) to derive that $V_i(x(t)) \leq V_i(x(t_k))$, $\forall t \in [t_k, t_{k+1})$, such that if $x(t_k) \in \Omega_{\rho_i} / \Omega_{\rho_{s,i}}$, then $x(t) \in \Omega_{\rho_i}$, $\forall t \in [t_k, t_{k+1})$. If $x(t_k) \in \Omega_{\rho_{s,i}}$, then when Equation (76) holds, we obtain that $x(t) \in \Omega_{\rho_{\min,i}}$, $t \in [t_k, t_{k+1})$, so that $x(t) \in \Omega_{\rho_i}$ for $t \in [t_k, t_{k+1})$. Since $\Omega_{\rho_i} \subseteq \Omega_{\rho_1}$, we again obtain that if $x(t_k) \in \Omega_{\rho_i}$ and $h_i(x(t_k))$ is applied for $t \in [t_k, t_{k+1})$, then $x(t) \in \Omega_{\rho_1}$, $\forall\, t \in [t_k, t_{k+1})$.

The above results indicate that throughout every sampling period, if the conditions of Theorem 1 hold and the implementation strategy in Section 5.3.1 is used, then the closed-loop state does not leave Ω_{ρ_1}, implying that it also holds throughout all time if $x(t_0) \in \Omega_{\rho_1}$. This completes the proof. □

Remark 7. *Theorem 1 only speaks to the closed-loop state remaining in a bounded region of operation. If the randomness is removed and the $i = 1$ controller is selected to be used with the constraint of Equation (67) activated for all subsequent times (i.e., Equations (14)–(20) with $t > t'$), the closed-loop state is guaranteed to be ultimately bounded in a neighborhood of the origin [60]. If the randomness is not removed but $t > t'$ in Equations (61)–(67), the i-th controller will cause $V_i(x(t)) < V_i(x(t_k))$, $t \in (t_k, t_{k+1}]$ as noted in Section 2.4. However, consider the case that $x(t_k) \in \Omega_{\rho_i}$ and $x(t_k) \in \Omega_{\rho_z}$, but the i-th controller is selected at t_k. The decrease in V_i throughout the sampling period as a result of using the i-th controller does not necessarily imply that $V_z(x(t)) < V_z(x(t_k))$, $\forall t \in (t_k, t_{k+1}]$. If the randomness is removed, however, and only the $i = 1$ controller is used with $t > t'$, $V_1(x(t)) < V_1(x(t_k))$, $t \in (t_k, t_{k+1}]$ in every sampling period (i.e., a continuous decrease of the same Lyapunov function is ensured so that the closed-loop state is guaranteed to move to lower level sets of this Lyapunov function and not to again leave them) until the closed-loop state reaches $\Omega_{\rho_{s,1}}$, after which point it remains ultimately bounded in $\Omega_{\rho_{\min,1}}$. Another idea for driving the closed-loop state to a neighborhood of the origin with a randomized LMPC implementation strategy would be to change the implementation strategy at t' to only allow controllers to be selected in Steps 1-2 for which V_1 and h_1 are used in their design (e.g., h_1 and the $i = 1$ LMPC) so that each of the potential controllers would cause a decrease in the same Lyapunov function value over time.*

Remark 8. *The stability analysis reveals that despite the intuitive nature of the approach for deterring cyberattackers, it suffers the same problem as the controller in Section 5.2; namely, it does not meet Definition 4, and once the controller learns the implementation strategy itself, he or she could develop an attack policy that is not guaranteed to maintain closed-loop stability according to the proof methodology above. We can see a potential for the lack of resilience by referring again to Figure 3 and noting that if the actual state measurement is at x_a, the closed-loop stability proof relies on the $i = 2$ controller not being an option; however, a false state* *measurement of x_b may cause the $i = 2$ controller to be chosen when $x(t_k) = x_a$, such that the conditions required for the closed-loop stability proof in Theorem 1 (i.e., that the implementation strategy in Section 5.3.1 is correctly followed) do not hold. However, the closed-loop stability issues with the proposed design in the presence of a cyberattack are deeper than this; the problem is not necessarily that the control action computed by a controller that would not otherwise have been selected is used, but rather that regardless of whether that controller should have been allowed to be used per the implementation strategy in Section 5.3.1 is used or not, the input applied to the process has no relationship to the state in the sense that, for example, the state constraints in Equations (66) and (67) are not necessarily met (or even close to being met) by the actual process state even if the controller used at t_k indicated feasibility of the control action with respect to these constraints. This is because the controller is using a different initial condition than the actual process initial condition and therefore will compute, potentially, a state trajectory under the input selected as optimal by the LMPC that is very different from the actual process state trajectory under that same input, even in the absence of disturbances/plant-model mismatch. Mismatch is introduced by the cyberattack at the initial condition for the model of Equation (62).*

5.3.2. Problems with Incorporating Randomness in LMPC Design

In this section, we demonstrate the use of the randomized LMPC for the CSTR example of Section 5.2.1 during routine operation and also in the case that false state measurements are provided to demonstrate that the randomized LMPC implementation strategy can maintain closed-loop stability under normal operation, but may at best in certain sensor cyberattack cases only delay an unsafe condition from being reached (i.e., randomness by itself, without giving the properties in Definition 4, does not create cyberattack resilience in control). We first develop the set of LMPC's to be used to control the process of Equations (46) and (47). We begin by developing seven (i.e., $n_p = 7$) potential combinations of V_i, h_i, Ω_{ρ_i}, and $\Omega_{\rho_{e,i}}$. The form of each V_i is $x^T P_i x$, where P_i is a symmetric positive definite matrix of the following form:

$$\begin{bmatrix} P_{11} & P_{12} \\ P_{12} & P_{22} \end{bmatrix} \tag{88}$$

Sontag's control law [80] was used to set the value of the component of every $h_i = [h_{i,1}\ h_{i,2}]^T$ corresponding to u_2 as follows:

$$h_{i,2}(x) = \begin{cases} -\frac{L_{\tilde{f}}V_i + \sqrt{L_{\tilde{f}}V_i^2 + L_{\tilde{g}_2}V_i^4}}{L_{\tilde{g}_2}V_i}, & \text{if } L_{\tilde{g}_2}V_i \neq 0 \\ 0, & \text{if } L_{\tilde{g}_2}V_i = 0 \end{cases} \tag{89}$$

where if $h_{i,2}$ fell below or exceeded the upper or lower bound on u_2, $h_{i,2}$ was saturated at the respective bound. $L_{\tilde{f}}V_i$ and $L_{\tilde{g}_k}V_i$ represent the Lie derivatives of V_i with respect to $\tilde{f}$ and $\tilde{g}_k$, $k = 1, 2$. For simplicity, $h_{i,1}$ was taken to be 0 kmol/m^3 for $i = 1, \ldots, 7$. Using the values of the entries of each P_i associated with each V_i in Table 3 and the associated h_i, $i = 1, \ldots, 7$, the stability regions in Table 3 were obtained by discretizing the state-space and choosing an upper bound on each Lyapunov function in a region of state-space where $\dot{V}_i$ was negative at the discretized points under the controller h_i, $i = 1, \ldots, 7$ (the discretization was performed in increments of 0.01 kmol/m^3 in C_A for C_A between 0 and 4 kmol/m^3, and in increments of 1 in T for T between 340 and 560 K). Subsets of the stability regions were selected to be $\Omega_{\rho_{e,i}}$ with the goal of allowing several different control laws to be developed. For $i = 2, \ldots, 7$, $\Omega_{\rho_i} \subseteq \Omega_{\rho_1}$. The value of $\rho_{e,i}$ was not more than 80% of ρ_i in each case.

Table 3. *i*-th controller parameters.

i	P_{11}	P_{12}	P_{22}	ρ_i	$\rho_{e,i}$
1	1200	5	0.1	180	144
2	2000	−20	1	180	144
3	1500	−20	10	180	144
4	0.2	0	2000	180	144
5	1200	5	0.1	180	100
6	1200	5	0.1	180	130
7	1200	5	0.1	180	30

Initially, we evaluate the closed-loop stability properties of the process of Equations (46) and (47) for normal operation under the randomized LMPC implementation strategy and, for comparison, under the $i = 1$ LMPC used for all times. The process was initialized from $x_{init} = [-0.4 \text{ kmol/m}^3 \; 20 \text{ K}]^T$. For the randomized LMPC design, the implementation strategy in Section 5.3.1 was followed with the exception that, for simplicity, δ was set to 0 at every sampling time, and only $h_1(x)$ was considered as a candidate controller at a given sampling time as an alternative to controllers in Table 3. Therefore, at every sampling time, both the LMPC of Equations (61)–(67) with $i = 1$ and $h_1(x)$ were allowable control actions, and the i-th controller in Table 3 was also allowable if $x(t_k) \in \Omega_{\rho_i}$. The simulations were implemented in MATLAB using fmincon and the seed rng(5) and random integer generation function randi when the randomized LMPC implementation strategy was used. The integration step for the model of Equations (46) and (47) was set to 10^{-4} h, $N = 10$, and $\Delta = 0.01$ h, with 1 h of operation used. The Lyapunov-based stability constraint activated when $x(t_k) \in \Omega_{\rho_{e,i}}$ was enforced at the end of every sampling period in the prediction horizon, and whenever the Lyapunov-based stability constraint involving the time-derivative of the Lyapunov function was enforced, the other Lyapunov-based constraint was implemented at the end of the sampling periods after the first. The initial guess provided to fmincon in both cases was the steady-state input vector. The maximum and minimum values of u_2 were multiplied by 10^{-5} in numerically solving the optimization problem.

Figures 4–6 show the state, input, and state-space trajectories resulting from controlling the process with one LMPC throughout the time period of operation, and Figures 7–9 show the results of controlling the LMPC with one of the eight potential control laws selected at every sampling time, but depending on the position of the state measurement in state-space. The figures indicate that both the single LMPC implemented over time and the randomized LMPC implementation strategy were able to maintain the closed-loop state within Ω_{ρ_1}. Figure 10 shows which controller (i in Table 3) was selected by the randomized LMPC implementation strategy at each sampling time. Notably, the control laws associated with $i = 2$, 3, and 4 in Table 3 were not chosen, which is consistent with the requirement that a control law can only be available to be selected if $x(t_k) \in \Omega_{\rho_i}$ (from Figure 9, we see that the closed-loop state did not enter, for example, Ω_{ρ_2} and Ω_{ρ_3}, and the results of the simulations indicate that though the closed-loop state sometimes entered Ω_{ρ_4} as shown in Figure 9, it was never in this region at a sampling time, which explains why these controllers were never selected by the randomized implementation strategy). The time-integral of Equation (48) was monitored for the process of Equations (46) and (47) under the inputs applied to the process, and also for steady-state operation. For the single LMPC implemented over time, it evaluated to 32.2187, while for the randomized LMPC implementation strategy, it evaluated to 27.7536. There is some profit loss due to the randomized LMPC implementation strategy, and also large variations in states and inputs shown in Figures 7 and 8. If the randomized LMPC implementation strategy was able to deter cyberattacks, one could consider whether that made the variations and profit loss acceptable. Despite the decrease in profits due to the randomization, both the single LMPC over time and the LMPC's implemented within the randomized implementation strategy significantly outperformed steady-state operation, which had a value of the time-integral of Equation (48) of 13.8847.

After analyzing normal operation for the LMPC and randomized LMPC implementation strategy, we look at differences in their response to the cyberattack policy determined in Section 5.2.1, where the attack on the sensors is simulated for 10 sampling periods and the process is initialized at x_{init}. The metric that we use for comparing the results in the two scenarios is the time until the closed-loop state exceeds its threshold of 55 K for x_2 (as $x_2 > 55$ K occurs outside the stability region, the closed-loop state exits the stability region before this unsafe condition is reached). For the single LMPC, x_2 first exceeds it threshold around 0.0142 h. In the case of the randomized LMPC, different input policies (i.e., different sequences of randomly selected control laws) give different behavior in the presence of the cyberattack. Therefore, in Table 4, we present the approximate time that x_2 exceeds its threshold for 10 different arguments provided to the MATLAB seeding function rng to create 10 different seeds for the random number generator that selects which control law to randomly select at each sampling time. The table indicates that the randomization may slightly delay the time at which x_2 first exceeds its threshold compared to the case that the single LMPC is used. However, in none of the cases simulated was it able to prevent the cyberattack from driving the value of x_2 above its threshold in 0.1 h of operation. If a cyberattacker believes that some delay in the attack may cause him or her to be caught, this strategy may help with deterring some types of attacks. However, the results indicate that it is not cyberattack-resilient according to Definition 4. Figure 11 shows the results of the simulations for 0.1 h with the randomized LMPC implementation strategy for different arguments of rng in state-space.

Figure 12 displays data on the inputs and value of $V_1(x)$ over time under both the randomized LMPC implementation strategy and the single LMPC, as well as the selected control law among the 8 possibilities at each sampling time in the case that the argument of rng is set to 20. This figure suggests that some of the difficulty with maintaining the closed-loop state in a bounded region under the attack is that for the falsified state measurement, the available controllers (the $i = 3$ and $i = 4$ controllers are not available because the false state measurement that the controller receives and uses in determining which control laws should be made available according to the randomized LMPC implementation strategy is outside of Ω_{ρ_3} and Ω_{ρ_4}) compute inputs with similarities to each other and to the inputs which the single LMPC would compute in the sense that they are either close in value or create similar effects on the closed-loop state (i.e., the fact that different control laws may be chosen to compute an input is not very effective in this case at obscuring the mapping between $x(t_k)$ and the inputs applied to the process). From Figure 12, we see that all of the available control laws were used at some point, but the inputs computed in every case except for the $i = 8$ controller were close to those of the single LMPC, and the $i = 8$ controller was also not effective at causing a direction change in the value of V_1, despite that it has some more noticeable differences compared to the trajectory computed by the single LMPC.

The attack policy chosen plays a role in the amount of delay in the success of an attack which the randomized LMPC implementation strategy of Section 5.3.1 may cause. For example, consider instead the falsified initial condition $x_1 = 0.0632$ kmol/m^3 and $x_2 = 21.2056$ K, which is also within the stability region (but not within the stability regions of the $i = 2$, 3, or 4 controllers). If used at each sampling time, it can cause $x_2 > 55$ K in 0.0319 h under the single LMPC. For this attack policy, the approximate time after which $x_2 > 55$ K for the randomized LMPC implementation strategy is reported in Table 5. Some of the delays in the success of the attack at driving $x_2 > 55$ K in this case are much more significant than in Table 4. The simulation results demonstrate that the lack of resiliency of the randomized LMPC policy can come from the lack of correlation between the inputs applied and the actual process state at each sampling time, as discussed in Remark 8. For example, for the case where the seed used is 5, the same inputs are applied to the process in both the case that the single LMPC is used and the case that the randomized LMPC implementation strategy is used at the sampling period beginning at $t_k = 0.02$ h, but because the initial condition at t_k in both cases is different (caused by the different input policies computed in the prior sampling period by the use of the different control laws), these same inputs in one case drive the closed-loop state out of the stability region in the sampling period, and in the other case they do not succeed in driving it out in the sampling period. Conversely, in the sampling periods between $t_k = 0.03$ h and 0.05 h, the inputs applied to the process under the randomized LMPC implementation strategy are not the values that would have been computed if the single LMPC had been used, but they drive the closed-loop state out of the stability region. Though the randomness may be beneficial at helping delay the success of attacks in some cases, it does not address the fundamental lack of correlation between the applied inputs and the actual process state that causes the cyberattack success.

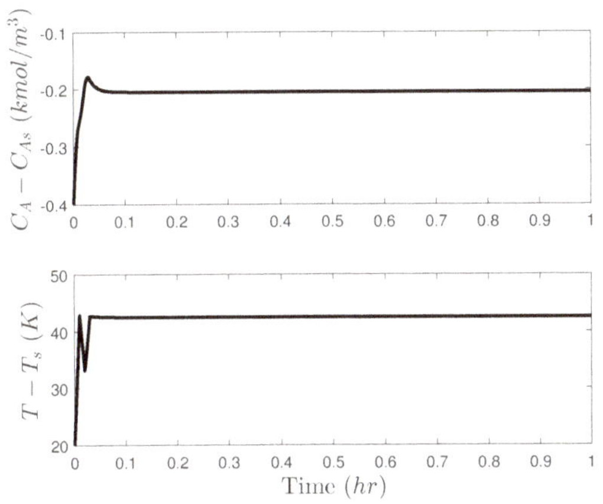

Figure 4. State trajectories under the single Lyapunov-based model predictive controller (LMPC) for the CSTR of Equations (46) and (47).

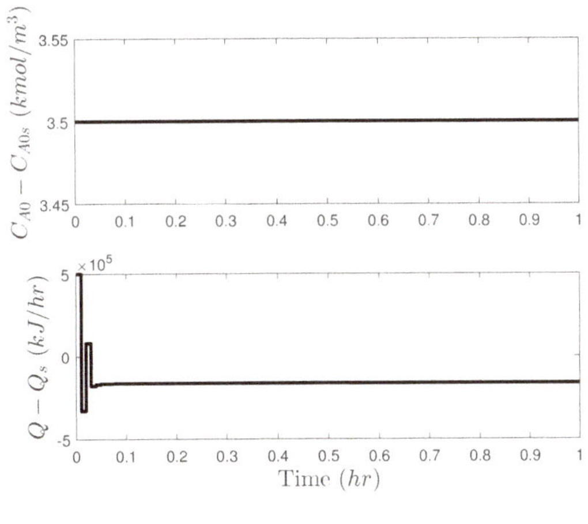

Figure 5. Input trajectories under the single LMPC for the CSTR of Equations (46) and (47).

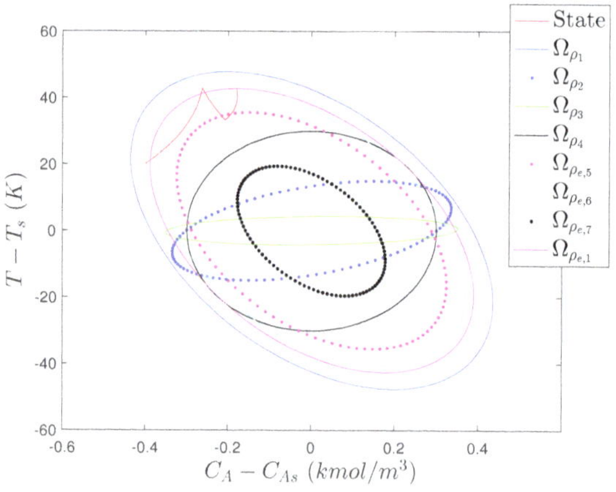

Figure 6. State-space trajectories under the single LMPC for the CSTR of Equations (46) and (47). The figure indicates that the closed-loop trajectory settled on the boundary of $\Omega_{\rho_{e,1}}$ to optimize the objective function while meeting the constraints. For simplicity, only one level set for each of the n_p potential LMPC's is shown (Ω_{ρ_i} is shown if $V_i \neq V_1$, and $\Omega_{\rho_{e,i}}$ is shown if $V_i = V_1$, $i > 1$).

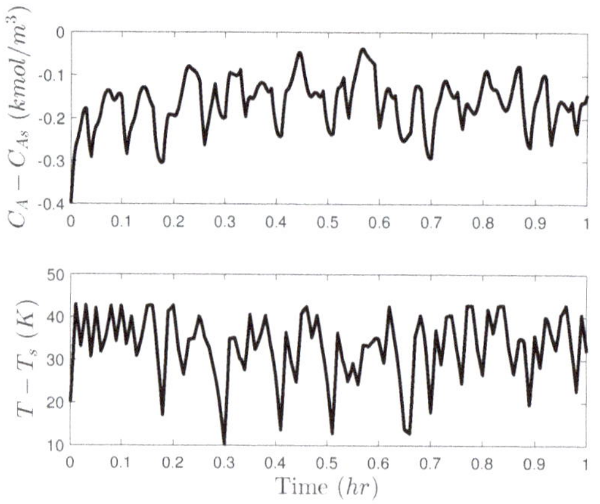

Figure 7. State trajectories under the randomized LMPC implementation strategy for the CSTR of Equations (46) and (47).

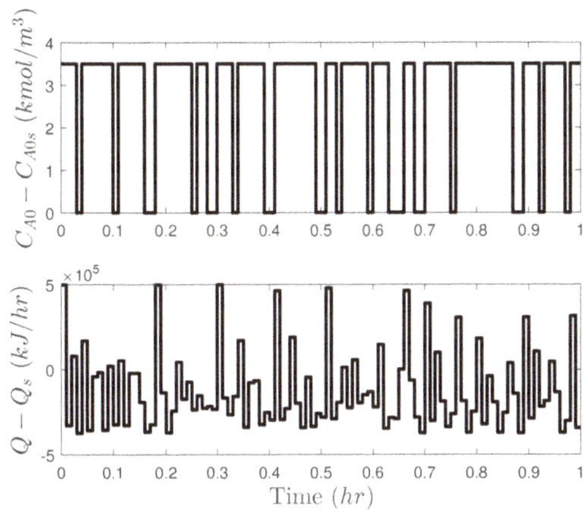

Figure 8. Input trajectories under the randomized LMPC implementation strategy for the CSTR of Equations (46) and (47).

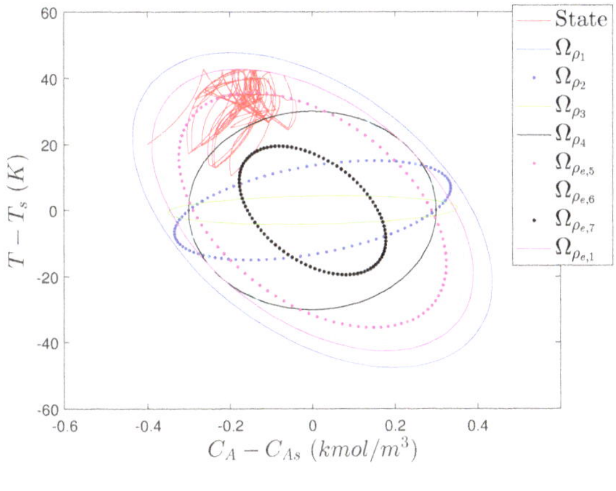

Figure 9. State-space trajectories under the randomized LMPC implementation strategy for the CSTR of Equations (46) and (47). For simplicity, only one level set for each of the n_p potential LMPC's is shown (Ω_{ρ_i} is shown if $V_i \neq V_1$, and $\Omega_{\rho_{e,i}}$ is shown if $V_i = V_1, i > 1$).

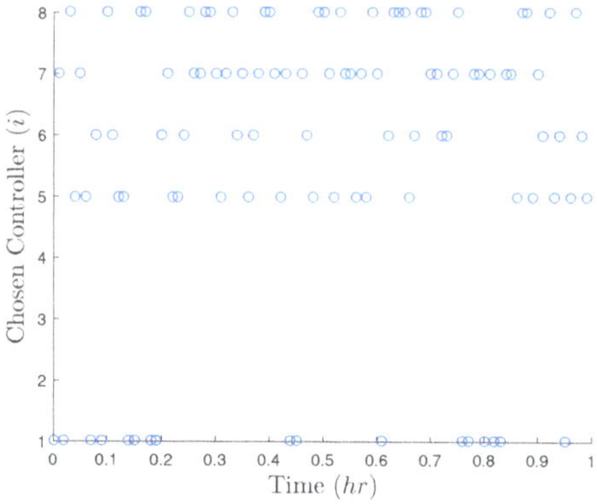

Figure 10. Scatter plot showing the control law chosen (i in Table 3) in each sampling period by the randomized LMPC implementation strategy.

Table 4. Approximate time after which $x_2 > 55$ K for various seed values of rng for the randomized LMPC design subjected to a cyberattack on the sensors determined in Section 5.2.1.

Seed	Time $x_2 > 55$ K (h)
5	0.0143
10	0.0148
15	0.0146
20	0.0324
25	0.0146
30	0.0142
35	0.0143
40	0.0147
45	0.0248
50	0.0231

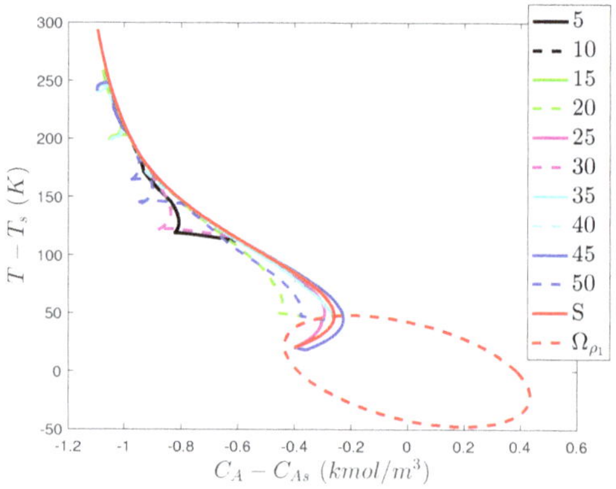

Figure 11. State-space trajectories for all of the situations in Table 4. The numbers in the caption represent the seed values for rng. 'S' represents the single LMPC.

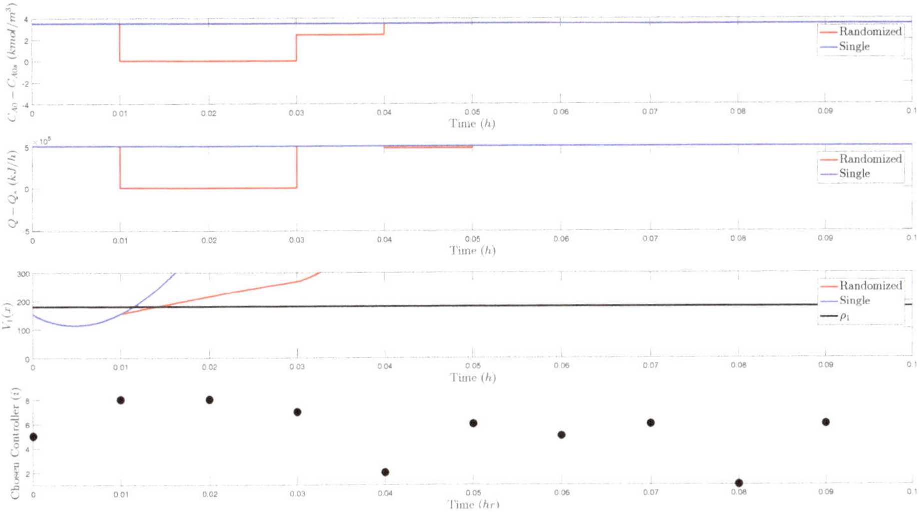

Figure 12. Trajectories of u_1, u_2, and V_1 under the randomized LMPC implementation strategy for rng(20) (denoted by 'Randomized' in the figure) and under the single LMPC (denoted by 'Single' in the figure). The value of ρ_1 is denoted by the horizontal line in the plot for the value of V_1. The bottom plot indicates the controller selected by the randomized LMPC implementation strategy at each of the 10 sampling times in the simulation.

Table 5. Approximate time after which $x_2 > 55$ K for various seed values of rng for the randomized LMPC design subjected to a cyberattack on the sensors with $x_1 = 0.0632$ kmol/m^3 and $x_2 = 21.2056$ K.

Seed	Time $x_2 > 55$ K (h)
5	0.0674
10	0.0458
15	0.0555
20	0.0767
25	0.0569
30	0.0418
35	0.0457
40	0.0874
45	0.0580
50	0.0950

Simulations were also performed in the case that it was attempted to operate the process at steady-state (instead of in a time-varying fashion) by removing the constraint of Equation (66) and using the following quadratic stage cost:

$$L_e = \tilde{x}^T Q \tilde{x} + u^T R u \tag{90}$$

where $Q = \text{diag}(10^4, 100)$ and $R = \text{diag}(10^4, 10^{-6})$. In this case, the LMPC and randomized LMPC implementation strategy with rng(5) drive the closed-loop state to a neighborhood of the origin in the absence of a cyberattack. If the falsified state measurement determined in Section 5.2.1 is applied (without attempting to see whether there may be a more problematic input policy for the tracking control design), $x_2 > 55$ K in 0.0834 h under the single LMPC and 0.1395 h under the randomized LMPC strategy with rng(5). This demonstrates that processes operated at steady-state are not immune to cyberattacks when operated under LMPC or a randomized LMPC implementation strategy because again in this case, the value of $x(t_k)$ becomes decoupled from the input being applied. In a coupled nonlinear system, this may result in state trajectories that do not drive the (actual) closed-loop state to the origin.

Remark 9. *The last result concerning steady-state operation indicates that the difficulties with the randomized LMPC design with respect to Definition 4 hold regardless of whether δ in Equations (61)–(67) is fixed at 0 or 1, as the issue does not stem from whether the controller is attempting to drive the state predictions it is making toward the steady-state through the inputs it computes or whether it is attempting to operate the process in a time-varying fashion, but rather stems from the disconnect between what the controller thinks it is achieving and what it is actually achieving due to the falsified state measurements. This also indicates that having the inputs computed by the different potential controllers be significantly different from one another to create significant randomness in what input would be applied to the process may help in some cases (particularly if it sometimes reverses the direction in which V_1 changes), but it cannot address the input-state disconnect unless the manner in which random control laws are selected or generated can be proven to cause Definition 4 to be met. The fact that an allowable input policy exists that can cause problems means that even random attack strategies may pose a problem. Therefore, while a cyberattacker who cannot afford any delay in an attack might be deterred by the randomized LMPC implementation strategy, it is unlikely that this policy would provide a sufficient barrier to attacks.*

5.3.3. Creating Unpredictable Controller Outputs: Other Types of Randomness in MPC Design

There are many other techniques besides the randomized LMPC design of the prior sections which could be used to create randomness in control selection/design. For example, the closed-loop stability proofs for LMPC in [60] are independent of the objective function; therefore, one method for introducing randomness in the operation of the process of Equation (1) under LMPC without losing closed-loop stability during normal operation would be to make random modifications to the objective function of Equations (14)–(20) at each sampling time by adding penalty terms which change/are randomly generated at every sampling time (e.g., in some sampling periods they are zero, in some sampling periods they may penalize the difference between the input values from randomly selected values within the input bounds). The LMPC could also seek to generate input policies that create significant input variation over time by using penalty terms in the objective function on the similarity between the input trajectory computed at t_k and that applied at t_{k-1} (through, for example, terms such as $\sum_{i=1}^{m}(u_i(t_k) - u_i(t_{k-1}))^2$ subtracted from the stage cost to minimize the objective function more strongly if the difference between the inputs is greater between two sampling periods; this is not a randomly generated penalty but it is one that can differ between sampling times as $u(t_{k-1})$ can be different at each sampling time). A potential disadvantage of this approach, however, is that it causes other terms in the objective function, which are chosen to be meaningful with respect to operating objectives such as profit or steady-state tracking, to compete with randomly generated terms.

Another idea for creating randomness within the control design that does not impact the objective function (and therefore does not require the difficult task of determining an appropriate tuning that can trade off meaningful terms against randomly generated terms, as in the policies of the prior paragraph) would be to randomly generate constraints for an MPC at every sampling time. For example, the state constraint of Equation (17) might be modified to become $\tilde{x}(t) \in \tilde{X}$, $t \in [t_k, t_{k+N})$, where $\tilde{X}$ is a state-space region that is randomly generated at every sampling time (but $\tilde{X} \subset X$ to ensure that the modified state constraint maintains the closed-loop state predictions in X). As an example, consider that $\tilde{x}(t) \in X$ represents a state constraint of the form $x_{\min} \leq \tilde{x}(t) \leq x_{\max}$, $t \in [t_k, t_{k+N})$. A constraint of the form $\tilde{x}(t) \in \tilde{X}$ might require that at every sampling time, $x_{rand,\min} \leq \tilde{x}(t) \leq x_{rand,\max}$, where $x_{rand,\min}$ and $x_{rand,\max}$ are two randomly selected real numbers (at every sampling time) with $x_{rand,\min} \geq x_{\min}$, $x_{rand,\max} \leq x_{\max}$, and $x_{rand,\min} \leq x_{rand,\max}$. However, these modified state constraints are hard constraints that are not guaranteed to be satisfied throughout

Ω_{ρ_1} ($\tilde{x} \in X$ can be guaranteed to be satisfied by defining Ω_{ρ_1} to be in X, but it is not guaranteed that $\tilde{x}$ can be maintained in randomly generated subsets of X that may only constitute subsets of the stability region that are not necessarily related to V_1 and therefore are not necessarily forward invariant). Therefore, the randomly generated hard constraints may impact feasibility of an LMPC. Methods for handling this could include reformulating the hard constraints as soft constraints in the objective function when the problem is determined to be infeasible at t_k, or generating multiple (i.e., up to $\bar{p}$) random subsets of X at t_k, and up to $\bar{p}$ LMPC's using these different subsets to form the state constraints of Equation (17), and then attempting to solve these LMPC's in order from 1 to $\bar{p}$ to see whether one is feasible and can be used to compute a control action before applying a backup control law that guarantees closed-loop stability such as $h_1(x)$. Closed-loop stability of the system of Equation (1) under the LMPC of Equations (14)–(20) with Equation (17) modified to allow for random state constraint generation would follow from the results in [60] if feasibility is maintained. One could also consider other methods for developing randomly generated state constraints, such as exploring the potential for randomly generating constraints on regions for the closed-loop state to avoid [9–11] at each sampling time. However, even if optimization-based control designs with randomly generated constraints are feasible at a sampling time, they may also have disadvantages with respect to profit. For example, if the objective function is related to process economics and subsets of the allowable operating region are disallowed by hard constraints, the inputs seek to optimize the economics with a more restricted constraint set than is actually available, which would be expected to negatively impact profits. This is because the goal of the randomization would be to cause the controller to compute inputs which it would not normally compute if the constraint set was less restrictive in order to prevent an attacker from mapping $x(t_k)$ to an input. If the global optimum of the objective function within the allowable constraint set is assumed to be achieved with the solution to the controller without the randomization, then any deviations of the solution from this optimal value for the purpose of making the input-state measurement mapping difficult to determine would result in a decrease in profit compared to the optimum condition. If the global optimum is achieved, however, this means that the randomization is not succeeding in computing inputs which are difficult to map to the state measurements. Therefore, the premise of the randomized constraint designs would cause a profit reduction in cases where the economics are being optimized in the objective function (though popular techniques for solving nonlinear optimization problems (e.g., [83]) may find local rather than global optima, making it less obvious whether the randomization strategy will result in a profit loss compared to the (local) solution which might be found without the randomization).

The results of the prior sections of this work indicate that cyberattack-deterring control policies incorporating randomness cannot rely on randomness alone to prevent cyberattacks from being successful or from being attempted; the inputs computed by any cyberattack-resilient policy according to Definition 4 must have a structure that prevents the fact that they are decoupled from the state measurements from driving the closed-loop state out of a set of safe operating conditions.

5.4. Deterring Sensor Measurement Falsification Cyberattacks on Safety: Using Open-Loop Controller Outputs

Whereas the "intuitive" approaches of the prior sections failed to be cyberattack-resilient, in this section, we show that it may be possible to develop operating policies for which sensor falsification cyberattacks intended to impact process safety cannot be successful. The policy to be examined is specific to a subset of the class of systems of Equation (1), specifically those which have an open-loop asymptotically stable equilibrium. For clarity of notation in the following, we will denote the set of nonlinear systems of the form of Equation (1) with an open-loop asymptotically stable equilibrium as follows:

$$\dot{x} = f_{as}(x, u, w) \tag{91}$$

where f_{as} is a locally Lipschitz vector function of its arguments and $f_{as}(0,0,0) = 0$. The following conditions hold for all $x \in D' \subset R^n$, where D' is a neighborhood of the origin:

$$\alpha_5(|x|) \le V'(x) \le \alpha_6(|x|) \tag{92}$$

$$\frac{\partial V'(x)}{\partial x} f_{as}(x, u_s, 0) \le -\alpha_7(|x|) \tag{93}$$

where $u_s = 0$ denotes the steady-state input, $V' : R^n \to R_+$ is a sufficiently smooth positive definite Lyapunov function, and the functions α_5, α_6 and α_7 are of class $\mathcal{K}$. We define a level set of V' within D' where $x \in X$ as a stability region $\Omega_{\rho'}$ of the nominal system of Equation (91) under u_s ($\Omega_{\rho'} := \{x \in X \cap D' : V'(x) \le \rho'\}$). In the remaining developments, we assume that V' can be chosen to be the same as V_1.

5.4.1. Using Open-Loop Controller Outputs: Integration with LMPC

For the system of Equation (91), u_s itself is a cyberattack-deterring input policy according to Definition 4 when $x(t_0) \in \Omega_{\rho_1} \subset \Omega_{\rho'} \subset X$ because it drives the closed-loop state to the origin and is independent of the sensor measurements. However, it does not use feedback of the process state to impact the speed with which the steady-state is approached. Furthermore, it cannot drive the closed-loop state off of the steady-state in a fashion that seeks to optimize process economics. It therefore lacks the desirable properties of feedback controllers for non-attack scenarios, but in the case of cyberattacks on sensors, it has advantages over feedback control in that it is not dependent on sensor readings. This indicates that u_s and feedback controllers complement one another; the former is beneficial for preventing cyberattack success, and the latter is beneficial for normal operation. Therefore, in this section, we explore integrating these two types of control in an implementation strategy that, as will be proven in the next section, is guaranteed under sufficient conditions to maintain closed-loop stability both in the presence and absence of cyberattacks (i.e., it meets Definition 4). For developing this implementation strategy, we again use LMPC because the *a priori* characterizable region Ω_{ρ_1} within which LMPC maintains the process state during normal operation can be beneficial for developing a controller implementation strategy that guarantees that Definition 4 is met (in general, the results of this work suggest that theory-based control designs may be important for allowing cyberattack-resilient control designs to be developed, indicating that an important direction of future research may be making theory-based control designs easier to use in an industrial setting). The implementation strategy proposed is as follows:

Step 1. Given $x(t_0) \in \Omega_{\rho_1} \subset \Omega_{\rho'} \subset X$, apply u_s for N_1 sampling periods. Go to Step 2.

Step 2. Utilize an LMPC with the form in Equations (14)–(20) to control the process of Equation (91) for N_2 sampling periods. Go to Step 3.

Step 3. Apply u_s for N_1 sampling periods. Return to Step 2.

Characterizations of N_1 and N_2 that allow closed-loop stability of the system of Equation (91) to be guaranteed, even in the presence of cyberattacks and sufficiently small disturbances, under this implementation strategy are presented in the next section.

Stability Analysis of Open-Loop Control Integrated with LMPC

This section presents the conditions under which closed-loop stability of the system of Equation (91) under the implementation strategy in Section 5.4.1 is guaranteed in both the presence of and absence of a cyberattack that provides false state measurements $x_f \in \Omega_{\rho_1}$ at every sampling time (where the notation x_f represents a falsified sensor signal that in general can be different at each sampling time). The results are presented in a theorem that relies on the following proposition.

Proposition 4. *Ref. [62] Consider u_s for the system of Equation (91) such that the inequalities of Equations (92) and (93) are met with Lyapunov function $V'(\cdot) = V_1(\cdot)$. If $\rho' > \rho'_{\min} > \rho'_s$, and $\theta > 0$, $\Delta > 0$, and $\epsilon'_w > 0$ satisfy:*

$$-\alpha_7(\alpha_6^{-1}(\rho'_s)) + L'_{w,1}\theta \leq -\epsilon'_w/\Delta \tag{94}$$

then $\forall x(t_k) \in \Omega_{\rho'}/\Omega_{\rho_{s'}}$,

$$V'(x(t)) \leq V'(x(t_k)) \tag{95}$$

for $t \in [t_k, t_{k+1})$ and $x(t) \in \Omega_{\rho'}$. Furthermore, if $\rho'_{\min}$ is defined as follows:

$$\rho'_{\min} = \max\{V'(x(t+\Delta)) : V'(x(t)) \leq \rho'_s\} \tag{96}$$

then the closed-loop state is ultimately bounded in $\Omega_{\rho'_{\min}}$ in the sense that:

$$\limsup_{t\to\infty} |x(t)| \in \Omega_{\rho'_{\min}} \tag{97}$$

Theorem 2. *Consider the system of Equation (91) under the implementation strategy of Section 5.4.1 based on controllers u_s and $h_1(x)$ that satisfy Equations (92) and (93) and (2)–(5), respectively, and consider that the conditions in Proposition 4 hold, as well as those in Proposition 3 and Equation (78) with $i = 1$. If $x(t_0) \in \Omega_{\rho_1}$, $\Omega_{\rho'_s} \subset \Omega_{\rho'_{\min}} \subset \Omega_{\rho_{e,1}} \subset \Omega_{\rho_1} \subset \Omega_{\rho'}$, $V'(\cdot) = V_1(\cdot)$, $N \geq 1$, $N_1 = \lceil \frac{(\rho_1 - \rho'_{\min})}{\epsilon'_w} \rceil$, and $N_2 = \lfloor \frac{(\rho_1 - \rho'_{\min})}{(\alpha_{4,1}(\alpha_5^{-1}(\rho_1)))M\Delta} \rfloor$, then the state $x(t)$ of the closed-loop system is always bounded in Ω_{ρ_1}, $\forall\, t \geq 0$, regardless of the value of $\tilde{x}(t_k)$ in Equation (16), $\forall\, k \geq 0$, if $\tilde{x}(t_k) \in \Omega_{\rho_1}$ when Equations (14)–(20) are used at a sampling time for computing the control action applied to the process according to the implementation strategy in Section 5.4.1.*

Proof. The proof consists of four parts. In the first part, feasibility of the LMPC of Equations (14)–(20) at every sampling time in which it is used according to the implementation strategy in Section 5.4.1 will be demonstrated, regardless of whether the state measurements provided to the LMPC in Equation (16) are accurate or falsified, if they are within Ω_{ρ_1}. The second part will demonstrate that for any $x(t_k) \in \Omega_{\rho_1}$, $x(t_{k+N_1}) \in \Omega_{\rho'_{\min}}$ when u_s is used for N_1 sampling periods. The third part demonstrates that if $x(t_k) \in \Omega_{\rho'_{\min}}$ and the LMPC of Equations (14)–(20) is used for the next N_2 sampling periods to control the system of Equation (91) with potentially falsified state measurements, then $x(t_{k+N_2}) \in \Omega_{\rho_1}$. The fourth part combines the results of the prior three parts to demonstrate that the implementation strategy of Section 5.4.1 guarantees that the closed-loop state remains in Ω_{ρ_1} at all times, whether or not cyberattacks which provide falsified state measurements occur.

Part 1. When the input u_s is applied to the system of Equation (91) according to the implementation strategy in Section 5.4.1, no optimization problem is solved, and therefore there is no feasibility issue with using u_s at t_k. However, if the LMPC of Equations (14)–(20) is used, then if the state measurement $\tilde{x}(t_k) \in \Omega_{\rho_1}$ (regardless of whether $\tilde{x}(t_k)$ equals the true state measurement $x(t_k)$ or a falsified state measurement $x_f(t_k) \in \Omega_{\rho_1}$), $h_1(\tilde{x}(t_q))$, $q = k, \ldots, k+N-1$, $t \in [t_q, t_{q+1})$, is a feasible solution to all constraints of the optimization problem when $\tilde{x}(t_k) \in \Omega_{\rho_{e,1}}$ or when $x(t_k) \in \Omega_{\rho}/\Omega_{\rho_{e,1}}$ for the reasons noted in the proof of Part 1 of Theorem 1. While x_f can always be chosen to be in Ω_{ρ_1} to guarantee feasibility when the LMPC is used in computing control actions, the proof that $x(t_k)$ is always in Ω_{ρ_1} when the LMPC is used so that the feasibility guarantees at each sampling time hold when no cyberattack occurs at t_k will be developed in subsequent parts of this proof.

Part 2. To demonstrate that for any $x(t_k) \in \Omega_{\rho_1}$, $x(t_{k+N_1}) \in \Omega_{\rho'_{\min}}$, we look at the change in the value of V' along the closed-loop state trajectory of the system of Equation (91) as follows:

$$\begin{aligned}\dot{V}'(x(t)) =& \frac{\partial V'(x(t))}{\partial x} f_{as}(x(t), u_s, w(t)) + \frac{\partial V'(x(t))}{\partial x} f_{as}(x(t), u_s, 0) - \frac{\partial V'(x(t))}{\partial x} f_{as}(x(t), u_s, 0) \\ &\leq -\alpha_7(|x(t)|) + \left| \frac{\partial V'(x(t))}{\partial x} f_{as}(x(t), u_s, w) - \frac{\partial V'(x(t))}{\partial x} f_{as}(x(t), u_s, 0) \right| \\ &\leq -\alpha_7(|x(t)|) + L'_{w,1}\theta\end{aligned} \tag{98}$$

which follows from Equations (93) and (7) (since $V' = V_1$ and systems of the form of Equation (91) are members of the class of Equation (1)), and the bound on w. If we consider that $x(t_k) \in \Omega_{\rho_1}/\Omega_{\rho'_s}$, then from Equation (92), $\alpha_6^{-1}(\rho'_s) \leq |x(t)|$ such that the upper bound on $\dot{V}'(x(t))$ is determined as follows:

$$\dot{V}'(x(t)) \leq -\alpha_7(\alpha_6^{-1}(\rho'_s)) + L'_{w,1}\theta \tag{99}$$

If Equation (94) holds, then $\frac{dV'}{dt} \leq -\epsilon'_w/\Delta$. Integrating this equation gives:

$$V'(x(t)) \leq V'(x(t_k)) - \frac{\epsilon'_w(t - t_k)}{\Delta} \tag{100}$$

for $t \geq t_k$ while $x(t) \in \Omega_{\rho_1}/\Omega_{\rho'_s}$.

We are interested in the amount of time that it would take to drive the closed-loop state from any $x(t_k) \in \Omega_{\rho_1}$ into $\Omega_{\rho'_{\min}}$ using u_s. In a worst case, $V'(x(t_k)) = V_1(x(t_k)) = \rho_1$, and we would like V' at t to be $\rho'_{\min}$. From Equation (100), the worst-case time t_{WC} that it would take to drive $x(t_k)$ from the boundary of Ω_{ρ_1} to the boundary of $\Omega_{\rho'_{\min}}$ using u_s is $t_{WC} = \frac{(\rho_1 - \rho'_{\min})\Delta}{\epsilon'_w}$. However, t_{WC} may not be an integer multiple of a sampling period; to guarantee that at least the worst-case amount of time passes after t_k during which u_s is applied to the process, $N_1 = \lceil \frac{t_{WC}}{\Delta} \rceil$ is the number of sampling periods throughout which u_s must be applied to guarantee that for any $x(t_k) \in \Omega_{\rho_1}$, $x(t_{k+N_1}) \in \Omega_{\rho'_{\min}}$.

Part 3. We next demonstrate that if $x(t_k) \in \Omega_{\rho'_{\min}}$, it will not exit Ω_{ρ_1} within N_2 sampling periods under any input within the input bounds (i.e., under any input which the LMPC of Equations (14)–(20) may compute in the presence or absence of cyberattacks). Specifically, the following inequality holds for the time derivative of V' along the closed-loop state trajectory of the system of Equation (91) for any $x \in \Omega_{\rho_1}$, $u \in U$, and $w \in W$:

$$\begin{aligned}\frac{\partial V'(x)}{\partial x} f_{as}(x, u, w) &\leq \left| \frac{\partial V'(x)}{\partial x} f_{as}(x, u, w) \right| \\ &\leq \left| \frac{\partial V'(x)}{\partial x} \right| |f_{as}(x, u, w)| \\ &\leq \alpha_{4,1}(|x|)M \\ &\leq \alpha_{4,1}(\alpha_5^{-1}(\rho_1))M\end{aligned} \tag{101}$$

which follows from Equations (4) and (8) (f_{as} is a member of the class of systems of Equation (1)), Equation (92), and $V' = V_1$. The result of Equation (101) can be integrated to give:

$$V'(x(t)) \leq V'(x(t_k)) + \alpha_{4,1}(\alpha_5^{-1}(\rho_1))M(t - t_k) \tag{102}$$

for $t \geq t_k$.

To find the shortest possible time that it would take for a sequence of inputs $u(t) \in U$ applied in sample-and-hold to drive the closed-loop state to the border of Ω_{ρ_1}, we compute t in Equation (102) if $V'(x(t_k)) = \rho'_{\min}$ and $V'(x(t_{ST})) = \rho_1$, where t_{ST} denotes the first possible time at which $V'(x(t)) = \rho_1$. This gives a shortest time of $t_{ST} = \frac{(\rho_1 - \rho'_{\min})}{(\alpha_{4,1}(\alpha_5^{-1}(\rho_1)))M}$. However, this may not be an integer multiple of a sampling period, so that the maximum number of sampling periods over which the LMPC of Equations (14)–(20) can be used in the implementation strategy of Section 5.4.1 while guaranteeing closed-loop stability even in the presence of cyberattacks on the sensor measurements is $N_2 = \lfloor \frac{t_{ST}}{\Delta} \rfloor$.

Part 4. Finally, we prove the results of Theorem 2 by combining the results of the prior parts of the proof. According to the implementation strategy of Section 5.4.1, for any $x(t_0) \in \Omega_{\rho_1}$, u_s will be applied for N_1 sampling periods. From Part 2 of this proof, this will drive the closed-loop state into $\Omega_{\rho'_{\min}}$ by t_{k+N_1} and also, from Proposition 4, will maintain the closed-loop state in Ω_{ρ_1} at all times from Equations (95)–(97). Subsequently, the LMPC of Equations (14)–(20) may be used for N_2 sampling periods. In this case, the closed-loop state will also remain bounded within Ω_{ρ_1} from Part 3 of this proof. Then, u_s will be used again for N_1 sampling periods, and will again maintain the closed-loop state in Ω_{ρ_1}. This sequence of steps will then continue according to the implementation strategy of Section 5.4.1 such that the closed-loop state will be maintained within Ω_{ρ_1} at all times. □

Remark 10. *Minimal assumptions are made on the trajectory of x_f over time in the above proof (only that $x_f(t_k) \in \Omega_{\rho_1}$, $\forall t_k \geq 0$). Therefore, the applied policy can handle attacks where x_f changes at each sampling time, regardless of the manner in which it changes as long as the assumptions are met (e.g., there is no need for separate implementation strategies for different types of sensor attack policies such as surge, bias, geometric, or replay attacks [20,84]). u_s is also an attack-resistant policy for denial-of-service attacks [46] of any length, and the implementation strategy can handle such attacks if an additional statement of what the LMPC should do when it is not provided a state measurement at t_k is added (the proof of Theorem 2 indicates that the controller could choose any $u \in U$ if no sensor signal is provided to it at t_k when the LMPC should be used and if the implementation strategy is followed, closed-loop stability is maintained). Furthermore, the implementation strategy can also be used with closed-loop stability guarantees if x_f is received at some sampling times and $x(t_k)$ at others (as both meet the requirement of Theorem 2 that the state measurement must be in Ω_{ρ_1}). The results also hold if only a partially falsified state measurement is received (i.e., only some components of the state vector are falsified due to only some sensors being compromised), as long as the full state measurement vector received by the controller at every sampling time is in Ω_{ρ_1} (if not, this may indicate that a cyberattack may be occurring and could trigger the use of u_s only so that closed-loop stability is still guaranteed but without the potential benefits of trading it off with a feedback controller).*

5.4.2. Problems with Integrating Open-Loop Control and LMPC

Despite the guarantees which are developed in the prior section for open-loop control integrated with LMPC, the fact that open-loop inputs are required and that both N_1 and N_2 depend on the process dynamics through, for example, ϵ'_w and $\alpha_{4,1}$, α_5, and M indicates that this method has fundamental limitations based on the process time constants. The open-loop policy removes the benefits of feedback control in terms of speeding up the process response. The values of N_1 and N_2 may be such that the process would essentially always have to operate in open-loop (i.e., N_1 is large and N_2 is zero) to guarantee that no cyberattack can impact closed-loop stability. Open-loop control is not a viable alternative for feedback control as an operating strategy at all times.

Another problem that may occur with the proposed approach is that the region $\Omega_{\rho'}$ within which u_s is guaranteed to drive the closed-loop state to the steady-state may be very small. V' might be adjusted to try to increase the size of $\Omega_{\rho'}$, but it is not guaranteed that the input u_s can drive the closed-loop state to the steady-state from a large region around the steady-state, as only local asymptotic stability is implied by Equations (92) and (93). Therefore, the fact that $\Omega_{\rho'}$ is small may be a fundamental limitation of the system for any V'. Because the results of Theorem 2 require $\Omega_{\rho_1} \subset \Omega_{\rho'}$, a small $\Omega_{\rho'}$ means that Ω_{ρ_1} must be small as well, which can significantly limit the potential of the LMPC to enforce a policy that is not steady-state operation or that is economically beneficial compared to steady-state operation. If steady-state operation is desired, a small Ω_{ρ_1} means that closed-loop stability is only guaranteed in a small region around the steady-state, requiring small sampling times and small disturbances to maintain the closed-loop state in the resulting small

$\Omega_{\rho_1'} \subset \Omega_{\rho_{min}'} \subset \Omega_{\rho_1} \subset \Omega_{\rho'}$ per Equations (94) and (74), which may not be practical for certain processes with larger disturbances or larger computation time requirements that restrict the minimum size of Δ. For this reason as well, the proposed technique, despite the guarantees of Theorem 2, is not likely to pose a viable solution to the cyberattack problem. Furthermore, the approach only holds for an open-loop stable steady-state; this is overly restrictive as there are many cases where it may be desirable to operate around an open-loop unstable steady-state. It may be necessary to utilize additional assumptions (e.g., that there is an alternative way to obtain a state measurement that is known to be accurate at certain times) to develop cyberattack-resilient controllers in general that meet Definition 4.

5.5. Deterring Sensor Measurement Falsification Cyberattacks on Safety: Perspectives

The prior sections demonstrated that due to the fundamental nonlinear dynamics considerations which define cyberattacks, concepts for deterring cyberattacks on chemical process control systems that at first seem intuitive may not be proper solutions to the problem. However, the characteristics of proper solutions can be explicitly defined mathematically. Some policies which meet the mathematical definition, however, such as the policy developed in Section 5.4, may be undesirable for some processes under normal operation. Though policies like that in Section 5.4 might be considered to be a reasonable policy if a cyberattack is detected (i.e., it becomes reasonable to give up the benefits of feedback control), the difficulty of predicting the responses of nonlinear systems to changes in the process inputs *a priori* makes it difficult to assess all cyberattack possibilities during the design of the detection policies to ensure that detection policies will not miss any attacks; therefore, there is value in continuing to search for control laws/implementation strategies which are resilient to any cyberattack of a certain type. The results of the prior sections suggest that cyberattack-resilient control designs may need to incorporate special features compared to techniques such as LMPC that do not account for cyberattack-resilience, potentially making them more conservative than control designs which do not account for cyberattacks in the sense that they may not achieve instantaneous profits as large as those with alternative controllers; however, a company could assess the potential for profit loss over time with a cyberattack-resilient controller compared to potential reductions in information technology-related security costs and the potential economic and human costs of accidents without cyberattack-resilient control when selecting a controller for a process.

The control designs presented in Sections 5.2–5.4 for investigating the nature of cyberattacks and of cyberattack-resilient control demonstrated several principles that can be used to guide future research. The design in Section 5.2 led to the presentation of a potential cyberattack-development methodology that uses optimization to attempt to systematically determine attack policies in terms of both inputs and false sensor measurements. Though only one potential computational technique for cyberattack development was explored, it suggests that cyberattack development for non-intuitive situations, such as large-scale processes under control laws with many constraints, may be able to be approached computationally, rather than requiring a trial-and-error approach, which is critical for enabling research on cyberattack-resilient control designs for the process industries to include simulation case studies. The developments in Section 5.3 demonstrate that randomness that impacts process operation may be able to be achieved with closed-loop stability guarantees as part of a cyberattack prevention policy, and therefore can be considered in developing future designs geared toward addressing Definition 4. Finally, in Section 5.4, we demonstrated that despite the strength of the conditions required to meet Definition 4, it may be possible to develop control laws with their implementation policies that do satisfy the definition, particularly by relying on the implementation strategy or potentially additional assumptions on the process dynamics or instrumentation setup/accurate measurement availability. For example, though it is not guaranteed in the strategy presented in Section 5.4 that if $V_1(x(t_0)) = \rho_1$, there is no input that could be computed by the LMPC of Equations (14)–(20) for any provided false state measurement in Ω_{ρ_1}, the implementation strategy that trades off the use of LMPC with the open-loop input policy prevents the state from ever reaching a condition where closed-loop stability would be compromised in the face of a cyberattack. It may also be beneficial to consider control designs such as LMPC that are based on theory that allow rigorous guarantees to be made even in the presence of disturbances, particularly from a set of initial conditions that can be characterized *a priori*, since cyberattack-resilience according to Definition 4 depends on the allowable set of initial conditions for the system.

A final outcome of the results in this work is that they indicate the utility of the recent theoretical developments resulting from the study of the stability properties of economic model predictive control (EMPC) [85–90], which have included notions of stability developed for processes operated in a time-varying fashion, in studying cybersecurity even for processes that would be operated at steady-state without cyberattacks. Closed-loop stability when analyzing cyberattacks requires characterizing the boundedness of the closed-loop state in operating regions in state-space under the attack (in a sense, the state is being manipulated in a time-varying fashion by the attacker) and not necessarily driving the state to the steady-state under the attack, as the attacker's goal for a process typically operated at steady-state would involve moving it off of that steady-state. As we consider more complex process [91,92] and control designs (in the sense of greater coupling between process states due to process designs and controllers intended to improve efficiency and enhance economics), it may become more difficult to predict all the potential methods by which a cyberattacker may attack a plant, enhancing the need for cyberattack-resilient systems by process and control design.

6. Conclusions

This work developed a comprehensive nonlinear systems characterization of cyberattacks of different kinds on chemical process control systems, which indicated that cyberattacks on control systems in the chemical process industries are first and foremost a chemical engineering problem which should be considered during process and control design. We subsequently focused on a specific type of cyberattack in which sensor measurements to feedback controllers are compromised with the goal of impacting process safety and discussed the nonlinear systems definition of a process system resilient to these types of cyberattacks. We used three control designs to explore the concept of cyberattack-resilience against sensor measurement attacks geared toward impacting process safety and to explore the properties required of controllers for making cyberattack-resilience guarantees. The results indicate that a control design/implementation strategy which can be effective at deterring sensor measurement falsification-based cyberattacks geared toward impacting process safety should: (1) maintain closed-loop stability under normal operating conditions and also guarantee closed-loop stability when inputs that have no relationship to the state measurement are applied to the process; and (2) result in a desirable operating policy (i.e., not open-loop) during normal operation (i.e., in the absence of cyberattacks).

Future work will explore cyberattack-resilient control design for larger-scale, more realistic and complex chemical process models. It will also seek to use the insights gained on cyberattack-resilient control for nonlinear systems as developed in this work to create cyberattack-resilient controllers, and to more thoroughly investigate a range of MPC designs which handle disturbances or measurement noise in control designs such as MPC (e.g., [93–97]) in the context of cyberattack-resilience. All future work will consider that a defining feature of cyberattacks is that they remove the association between the input physically implemented on the process and the process state, attempting to make the controller a vehicle for computing a problematic process input (i.e., misusing the controller) rather than using the controller formulation to maintain closed-loop stability in the case that state measurements are falsified.

Funding: Financial support from Wayne State University is gratefully acknowledged.

Conflicts of Interest: The author declares no conflict of interest.

References

1. Leveson, N.G.; Stephanopoulos, G. A system-theoretic, control-inspired view and approach to process safety. *AIChE J.* **2014**, *60*, 2–14.
2. Mannan, M.S.; Sachdeva, S.; Chen, H.; Reyes-Valdes, O.; Liu, Y.; Laboureur, D.M. Trends and challenges in process safety. *AIChE J.* **2015**, *61*, 3558–3569.
3. Venkatasubramanian, V. Systemic failures: Challenges and opportunities in risk management in complex systems. *AIChE J.* **2011**, *57*, 2–9.
4. Albalawi, F.; Durand, H.; Christofides, P.D. Process operational safety via model predictive control: Recent results and future research directions. *Comput. Chem. Eng.* **2018**, *114*, 171–190.
5. Albalawi, F.; Durand, H.; Alanqar, A.; Christofides, P.D. Achieving operational process safety via model predictive control. *J. Loss Prev. Process Ind.* **2018**, *53*, 74–88.
6. Albalawi, F.; Durand, H.; Christofides, P.D. Process operational safety using model predictive control based on a process Safeness Index. *Comput. Chem. Eng.* **2017**, *104*, 76–88.
7. Zhang, Z.; Wu, Z.; Durand, H.; Albalawi, F.; Christofides, P.D. On integration of feedback control and safety systems: Analyzing two chemical process applications. *Chem. Eng. Res. Des.* **2018**, *132*, 616–626.
8. Carson, J.M.; Açıkmeşe, B.; Murray, R.M.; MacMartin, D.G. A robust model predictive control algorithm augmented with a reactive safety mode. *Automatica* **2013**, *49*, 1251–1260.
9. Wu, Z.; Durand, H.; Christofides, P.D. Safe economic model predictive control of nonlinear systems. *Syst. Control Lett.* **2018**, *118*, 69–76.
10. Wieland, P.; Allgöwer, F. Constructive Safety Using Control Barrier Functions. *IFAC Proc. Vol.* **2007**, *40*, 462–467.
11. Braun, P.; Kellett, C.M. On (the existence of) Control Lyapunov Barrier Functions. 2017. Available online: https://epub.uni-bayreuth.de/3522/ (accessed on 10 August 2018).
12. Shahnazari, H.; Mhaskar, P. Distributed fault diagnosis for networked nonlinear uncertain systems. *Comput. Chem. Eng.* **2018**, *115*, 22–33.
13. Shahnazari, H.; Mhaskar, P. Actuator and sensor fault detection and isolation for nonlinear systems subject to uncertainty. *Int. J. Robust Nonlinear Control* **2018**, *28*, 1996–2013.
14. Yin, X.; Liu, J. Distributed output-feedback fault detection and isolation of cascade process networks. *AIChE J.* **2017**, *63*, 4329–4342.
15. Alanqar, A.; Durand, H.; Christofides, P.D. Fault-Tolerant Economic Model Predictive Control Using Error-Triggered Online Model Identification. *Ind. Eng. Chem. Res.* **2017**, *56*, 5652–5667.
16. Demetriou, M.A.; Armaou, A. Dynamic online nonlinear robust detection and accommodation of incipient component faults for nonlinear dissipative distributed processes. *Int. J. Robust Nonlinear Control* **2012**, *22*, 3–23.
17. Xue, D.; El-Farra, N.H. Resource-aware fault accommodation in spatially-distributed processes with sampled-data networked control systems. In Proceedings of the American Control Conference, Seattle, WA, USA, 24–26 May 2017; pp. 1809–1814.
18. Xue, D.; El-Farra, N.H. Actuator fault-tolerant control of networked distributed processes with event-triggered sensor-controller communication. In Proceedings of the American Control Conference, Boston, MA, USA, 6–8 July 2016; pp. 1661–1666.
19. Smith, R.E. *Elementary Information Security*; Jones & Bartlett Learning, LLC: Burlington, MA, USA, 2016.
20. Cárdenas, A.A.; Amin, S.; Lin, Z.S.; Huang, Y.L.; Huang, C.Y.; Sastry, S. Attacks against process control systems: Risk assessment, detection, and response. In Proceedings of the ACM Asia Conference on Computer & Communications Security, Hong Kong, China, 22–24 March 2011.
21. Greenberg, A. How an Entire Nation Became Russia's Test Lab for Cyberwar. 2017. Available online: https://www.wired.com/story/russian-hackers-attack-ukraine/ (accessed on 11 July 2018).
22. Clark, R.M.; Panguluri, S.; Nelson, T.D.; Wyman, R.P. Protecting drinking water utilities from cyberthreats. *J. Am. Water Works Assoc.* **2017**, *109*, 50–58.
23. Langner, R. Stuxnet: Dissecting a Cyberwarfare Weapon. *IEEE Secur. Priv.* **2011**, *9*, 49–51.
24. Perlroth, N.; Krauss, C. A Cyberattack in Saudi Arabia Had a Deadly Goal. Experts Fear Another Try. 2018. Available online: https://www.nytimes.com/2018/03/15/technology/saudi-arabia-hacks-cyberattacks.html (accessed on 11 March 2018).

25. Groll, E. Cyberattack Targets Safety System at Saudi Aramco. 2017. Available online: https://foreignpolicy.com/2017/12/21/cyber-attack-targets-safety-system-at-saudi-aramco/ (accessed on 11 July 2018).
26. Liu, Y.; Sarabi, A.; Zhang, J.; Naghizadeh, P.; Karir, M.; Bailey, M.; Liu, M. Cloudy with a Chance of Breach: Forecasting Cyber Security Incidents. In Proceedings of the USENIX Security Symposium, Washington, DC, USA, 12–14 August 2015; pp. 1009–1024.
27. Solomon, M.G.; Kim, D.; Carrell, J.L. *Fundamentals of Communications and Networking*; Jones & Bartlett Publishers: Burlington, MA, USA, 2014.
28. McLaughlin, S.; Konstantinou, C.; Wang, X.; Davi, L.; Sadeghi, A.R.; Maniatakos, M.; Karri, R. The Cybersecurity Landscape in Industrial Control Systems. *Proc. IEEE* **2016**, *104*, 1039–1057.
29. Hull, J.; Khurana, H.; Markham, T.; Staggs, K. Staying in control: Cybersecurity and the modern electric grid. *IEEE Power Energy Mag.* **2012**, *10*, 41–48.
30. Ginter, A. Unidirectional Security Gateways: Stronger than Firewalls. In Proceedings of the ICALEPCS, San Francisco, CA, USA, 6–11 October 2013; pp. 1412–1414.
31. Khorrami, F.; Krishnamurthy, P.; Karri, R. Cybersecurity for Control Systems: A Process-Aware Perspective. *IEEE Des. Test* **2016**, *33*, 75–83.
32. He, D.; Chan, S.; Zhang, Y.; Wu, C.; Wang, B. How Effective Are the Prevailing Attack-Defense Models for Cybersecurity Anyway? *IEEE Intel. Syst.* **2014**, *29*, 14–21.
33. Ten, C.W.; Liu, C.C.; Manimaran, G. Vulnerability Assessment of Cybersecurity for SCADA Systems. *IEEE Trans. Power Syst.* **2008**, *23*, 1836–1846.
34. Pang, Z.H.; Liu, G.P. Design and implementation of secure networked predictive control systems under deception attacks. *IEEE Trans. Control Syst. Technol.* **2012**, *20*, 1334–1342.
35. Rieger, C.; Zhu, Q.; Başar, T. Agent-based cyber control strategy design for resilient control systems: Concepts, architecture and methodologies. In Proceedings of the 5th International Symposium on Resilient Control Systems, Salt Lake City, UT, USA, 14–16 August 2012; pp. 40–47.
36. Chavez, A.R.; Stout, W.M.S.; Peisert, S. Techniques for the dynamic randomization of network attributes. In Proceedings of the IEEE International Carnahan Conference on Security Technology, Taipei, Taiwan, 21–24 September 2015; pp. 1–6.
37. Linda, O.; Manic, M.; McQueen, M. Improving control system cyber-state awareness using known secure sensor measurements. In *Critical Information Infrastructures Security. CIRITIS 2012*; Hämmerli, B.M., Kalstad Svendsen, N., Lopez, J., Eds.; Lecture Notes in Computer Science; Springer: Berlin/Heidelberg, Germany, 2013; Volume 7722, pp. 46–58.
38. Plosz, S.; Farshad, A.; Tauber, M.; Lesjak, C.; Ruprechter, T.; Pereira, N. Security vulnerabilities and risks in industrial usage of wireless communication. In Proceedings of the IEEE International Conference on Emerging Technology and Factory Automation, Barcelona, Spain, 6–19 September 2014; pp. 1–8.
39. Lopez, J.; Zhou, J. (Eds.) *Wireless Sensor Network Security*; IOS Press: Amsterdam, The Netherlands, 2008.
40. Xu, L.D.; He, W.; Li, S. Internet of Things in Industries: A Survey. *IEEE Trans. Ind. Inform.* **2014**, *10*, 2233–2243.
41. Almorsy, M.; Grundy, J.; Müller, I. An analysis of the cloud computing security problem. *arXiv* **2016**, arXiv:1609.01107.
42. Rieger, C.G. Notional examples and benchmark aspects of a resilient control system. In Proceedings of the 2010 3rd International Symposium on Resilient Control Systems, Idaho Falls, ID, USA, 10–12 August 2010; pp. 64–71.
43. Rieger, C.G.; Gertman, D.I.; McQueen, M.A. Resilient control systems: Next generation design research. In Proceedings of the 2009 2nd Conference on Human System Interactions, Catania, Italy, 21–23 May 2009; pp. 632–636.
44. Wakaiki, M.; Tabuada, P.; Hespanha, J.P. Supervisory control of discrete-event systems under attacks. *arXiv* **2017**, arXiv:1701.00881.
45. Bopardikar, S.D.; Speranzon, A.; Hespanha, J.P. An H-infinity approach to stealth-resilient control design. In Proceedings of the 2016 Resilience Week, Chicago, IL, USA, 16–18 August 2016; pp. 56–61.
46. Amin, S.; Cárdenas, A.A.; Sastry, S.S. Safe and secure networked control systems under denial-of-service attacks. In *Hybrid Systems: Computation and Control. HSCC 2009*; Majumdar, R., Tabuada, P., Eds.; Lecture Notes in Computer Science; Springer: Berlin/Heidelberg, Germany, 2009; Volume 5469, pp. 31–45.
47. Fawzi, H.; Tabuada, P.; Diggavi, S. Secure Estimation and Control for Cyber-Physical Systems Under Adversarial Attacks. *IEEE Trans. Autom. Control* **2014**, *59*, 1454–1467.

48. Zhu, Q.; Başar, T. Game-Theoretic Methods for Robustness, Security, and Resilience of Cyberphysical Control Systems: Games-in-Games Principle for Optimal Cross-Layer Resilient Control Systems. *IEEE Control Syst.* **2015**, *35*, 46–65.
49. Zhu, Q.; Başar, T. Robust and resilient control design for cyber-physical systems with an application to power systems. In Proceedings of the 2011 50th IEEE Conference on Decision and Control and European Control Conference, Orlando, FL, USA, 12–15 December 2011; pp. 4066–4071.
50. Zhu, Q.; Bushnell, L.; Başar, T. Resilient distributed control of multi-agent cyber-physical systems. In *Control of Cyber-Physical Systems*; Tarraf, D., Ed.; Lecture Notes in Control and Information Sciences; Springer: Berlin/Heidelberg, Germany, 2013; Volume 449, pp. 301–316.
51. Zonouz, S.; Rogers, K.M.; Berthier, R.; Bobba, R.B.; Sanders, W.H.; Overbye, T.J. SCPSE: Security-Oriented Cyber-Physical State Estimation for Power Grid Critical Infrastructures. *IEEE Trans. Smart Grid* **2012**, *3*, 1790–1799.
52. Zheng, S.; Jiang, T.; Baras, J.S. Robust State Estimation under False Data Injection in Distributed Sensor Networks. In Proceedings of the 2010 IEEE Global Telecommunications Conference, Miami, FL, USA, 6–10 December 2010; pp. 1–5.
53. Pasqualetti, F.; Dorfler, F.; Bullo, F. Control-Theoretic Methods for Cyberphysical Security: Geometric Principles for Optimal Cross-Layer Resilient Control Systems. *IEEE Control Syst.* **2015**, *35*, 110–127.
54. Pasqualetti, F.; Dörfler, F.; Bullo, F. Attack Detection and Identification in Cyber-Physical Systems. *IEEE Trans. Autom. Control* **2013**, *58*, 2715–2729.
55. McLaughlin, S. CPS: Stateful policy enforcement for control system device usage. In Proceedings of the 29th Annual Computer Security Applications Conference, New Orleans, LA, USA, 9–13 December 2013; pp. 109–118.
56. Melin, A.; Kisner, R.; Fugate, D.; McIntyre, T. Minimum state awareness for resilient control systems under cyber-attack. In Proceedings of the 2012 Future of Instrumentation International Workshop, Gatlinburg, TN, USA, 8–9 October 2012; pp. 1–4.
57. Qin, S.J.; Badgwell, T.A. A survey of industrial model predictive control technology. *Control Eng. Pract.* **2003**, *11*, 733–764.
58. Rawlings, J.B. Tutorial overview of model predictive control. *IEEE Control Syst.* **2000**, *20*, 38–52.
59. Durand, H. State Measurement Spoofing Prevention through Model Predictive Control Design. In Proceedings of the IFAC NMPC-2018, Madison, WI, USA, 19–22 August 2018; pp. 643–648.
60. Heidarinejad, M.; Liu, J.; Christofides, P.D. Economic model predictive control of nonlinear process systems using Lyapunov techniques. *AIChE J.* **2012**, *58*, 855–870.
61. Mhaskar, P.; El-Farra, N.H.; Christofides, P.D. Stabilization of nonlinear systems with state and control constraints using Lyapunov-based predictive control. *Syst. Control Lett.* **2006**, *55*, 650–659.
62. Muñoz de la Peña, D.; Christofides, P.D. Lyapunov-Based Model Predictive Control of Nonlinear Systems Subject to Data Losses. *IEEE Trans. Autom. Control* **2008**, *53*, 2076–2089.
63. Zhu, B.; Joseph, A.; Sastry, S. A taxonomy of cyber attacks on SCADA systems. In Proceedings of the 2011 IEEE International Conferences on Internet of Things, and Cyber, Physical and Social Computing, Dalian, China, 19–22 October 2011; pp. 380–388.
64. Krotofil, M.; Cárdenas, A.A. Resilience of process control systems to cyber-physical attacks. In Proceedings of the Nordic Conference on Secure IT Systems, Ilulissat, Greenland, 18–21 October 2013; pp. 166–182.
65. Gentile, M.; Rogers, W.J.; Mannan, M.S. Development of an inherent safety index based on fuzzy logic. *AIChE J.* **2003**, *49*, 959–968.
66. Heikkilä, A.M.; Hurme, M.; Järveläinen, M. Safety considerations in process synthesis. *Comput. Chem. Eng.* **1996**, *20*, S115–S120.
67. Khan, F.I.; Amyotte, P.R. How to Make Inherent Safety Practice a Reality. *Can. J. Chem. Eng.* **2003**, *81*, 2–16.
68. Gupta, J.P.; Edwards, D.W. Inherently Safer Design—Present and Future. *Process Saf. Environ. Prot.* **2002**, *80*, 115–125.
69. Kletz, T.A. Inherently safer plants. *Plant/Oper. Prog.* **1985**, *4*, 164–167.
70. Li, L.; Hu, B.; Lemmon, M. Resilient event triggered systems with limited communication. In Proceedings of the 2012 51st IEEE Conference on Decision and Control, Maui, HI, USA, 10–13 December 2012; pp. 6577–6582.

71. Melin, A.M.; Ferragut, E.M.; Laska, J.A.; Fugate, D.L.; Kisner, R. A mathematical framework for the analysis of cyber-resilient control systems. In Proceedings of the 2013 6th International Symposium on Resilient Control Systems, San Francisco, CA, USA, 13–15 August 2013; pp. 13–18.
72. Chandy, S.E.; Rasekh, A.; Barker, Z.A.; Shafiee, M.E. Cyberattack Detection using Deep Generative Models with Variational Inference. *arXiv* **2018**, arXiv:1805.12511.
73. Rosich, A.; Voos, H.; Li, Y.; Darouach, M. A model predictive approach for cyber-attack detection and mitigation in control systems. In Proceedings of the IEEE Conference on Decision and Control, Florence, Italy, 10–13 December 2013; pp. 6621–6626.
74. Tajer, A.; Kar, S.; Poor, H.V.; Cui, S. Distributed joint cyber attack detection and state recovery in smart grids. In Proceedings of the IEEE International Conference on Smart Grid Communications, Brussels, Belgium, 17–20 October 2011; pp. 202–207.
75. Kiss, I.; Genge, B.; Haller, P. A clustering-based approach to detect cyber attacks in process control systems. In Proceedings of the IEEE 13th International Conference on Industrial Informatics, Cambridge, UK, 22–24 July 2015; pp. 142–148.
76. Valdes, A.; Cheung, S. Intrusion Monitoring in Process Control Systems. In Proceedings of the 42nd Hawaii International Conference on System Sciences, Big Island, HI, USA, 5–8 January 2009; pp. 1–7.
77. Wu, Z.; Albalawi, F.; Zhang, J.; Zhang, Z.; Durand, H.; Christofides, P.D. Detecting and Handling Cyber-attacks in Model Predictive Control of Chemical Processes. *Mathematics* **2018**, accepted.
78. Ricker, N.L. Model predictive control of a continuous, nonlinear, two-phase reactor. *J. Process Control* **1993**, *3*, 109–123.
79. Alanqar, A.; Ellis, M.; Christofides, P.D. Economic model predictive control of nonlinear process systems using empirical models. *AIChE J.* **2015**, *61*, 816–830.
80. Lin, Y.; Sontag, E.D. A universal formula for stabilization with bounded controls. *Syst. Control Lett.* **1991**, *16*, 393–397.
81. Grossmann, I.E. Review of nonlinear mixed-integer and disjunctive programming techniques. *Optim. Eng.* **2002**, *3*, 227–252.
82. Mhaskar, P.; Liu, J.; Christofides, P.D. *Fault-Tolerant Process Control: Methods and Applications*; Springer: London, UK, 2013.
83. Wächter, A.; Biegler, L.T. On the implementation of an interior-point filter line-search algorithm for large-scale nonlinear programming. *Math. Program.* **2006**, *106*, 25–57.
84. Mo, Y.; Sinopoli, B. Secure control against replay attacks. In Proceedings of the 2009 47th Annual Allerton Conference on Communication, Control, and Computing, Monticello, IL, USA, 30 September–2 October 2009; pp. 911–918.
85. Ellis, M.; Durand, H.; Christofides, P.D. A tutorial review of economic model predictive control methods. *J. Process Control* **2014**, *24*, 1156–1178.
86. Rawlings, J.B.; Angeli, D.; Bates, C.N. Fundamentals of economic model predictive control. In Proceedings of the Conference on Decision and Control, Maui, HI, USA, 10–13 December 2012; pp. 3851–3861.
87. Faulwasser, T.; Korda, M.; Jones, C.N.; Bonvin, D. Turnpike and dissipativity properties in dynamic real-time optimization and economic MPC. In Proceedings of the IEEE 53rd Annual Conference on Decision and Control, Los Angeles, CA, USA, 15–17 December 2014; pp. 2734–2739.
88. Müller, M.A.; Grüne, L.; Allgöwer, F. On the role of dissipativity in economic model predictive control. *IFAC-PapersOnLine* **2015**, *48*, 110–116.
89. Huang, R.; Harinath, E.; Biegler, L.T. Lyapunov stability of economically oriented NMPC for cyclic processes. *J. Process Control* **2011**, *21*, 501–509.
90. Omell, B.P.; Chmielewski, D.J. IGCC power plant dispatch using infinite-horizon economic model predictive control. *Ind. Eng. Chem. Res.* **2013**, *52*, 3151–3164.
91. Amini-Rankouhi, A.; Huang, Y. Prediction of maximum recoverable mechanical energy via work integration: A thermodynamic modeling and analysis approach. *AIChE J.* **2017**, *63*, 4814–4826.
92. Tula, A.K.; Babi, D.K.; Bottlaender, J.; Eden, M.R.; Gani, R. A computer-aided software-tool for sustainable process synthesis-intensification. *Comput. Chem. Eng.* **2017**, *105*, 74–95.
93. Limon, D.; Alamo, T.; Salas, F.; Camacho, E. Input to state stability of min–max MPC controllers for nonlinear systems with bounded uncertainties. *Automatica* **2006**, *42*, 797–803.

94. Campo, P.J.; Morari, M. Robust Model Predictive Control. In Proceedings of the American Control Conference, Minneapolis, MN, USA, 10–12 June 1987; pp. 1021–1026.
95. Pannocchia, G.; Gabiccini, M.; Artoni, A. Offset-free MPC explained: Novelties, subtleties, and applications. *IFAC-PapersOnLine* **2015**, *48*, 342–351.
96. Ellis, M.; Zhang, J.; Liu, J.; Christofides, P.D. Robust moving horizon estimation based output feedback economic model predictive control. *Syst. Control Lett.* **2014**, *68*, 101–109.
97. Das, B.; Mhaskar, P. Lyapunov-based offset-free model predictive control of nonlinear process systems. *Can. J. Chem. Eng.* **2015**, *93*, 471–478.

mathematics

MDPI

Article

Recurrent Neural Network-Based Model Predictive Control for Continuous Pharmaceutical Manufacturing

Wee Chin Wong, Ewan Chee, Jiali Li and Xiaonan Wang *

Department of Chemical & Biomolecular Engineering, Faculty of Engineering, National University of Singapore, 4 Engineering Drive 4, Singapore 117585, Singapore; weechin.wong@gmail.com (W.C.W.); ewan-jun-xian.chee@student.ecp.fr (E.C.); e0276496@u.nus.edu (J.L.)

* Correspondence: chewxia@nus.edu.sg; Tel.: +65-6601-6221

Received: 5 August 2018; Accepted: 26 October 2018; Published: 7 November 2018

Abstract: The pharmaceutical industry has witnessed exponential growth in transforming operations towards continuous manufacturing to increase profitability, reduce waste and extend product ranges. Model predictive control (MPC) can be applied to enable this vision by providing superior regulation of critical quality attributes (CQAs). For MPC, obtaining a workable system model is of fundamental importance, especially if complex process dynamics and reaction kinetics are present. Whilst physics-based models are desirable, obtaining models that are effective and fit-for-purpose may not always be practical, and industries have often relied on data-driven approaches for system identification instead. In this work, we demonstrate the applicability of recurrent neural networks (RNNs) in MPC applications in continuous pharmaceutical manufacturing. RNNs were shown to be especially well-suited for modelling dynamical systems due to their mathematical structure, and their use in system identification has enabled satisfactory closed-loop performance for MPC of a complex reaction in a single continuous-stirred tank reactor (CSTR) for pharmaceutical manufacturing.

Keywords: pharmaceuticals; critical quality attributes (CQAs); recurrent neural networks; model predictive control (MPC); system identification

1. Introduction

The pharmaceutical industry has shown a growing interest in adopting the concept of continuous manufacturing [1,2] to reduce waste, cost, environmental footprints and lead-times. Coupled with developments in continuous manufacturing technologies and quality by design (QbD) paradigms, there has thus recently been an increasing demand for more advanced system identification and process control strategies for continuous pharmaceutical manufacturing [1].

Continuous manufacturing has the potential of extending the palette of permissible reaction conditions and enabling reaction outcomes that are quite challenging when performed under batch conditions [3–5], and its application to the production of complex Active Pharmaceutical Ingredients (APIs) is an emerging research domain. However, the reactions involved are generally both reversible and coupled with many side reactions, which leads to highly non-linear systems that impose difficulties for system identification. The need to operate in tight and extreme conditions also necessitates stricter control requirements [6–8]. The complexity of the system, as well as its potential impact on reaction yields, downstream processing requirements and critical quality attributes (CQAs) of final products, render model identification and control of continuous API reactions particularly challenging.

Advanced model-based control technologies promise to improve control of CQAs and are seen as vital to enabling continuous pharmaceutical manufacturing, and their performance hinges on the quality of the model obtained from system identification. Many rigorous physics-based models have been proposed to describe different API reactions [6,9,10] and they have clear physical interpretations. However, these models can suffer from complicated structures and may incur excessive computational costs during online control. The need for models that evaluate quickly for online control purposes is not unique to the pharmaceutical industry, and extensive work has also been done for thin-film deposition in the microelectronics industry [11,12]. Moreover, it may be extremely difficult and expensive to derive accurate models of non-linear systems from first principles in the first place [13]. In contrast, experimental data-driven heuristic models aim to describe underlying processes with relatively simpler mathematical expressions and are thus generally easier to obtain [14]. Neural networks belong to this class of data-driven models, and they have been extensively applied in the chemical and biochemical processing industries, particularly in modelling complex non-linear processes whose process understanding is limited to begin with [15–18]. The goal of system identification is to create models that capture the relationships between the process state variables and control variables, and data-based empirical models for system identification have been widely studied in previous work [19,20]. Artificial Neural Networks (ANNs) are interesting candidate models for these purposes due to their ability to approximate any continuous function, subject to assumptions on the activation function. RNNs constitute a sub-class of ANNs, and they are structured to use hidden variables as a memory for capturing temporal dependencies between system and control variables. These networks have been extensively used in sequence learning problems like scene labeling [21] and language processing [22], demonstrating good performance. Ref. [23] provides an overview of the neural network applications, particularly from a process systems engineering perspective. In this work, the use of RNNs for system identification of a complex reaction in a single CSTR for pharmaceutical API production will be demonstrated.

The design of effective model-based control algorithms for continuous manufacturing is also a key research focus in academia and industry [24], and the model predictive control (MPC) method is a prominent example that has been widely employed in industrial applications [25–27]. It has also been applied in the control of CQAs in continuous pharmaceutical manufacturing [10,24], and an example can be found in [10] where the authors presented two plant-wide end-to-end MPC designs for a continuous pharmaceutical manufacturing pilot plant. This plant covered operations from chemical synthesis to tablet formation, and the MPC designs were obtained using the quadratic dynamic matrix control (QDMC) algorithm. The results showed that, by monitoring the CQAs in real time, critical process parameters (CPPs) could be actively manipulated by feedback control to enable process operation that is more robust and flexible. The advantages that MPC has over conventional proportional-integral-derivative (PID) control are highlighted in [24], which compared their control performance for a feeding blending unit (FBU) used in continuous pharmaceutical manufacturing.

Previous studies have explored the potential of using RNNs with MPC. For example, ref. [28] investigated the use of RNNs in controlling non-linear dynamical systems. RNNs have also been used for predictive control of reactions in CSTRs [29]. However, this CSTR study was limited to relatively simple kinetics like single-step irreversible reactions, i.e., $A \rightarrow B$. This paper illustrates therefore how RNNs can be used with MPC of reactions with more complex kinetics.

The paper is organized as follows. Section 2 describes the plant model, which is the subject of system identification, and the control scenarios that will be used to assess closed-loop control performance. Section 3.1 articulates the RNN-based system identification methodology and Section 3.2 describes the MPC problem formulation. Results and discussions are shown in Section 4.

2. Plant Model and Control Scenarios

The following sub-sections describe (i) the true plant model for which system identification will be performed and (ii) the two control scenarios that will be used to assess closed-loop control performance.

2.1. Plant Model

The reaction system comprising of a single CSTR, as was also analysed in [30,31], will be the object of this study. For this work, the focus is on system identification using RNNs with a view towards closed-loop control with MPC, thereby marking a difference from how this reaction system was studied in previous literature. The reaction is described by Equation (1), where A is the feed species, R is the reaction intermediate and the desired product, and S is the undesired by-product:

$$A \underset{k_4}{\overset{k_1}{\rightleftarrows}} R \underset{k_3}{\overset{k_2}{\rightleftarrows}} S \tag{1}$$

The equations used to represent the dynamical behavior of this system are based on normalized dimensionless quantities and are as shown in Equations (2)–(4) below:

$$\frac{dC_A}{dt} = q[C_{A0} - C_A] - k_1 C_A + k_4 C_R \tag{2}$$

$$\frac{dC_R}{dt} = q[1 - C_{A0} - C_R] + k_1 C_A + k_3[1 - C_A - C_R] - [k_2 + k_4]C_R \tag{3}$$

$$k_j = k_{0,j} \exp\left\{\left[-\frac{E}{RT_0}\right]_j \left[\frac{1}{T} - 1\right]\right\}, j \in \{1,2,3,4\} \tag{4}$$

where, $C_{i\in\{A,R,S\}} \in \mathbb{R}_+$ refers to the species concentration in the reactor, $C_{A0} \in \mathbb{R}_+$ the feed concentration of A, $q \in \mathbb{R}_+$ the feed flow rate, and $k_{j\in\{1,2,3,4\}} \in \mathbb{R}_+$ the rate constants for the respective reaction. For each rate constant, $k_{0,j\in\{1,2,3,4\}} \in \mathbb{R}_+$ is its associated Arrhenius pre-exponential constant, and $[\frac{E}{RT_0}]_{j\in\{1,2,3,4\}} \in \mathbb{R}$ its associated normalised activation energy.

The reactions are first-order and reversible. The reactor is operated isothermally and perfect mixing is assumed. All reactor concentrations are measured in this example. The manipulated variables are q and T, and their values are also measured. It is also assumed that the only species entering the CSTR are A and R, such that their feed concentrations sum up to unity. As such, C_S is easily calculable as a function of time. The values of $k_{0,j\in\{1,2,3,4\}}$ and $[\frac{E}{RT_0}]_{j\in\{1,2,3,4\}}$ can be found in [31] and are reproduced in the Appendix A for ease of reference.

Figure 1 shows that, at steady-state conditions for a given q, C_R reaches a maximum at a certain temperature, beyond which the reaction gets driven back to the left of Equation (1) to eventually yield only A and no R. This system possesses rich and relevant dynamics for continuous pharmaceutical manufacturing and is therefore worth detailed investigation.

The control problem for this system is challenging due to the existence of input multiplicities [30,32], which occur when a set of outputs can be reached by more than one set of inputs. This implies the existence of sign changes in the determinant of the steady-state gain matrix within the operating region. Linear controllers with integral action therefore become unstable for this problem.

In what follows, as is common in digital control, the discrete-time domain with time-steps denoted by $k \in \mathbb{Z}_+$ will be used. The state vector, x, and manipulated vector (MV), u, are also defined as follows: $x \triangleq [C_A, C_R]'$, $u \triangleq [q, T]'$. The underlying plant model can therefore be described by a non-linear discrete-time difference equation, as Equations (5) and (6) show:

$$x_{k+1} = \Phi(x_k, u_k) \tag{5}$$

$$y_k = x_k \tag{6}$$

where $k \in \mathbb{Z}_+$ is the discrete-time index, and x_k and u_k the state vector and manipulated vector at time-step k respectively. $\Phi(.)$ is the state-transition equation consistent with Equations (2)–(4), and full state feedback is assumed, such that the measured output y_k equals x_k for all time-steps.

The system identification problem consists of finding an approximation of Equations (2)–(4), and forms the subject of Section 3.1 in this paper. It will be supposed that the $p-$step ahead prediction problem is of vital interest for MPC. Throughout this paper, all problems are considered from time-step zero, and a sampling time Δt, of 0.1 time units is used.

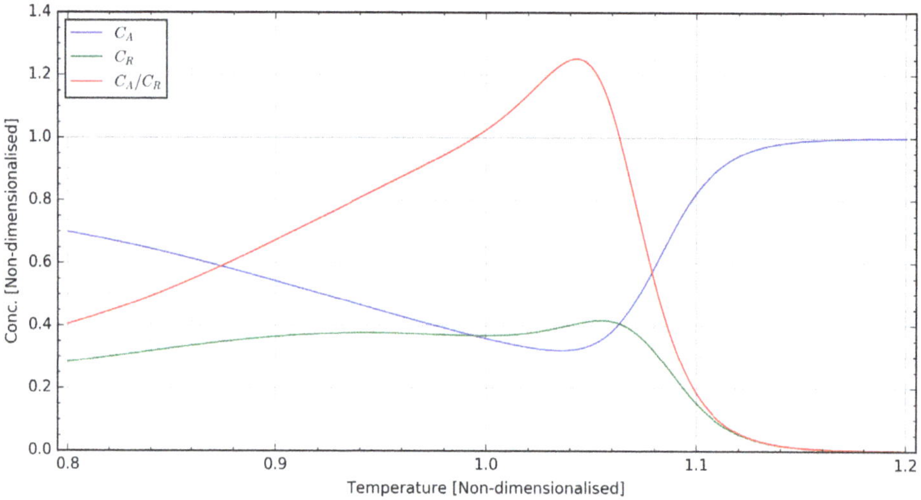

Figure 1. Steady-state conditions as a function of T with $q = 0.8$.

2.2. Closed-Loop Control Scenarios

The approach taken to construct the RNN for MPC is to train the RNN with data from the training set, then evaluate the model performance using previously unseen data from the test set. However, since the main focus of this study is process control, the final evaluation will be performed by comparing the closed-loop performance of the RNN-based MPC (RNN-MPC) against a benchmark non-linear MPC (NMPC) controller that directly uses the true plant model, as described in Equations (2)–(4), as its internal model.

Two control scenarios were selected for performance evaluation and the operating conditions for each scenario are reflected in the first two rows of Table 1. These scenarios are namely cases I and II, and they correspond respectively to a reactor system start-up case and an upset-recovery case. In case I, the system is assumed to initially be at a low temperature state with a relatively low product concentration. In case II, the initial conditions correspond instead to a high-temperature and low-yield state. Figure 2 shows the locations of the initial conditions for these control scenarios with respect to the set-point, with case I represented by the purple point to the left of the peak in the green C_R plot, and case II represented by the cyan point to the right of the same peak.

Table 1. Initial conditions and set-points for control scenarios.

Control Scenario	C_A	C_R	q	T
I: Start-up	0.692	0.287	0.800	0.800
II: Upset-recovery	0.822	0.152	0.800	1.100
Set-point (maximum C_R/C_A)	0.324	0.406	0.800	1.043

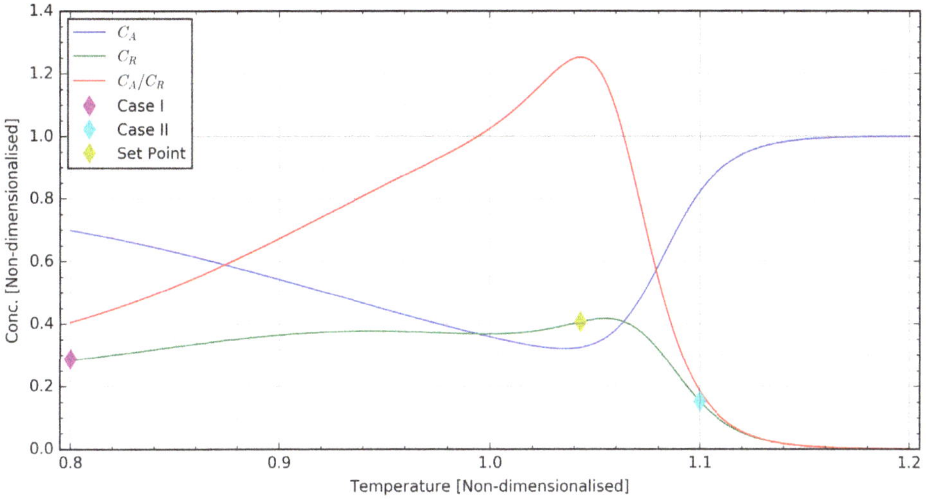

Figure 2. Initial conditions for cases I (start-up) and II (upset-recovery) relative to the set-point.

The set-point is judiciously located where the ratio of the product concentration to the feed concentration, C_R/C_A, is maximised. This set-point corresponds to a product concentration that is slightly lower than what is maximally achievable. The rationale for choosing this set-point over the point for maximum C_R is that it maximizes yield whilst minimising downstream separations cost. The operating conditions corresponding to the target set-point are shown in the last row of Table 1.

3. Methodology

3.1. Non-Linear Time-Series System Identification via Recurrent Neural Networks

ANNs serve as a potential source of models that capture the dynamics of highly non-linear systems sufficiently well while being relatively easy to obtain and evaluate online, and RNNs are a sub-class of ANNs that are structured to better capture temporal dependencies in the system. RNNs are particularly useful for making p-step ahead predictions of state variables for predictive control, because the prediction for time-step p depends on the present state and all control actions in time-step $k \in \{0, ..., p-1\}$. The prediction for time-step $p-1$ used in the above prediction for time-step p depends similarly on the present state and all control actions in time-step $k \in \{0, ..., p-2\}$, and so on.

Figure 3 illustrates the structure of an RNN layer in its compact and unfolded forms. The unfolded form reveals the repeating cells forming each layer, and each cell is taken in this study to represent a time-step, such that the state of the cell representing time-step $k \in \{0, ..., p-1\}$ serves as the input for a cell representing time-step $k+1$. Each cell contains N number of hidden nodes that encode the state representation, where N is user-specifiable. The equations that mathematically characterise a single RNN cell in a single-layer RNN are as shown in Equations (7) and (8) below:

$$h_k = W_{h,h}h_{k-1} + W_{u,h}u_{k-1} + b_1 \quad (7)$$

$$\hat{y}_k = \Phi(h_k + b_2) \quad (8)$$

where $k \in \{1, ..., p\}$ is the discrete-time index with p the prediction horizon, $h_k \in \mathbb{R}^N$ the (hidden) state of the cell representing time-step k, $u_{k-1} \in \mathbb{R}^2$ the cell input, $\hat{y}_k \in \mathbb{R}^2$ the cell output which corresponds to the state vector prediction for time-step k, $(b_1, b_2) \in (\mathbb{R}^N)^2$ the offset vectors, $W_{u,h} \in \mathbb{R}^{N\times 2}$ and $W_{h,h} \in \mathbb{R}^{N\times N}$ the input-to-hidden-state weight matrix and the hidden-state-to-hidden-state weight matrix respectively, and $\Phi : \mathbb{R}^N \rightarrow \mathbb{R}^2$ an activation function that is applied element-wise. h_0 is initialised in this study by using y_0.

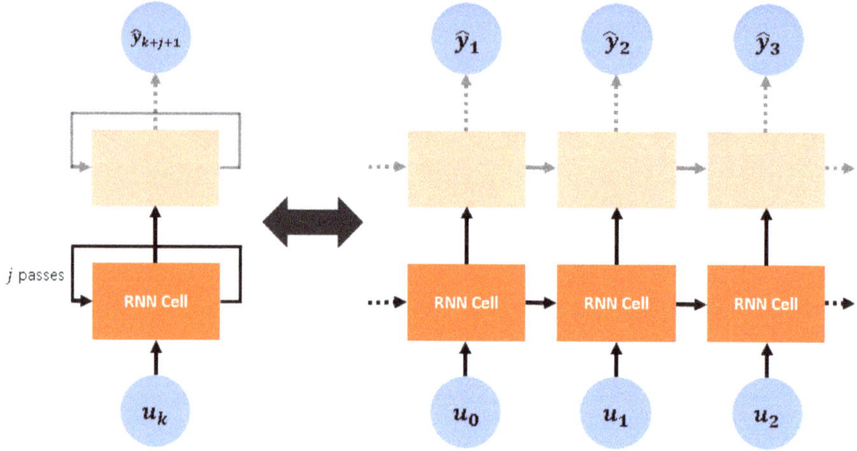

Figure 3. Illustration of a recurrent neural network (RNN) layer in compact form (**left**) and unfolded form (**right**).

Equations (7) and (8) are generalisable to deep RNNs containing more than one layer. For an l-layer RNN, the activation functions for layers $j \in \{2, ..., l\}$ take the form $\Phi : \mathbb{R}^N \rightarrow \mathbb{R}^N$, and the input-to-hidden-state weight matrices have dimensions $N \times N$ instead. Deep RNNs may be preferred to single-layer RNNs for their enhanced ability to learn features of the dynamical system that occur "hierarchically", but this comes at the expense of longer RNN training times stemming from more parameters to train.

The regressors required to predict $\hat{y}_{k \in \{1,...,p\}}$ are henceforth represented by $\phi_k := \{y_0, u_0, \ldots, u_{k-1}\}$, and they are introduced into the RNN in a fashion illustrated in Figure 4. Equation (9) below serves as a shorthand to describe the RNN:

$$\hat{y}_k = \hat{f}_{RNN}(\phi_k), k \in \{1, ..., p\} \tag{9}$$

An RNN is characterised by the values of $W_{h,h}$, $W_{u,h}$, b_1 and b_2 for all layers, and these values constitute the set of parameters. These parameters are learnt from training data by minimising the predictive error of the model on the training set as determined through a user-specified loss function. The learning process is performed through the back-propagation through time (BPTT) algorithm that estimates the gradient of the loss function as a function of the weights, and an optimisation algorithm that uses the calculated gradient to adjust the existing weights. The adaptive moment estimation algorithm (Adam) [33] is an example of an optimisation algorithm that is widely used.

The RNN parameters may be difficult to train in some cases due to problems with vanishing and exploding gradients, and RNN cells with different structures can be used to circumvent this issue. Long Short-Term Memory cells (LSTMs) were developed to address these problems, and they are used as the repeating cell in this study. Further details on LSTM cells may be found in the Appendix B.

To generate the data set for training the RNNs in this study, a numerical model of the CSTR system based on Equations (2)–(4) is needed, because this data consists precisely of this system's response to MV perturbations. This numerical model was implemented in Python version 3.6.5, and its temporal evolution between sampling points was simulated using the explicit Runge-Kutta method of order 5(4) through the `scipy.integrate.solve_ivp` function.

MV perturbations are then introduced into the system to obtain its dynamic response. This procedure mimics the manual perturbation of the system through experiments, which become necessary in more complicated reaction systems whose reaction kinetics and overall dynamics resist expression in neat analytical forms. These perturbations take the form of pre-defined sequences of control actions, $\{u_{\text{exp},k}\}_{k\in\{0,\ldots,T_e-1\}}$, where the "exp" subscript refers to "experiment" and T_e represents the final time-step for the experiment. These control actions are separated by Δt.

To simulate the system's dynamic response to the experimental sequence, the control actions associated to the element in this sequence, $u_{\text{exp},k} = [q_{\text{exp},k}, T_{\text{exp},k}]'$, $k \in \{0,\ldots,T_e-1\}$, are applied at time-step k for a period of Δt, during which the system's evolution with this MV is simulated using the Runge-Kutta method of order 5(4). Once Δt elapses, $u_{\text{exp},k+1}$ is applied with the system evolution proceeding in the same manner. This procedure repeats until the final time-step, T_e, is completed.

A history of the system's experimental dynamic response is associated to each experimental sequence, which takes the form of $\{y_{\text{exp},k}\}_{k\in\{1,\ldots,T_e\}}$. $y_{\text{exp},k}$ is the measured system output at time-step k after $u_{\text{exp},k-1}$ has been applied to the system for a period of Δt. This history corresponds to the labels in machine learning terminology, and the data set is thereafter constructed from both the experimental sequences and their associated labels. For the p-step ahead prediction problem, each data point thus takes the form $\{y_k, u_k, u_{k+1}, \ldots, u_{k+p-1}\}$ with the associated label $\{y_{k+1}, \ldots, y_{k+p}\}$, $k \in \{0,\ldots,T_e-p\}$. $T_e - p$ data points can thus be extracted from each experimental sequence. Multiple experimental sequences can be constructed by perturbing the MVs in different ways, and such a sequence is shown in Figure 5 with its associated system dynamical response, with this sequence showing variations of q in the range $[0.70, 1.05]$ for fixed values of T. For this study, these sequences were generated in a fashion similar to Figure 5, by introducing linear changes of different frequencies to one component of the MV spanning its experimental range while keeping the other MV component constant.

Before training the RNNs, it is necessary to split the data set into the training set and the test set. The training set contains the data with which the RNN trains its weight matrices, and the test set will be used to determine the models with the best hyperparameters. The two hyperparameters considered in this study are the number of hidden nodes, which also correspond to the dimensionality of the hidden states, and the number of hidden layers.

The experimental simulations and the extraction of training data were performed using Python version 3.6.5 for this study. Keras version 2.1.5, an open-source neural network library written in Python, was used for RNN training with TensorFlow version 1.7.0 as its backend.

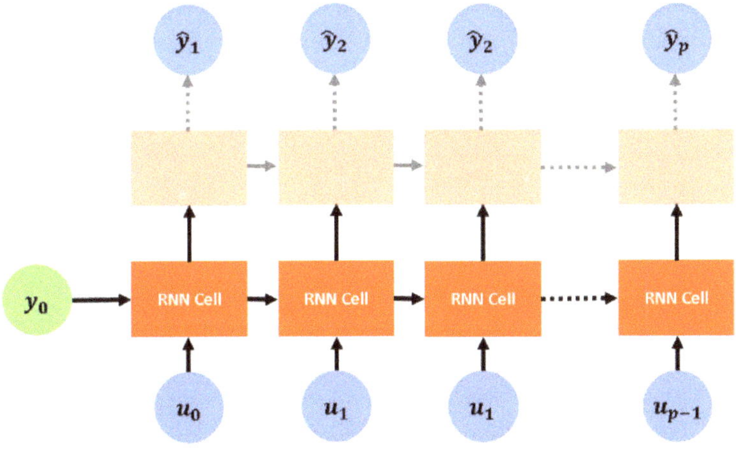

Figure 4. RNN structure for the p-step ahead prediction problem.

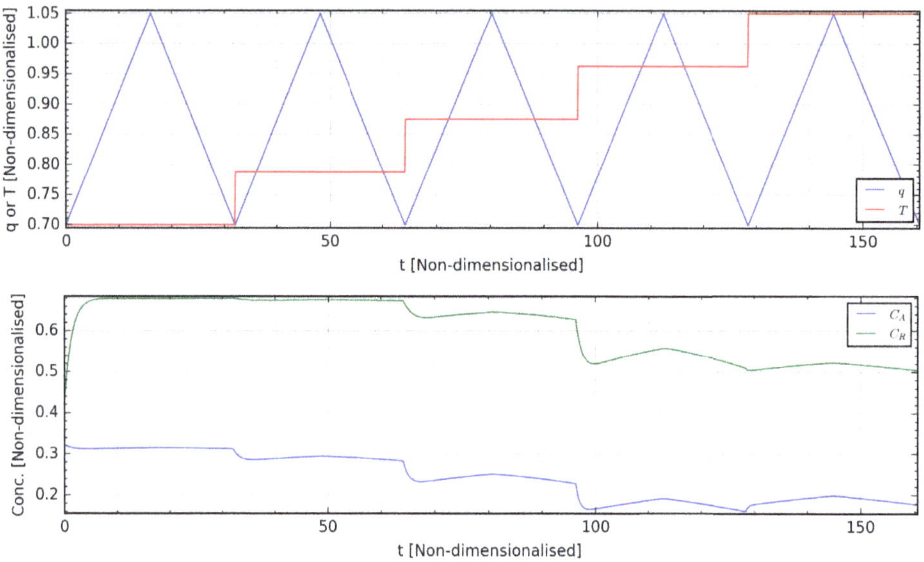

Figure 5. Example of an experimental sequence and the dynamic response with $\Delta t = 0.1$ time units.

3.2. Control Problem Formulation

MPC is a control strategy that selects a set of m future control moves, $\{u_0, ...u_{m-1}\}$, that minimises a cost function over a prediction of p steps by incorporating predictions of the dynamical system for these p steps, $\{\hat{y}_1, ..., \hat{y}_p\}$. The cost function is typically chosen to punish large control actions, which implies greater actuator power consumption, and differences between the state vector and the set-point at each time-step. Input and output constraints can also be factored into the MPC formulation. Since MPC performance depends on the quality of the predictions of the system, obtaining a reasonably accurate model through system identification is crucial.

The MPC problem can be formulated as shown in Equations (10)–(13) below:

$$\min_{\{\Delta u_0, \Delta u_1, ..., \Delta u_{m-1}\}} \left\{ \sum_{k=1}^{p} (\hat{y}_k - y_k^*)' Q_y (\hat{y}_k - y_k^*) + \sum_{k=0}^{m-1} \Delta u_k' Q_u \Delta u_k \right\} \text{ s.t.} \tag{10}$$

$$\hat{y}_k = \hat{f}_{RNN}(\phi_k), k \in \{1, \ldots, p\} \tag{11}$$

$$u_k \in [u_{\min,k}, u_{\max,k}], k \in \{0, \ldots, m-1\} \tag{12}$$

$$\Delta u_k \in [\Delta u_{\min,k}, \Delta u_{\max,k}], k \in \{0, \ldots, m-1\} \tag{13}$$

where $p \in \mathbb{Z}_+$ is the prediction horizon, $m \in \{1, \ldots, p\}$ the control horizon, $\hat{y}_k \in \mathbb{R}^2$ the prediction of the state vector for the discrete-time step k obtained from the RNN, $\hat{f}_{RNN}$ described in Equation (9), $y_k^* \in \mathbb{R}^2$ the set-point at time-step k, $u_k \in \mathbb{R}^2$ the MV for time-step k, $\Delta u_k \triangleq u_k - u_{k-1}$ the discrete-time rate of change of the MV which corresponds to the control action size at time-step k, $(Q_y, Q_u) \in (\mathbb{R}^{2\times 2})^2$ symmetric positive semi-definite weight matrices, and $(u_{\min,k}, u_{\max,k}, \Delta u_{\min,k}, \Delta u_{\max,k}) \in (\mathbb{R}^2)^4$ the lower and upper bounds for the control action and the rate of change of the control action at time-step k. In the above formulation, it is assumed that there are no changes in actuator position beyond time-step $m-1$, i.e., $\Delta u_{m-1} = \Delta u_{m+k} = 0, k \in \{0, \ldots, p-m-1\}$.

This optimization problem is not convex in general and thus does not possess special structures amenable to global optimality. This is therefore a Non-Linear Programming (NLP) problem, and modern off-the-shelf solvers can be used to solve it. This problem is solved at every time-step to yield the optimal control sequence for that time-step, $\{\Delta u_0^*, \ldots, \Delta u_{m-1}^*\}$. The first element, Δu_0^*, is then applied to the system until the next sampling instant, where the problem is solved again to yield another optimal control sequence. This procedure is then repeated in a moving horizon fashion.

The MPC controller in this study was implemented in Python version 3.6.5 through the `scipy.optimize.minimize` function, and the sequential least squares quadratic programming (SLSQP) algorithm was selected as the option for this solver.

4. Results and Discussion

The RNN-MPC workflow consists of (i) gathering experimental sequence data with perturbations of the MV to generate the data set; (ii) learning the RNNs from the training set and validating their performance with test data, and selecting the RNNs with the best test performance; and (iii) integrating the chosen RNN with the MPC and finally evaluating its control performance.

4.1. System Identification

RNNs with 250, 500 and 1000 hidden nodes were trained for this study, and the number of RNN layers for these models was also varied from 1 to 3. The validation performance of these models on the test set was quantified through the root-mean-square error (RMSE) metric, which is defined as shown in Equation (14) below:

$$\text{RMSE} = \sqrt{\left(\frac{1}{N_{\text{test}}}\sum_{i=1}^{N_{\text{test}}}\sum_{j=1}^{p}\left((\hat{y}_j)_i - (y_{\text{exp},j})_i\right)^2\right)} \tag{14}$$

where $N_{\text{test}} \in \mathbb{Z}_+$ is the number of data points in the test set, $p \in \mathbb{Z}_+$ the prediction horizon, $(\hat{y}_j)_i \in \mathbb{R}^2$ the j-ahead output prediction for data point i, and $(y_{\text{exp},j})_i \in \mathbb{R}^2$ the corresponding experimentally-determined output which also represents the ground truth that predictions are tested against. The smaller the RMSE of the model on the test data, the better its general predictive power.

Table 2 tabulates the validation performance of the RNNs on the test data, and it can be observed that prediction performance improves with an increasing number of hidden nodes. This is because hidden states of greater dimensionalities offer greater latitude for capturing more features of the underlying system, contributing to its temporal dynamics. An improvement of the validation performance can also be observed with an increase of the number of hidden layers from 1 to 2. This observation is attributable to the potential of deep hierarchical models to better represent some functions than shallower models [34].

Table 2. Root-mean-square error (RMSE) over test data of RNNs trained over 1000 epochs.

No. Layers / No. Nodes	250	500	1000
1	0.0299	0.0268	0.0206
2	0.0238	0.0118	0.0083
3	0.0262	0.0119	0.0125

However, performance deteriorated when the number of layers was increased from 2 to 3, and this may be due to the additional layers introducing more model parameters that needed to be tuned, such that the training data set became too small to tune all of them correctly.

Table 2 reveals that RNNs with two hidden layers and 1000 hidden nodes for each RNN cell gave the best validation performance over the test set, with this representing the optimised hyperparameter configuration. An RNN with two hidden layers and 2000 hidden nodes was also trained and it was observed that performance deteriorated when even more hidden nodes were used. This result is tabulated in Table 3, and possible explanations may be that the 2000-node network overfitted the training data, or that the use of 2000-node cells introduced more parameters to the model, rendering the existing training data set too small to tune the additional parameters correctly. RNN with two hidden layers will be used for subsequent closed-loop control studies.

Table 3. RMSE over test data of RNNs with hidden layers and either 1000 or 2000 nodes trained over 1000 epochs.

No. Layers	No. Nodes	RMSE
2	1000	0.0083
2	2000	0.0177

It is also worth noting that validation performance is also a function of the quality of the training set. This performance can be further improved by using a larger training set reflecting the system dynamics within the control range, particularly for any control scenarios earmarked as important.

The prediction performances of this optimised RNN on the training and test sets are respectively shown in Figure 6 below for completeness.

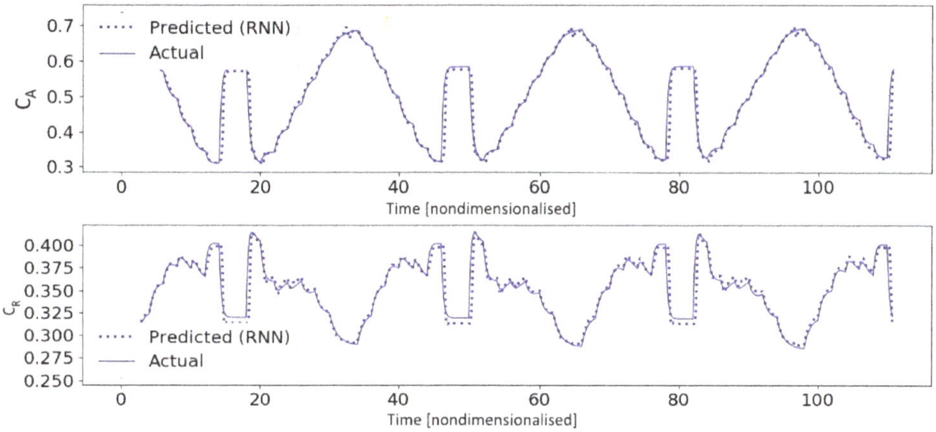

(**a**)

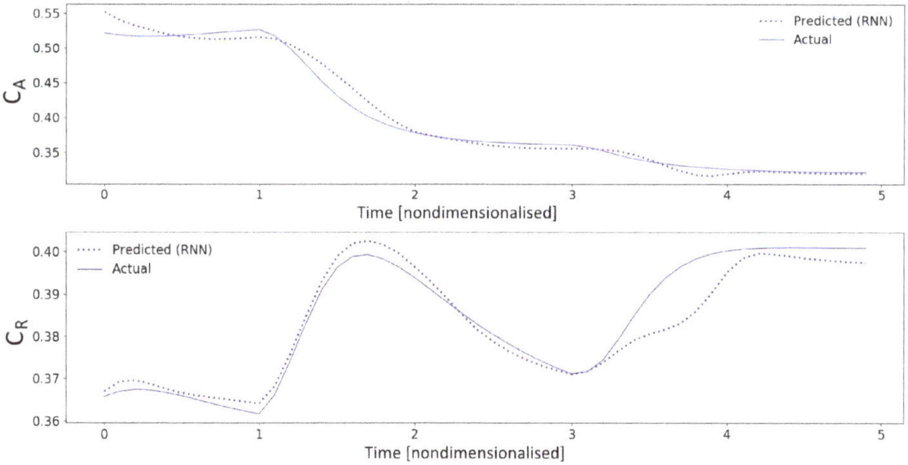

(**b**)

Figure 6. System identification performance of the optimised RNN. (**a**) Training performance of the optimised RNN. (**b**) Validation performance of the optimised RNN on test data.

Figure 6a reveals a good fit of the RNN to the training data, and testifies to the model's ability to reflect highly dynamic outputs from highly dynamic training data. Figure 6b shows that the model succeeded in capturing the general trends for previously unseen test data. Even though this fit was not perfect, particularly for C_R between 3 and 4 time units, it will be demonstrated in a later section that satisfactory closed-loop performance can still be achieved with predictions from this RNN.

4.2. RNN-MPC Closed-Loop Control Performance

As described in Section 2.2, the objective in both control scenarios is to bring the system quickly and smoothly to a target set-point where the ratio of C_R to C_A is maximised, because this maximises yield while minimising downstream separations cost. Since MPC strategies based on single linear models do not perform well for this example [32], non-linear MPC solutions will be required.

The MPC formulation was shown in Equations (10)–(13), and the prediction and control horizons for the MPC controller implemented for this study were set to 10 for both control scenarios, i.e., $p = m = 10$. Since the sampling rate, Δt, was set as 0.1 time units, p and m correspond to 1.0 time unit. The total simulation time, N_{sim}, was set to 40, corresponding to 4.0 time units. The weight matrices for the controller, Q_y and Q_u, were defined as diag[2.4, 5.67] and diag[25, 25] respectively. The coefficients selected for Q_y reflect that controlling C_R is more important than controlling C_A, and Q_u contain coefficients that are even larger to avoid control actions that are over-aggressive, since these control actions cause system instability. For the constraints, $[u_{\text{min},k}, u_{\text{max},k}] = [[0.75, 0.5]', [0.85, 1.1]']$, $\forall k \in \{0, N_{\text{sim}-1}\}$ and $[\Delta u_{\text{min},k}, \Delta u_{\text{max},k}] = [[-0.1, -0.1]', [0.1, 0.1]']$, $\forall k \in \{0, N_{\text{sim}-1}\}$.

In what follows, the closed-loop control performance of the RNN-MPC controller will be benchmarked against the NMPC controller whose internal model is the actual full plant model as described in Equations (2)–(4). This NMPC controller was implemented using the same software as that of the RNN-MPC controller, whose implementation was described in Section 3.2.

The indices used to evaluate control performance are defined as shown in Equations (15) and (16) below:

$$J = \sum_{k=1}^{N_{\text{sim}}} \left((y_k - y_k^*)' Q_y (y_k - y_k^*) + \Delta u_k' Q_u \Delta u_k \right) \tag{15}$$

$$\mathcal{I} = \left(1 - \frac{J_{\text{RNN}} - J_{\text{NMPC}}}{J_{\text{NMPC}}} \right) \times 100\% \tag{16}$$

where $y_{k \in \{1,\ldots,N_{\text{sim}}\}} \in \mathbb{R}^2$ and $y^*_{k \in \{1,\ldots,N_{\text{sim}}\}} \in \mathbb{R}^2$ are the measured output and the set-point at time-step k respectively, $\Delta u_k \in \mathbb{R}^2$ the size of the control action at time-step k, and $(Q_y, Q_u) \in (\mathbb{R}^{2\times 2})^2$ the symmetric positive semi-definite weight matrices.

Figures 7–10 below show the performance of the RNN-MPC and NMPC controllers on both control scenarios for 250, 500, 1000 and 2000 hidden nodes respectively. Table 4 tabulates the performance index averaged over both control scenarios for each RNN-MPC controller for the same set of values for the number of hidden nodes.

Table 4. Performance of closed-loop RNN-MPC as a function of RNN nodes (two layers).

No. Nodes	Average Performance Index, $\mathcal{I}_{\text{avg}}$	Comments
250	93.7	Steady-state Offset
500	95.8	Steady-state Offset
1000	100.0	Desired Performance
2000	98.6	Steady-state Offset

The results show that the RNN-MPC controllers exhibited stability even with only 250 hidden nodes for both control scenarios, suggesting a robustness associated with the RNN-MPC combination.

For the RNN-MPC controller with 250 hidden nodes, while the NMPC controller exhibited no steady-state offset, this RNN-MPC controller showed significant steady-state offsets for both control scenarios. However, the initial transient response of the system under RNN-MPC control was similar to its NMPC benchmark for both scenarios, suggesting that the RNN succeeded in capturing the global dynamics of the system, and that 250 nodes may have not been sufficient for the model to learn the subtler dynamics that would have allowed the controller to bring the system precisely back to the set-point. A notable improvement in performance was observed with 500 nodes, although steady-state offsets remained for both scenarios.

The best closed-loop control performance was obtained with 1000 nodes, and this is consistent with the finding that this RNN had the best validation performance on the test data. The control performance of this RNN-MPC was in fact comparable to its NMPC counterpart, with an $\mathcal{I}_{\text{avg}}$ of 100.0. For 2000 nodes, a deterioration of control performance could be seen with the reappearance of the steady-state offsets for both scenarios, and this may be attributable to the RNN's poorer validation performance on the test data.

To assess the temporal performance of the RNN-MPC controller, 1000 model inputs were generated randomly and the average computational times required by the NMPC and the RNN-MPC controllers to generate a single prediction for use in the optimiser are tabulated in Table 5. The RNN for this temporal benchmarking has 1000 hidden nodes and two layers, and the NMPC generates predictions by using the Runge-Kutta method of order 5(4) to march the system forward in time. The predictions for both controllers were generated on an Intel(R) Xeon(R) CPU E5-1650 with 16 GB RAM.

Table 5. Time required to generate a single prediction averaged over 1000 predictions.

	NMPC	RNN-MPC
Time required	1.55 ms	1.17 ms

The results in Table 5 show that, for the simulation context described in the previous paragraph, the times required by the internal models of both controllers to generate a single prediction are of the same order of magnitude, with the RNN even outperforming its NMPC counterpart slightly. This testifies to the RNNs being a model that can generate predictions for MPC control at least as fast as the true plant model for this CSTR system.

This CSTR system is a comparatively simple dynamical system, and other systems can have more complex exact representations whose evaluation for prediction purposes may take a considerably longer time. Even though these models generate very good predictions, they would be too slow to evaluate to be useful for online control. Having observed from earlier discussions that stable control performance can be achieved if the RNN succeeds in capturing the global dynamics, imperfect predictions may in fact be tolerable if it entails an RNN that is easier to train and faster to evaluate. RNNs can therefore serve to balance the needs of MPC for good and quick predictions.

To summarise, even if the system identification by the RNNs does not perfectly describe the underlying plant dynamics, stable closed-loop control performance can be achieved as long as the global dynamics of the system are captured. Further refinements to the RNN can be made by optimising its hyperparameters, which in this study have been selected to be the number of hidden nodes and hidden layers, to obtain a configuration that is capable of capturing the subtler dynamics of the system while avoiding overfitting on the training set.

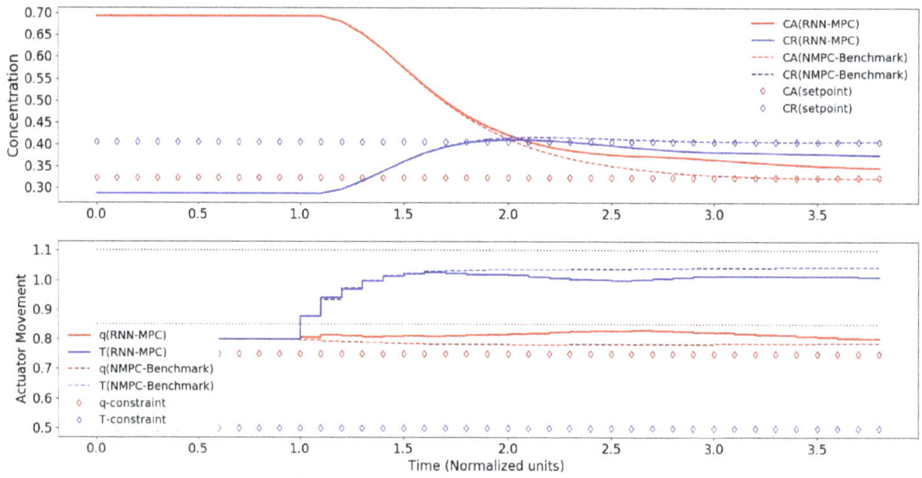

(**a**)

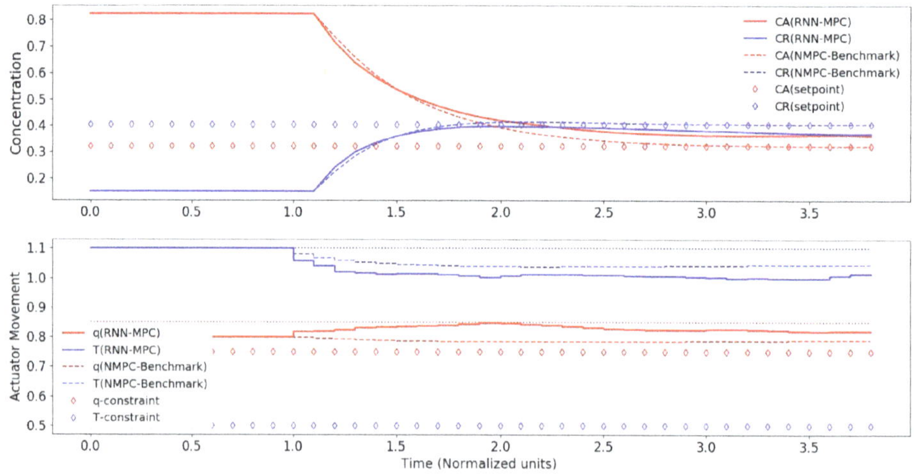

(**b**)

Figure 7. Control performance of RNN-MPC with 250 hidden nodes and two RNN layers. (**a**) Control performance for case I (start-up). (**b**) Control performance for case II (upset-recovery).

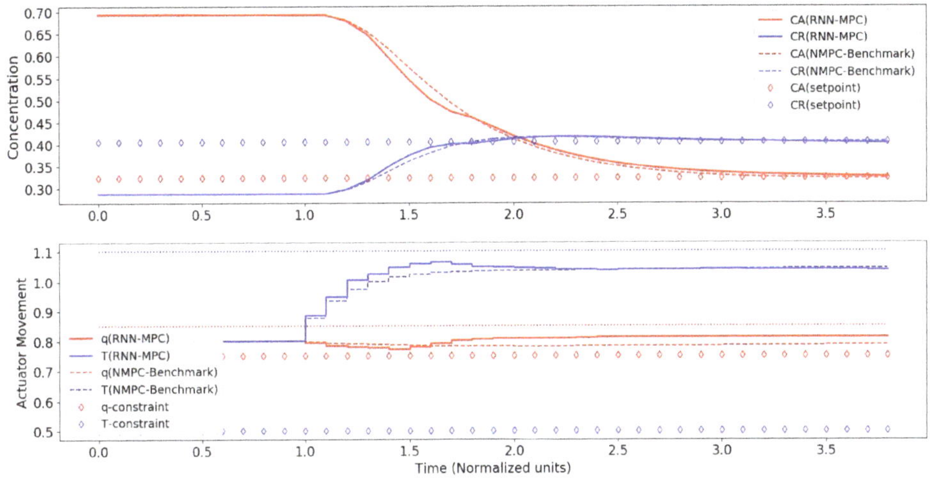

(**a**)

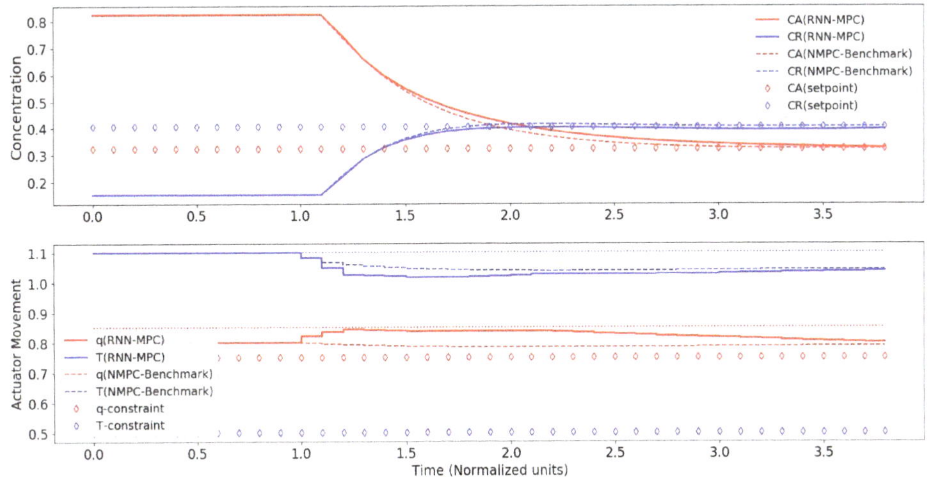

(**b**)

Figure 8. Control performance of RNN-MPC with 500 hidden nodes and two RNN layers. (**a**) Control performance for case I (start-up). (**b**) Control performance for case II (upset-recovery).

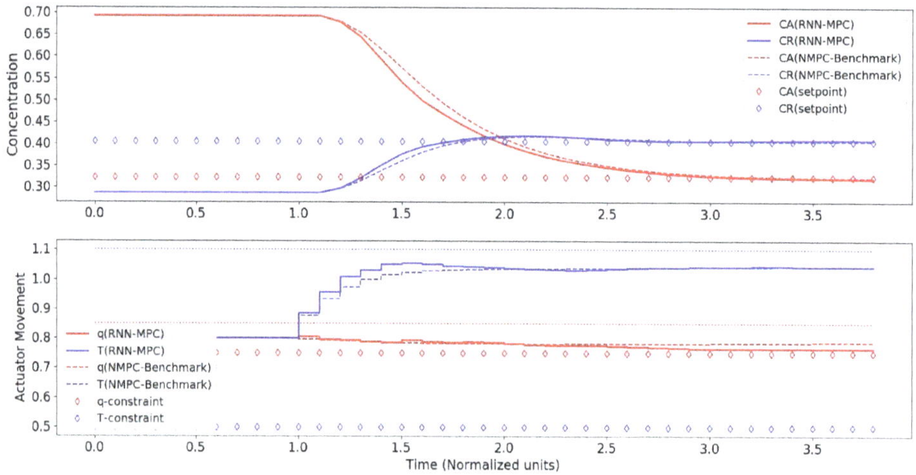

(**a**)

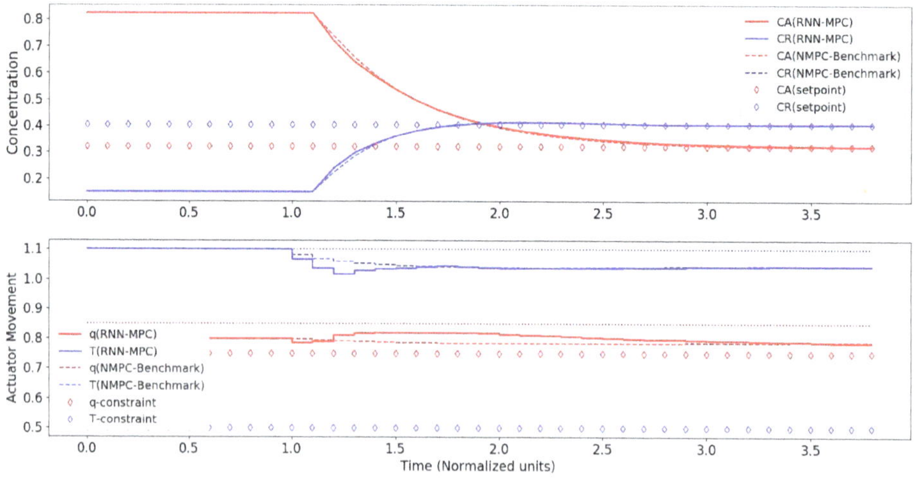

(**b**)

Figure 9. Control performance of RNN-MPC with 1000 hidden nodes and two RNN layers. (**a**) Control performance for case I (start-up). (**b**) Control performance for case II (upset-recovery).

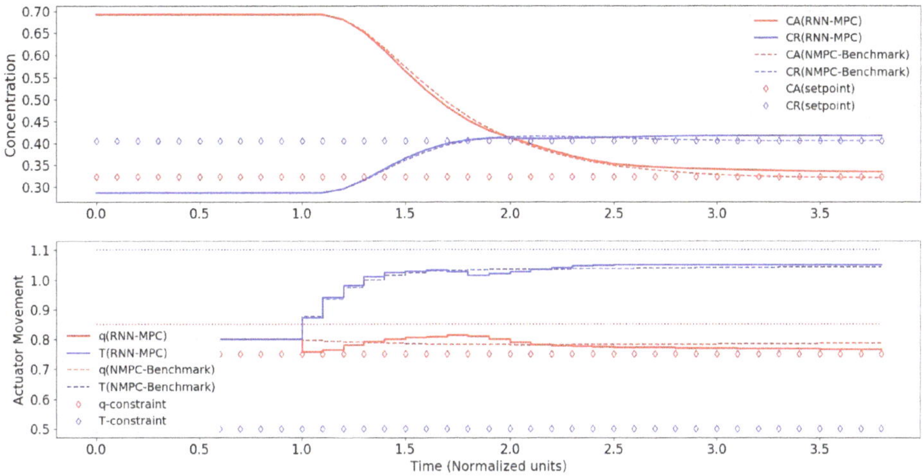

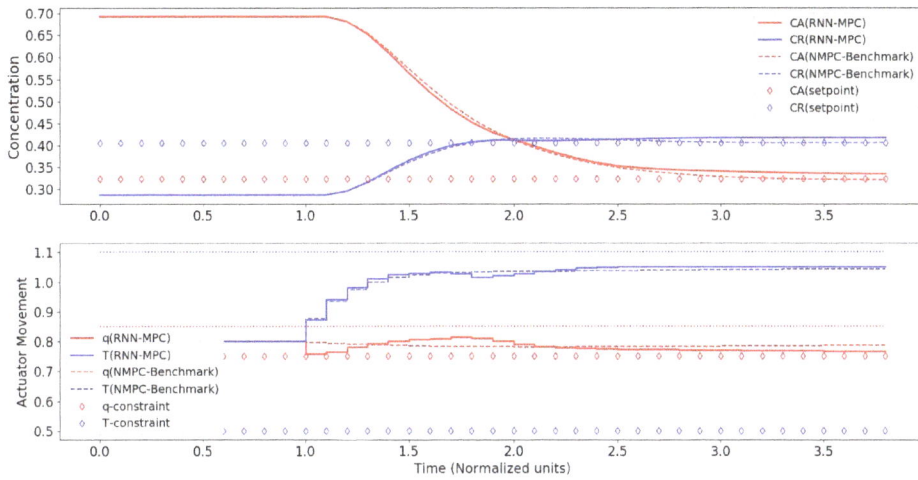

(**a**)

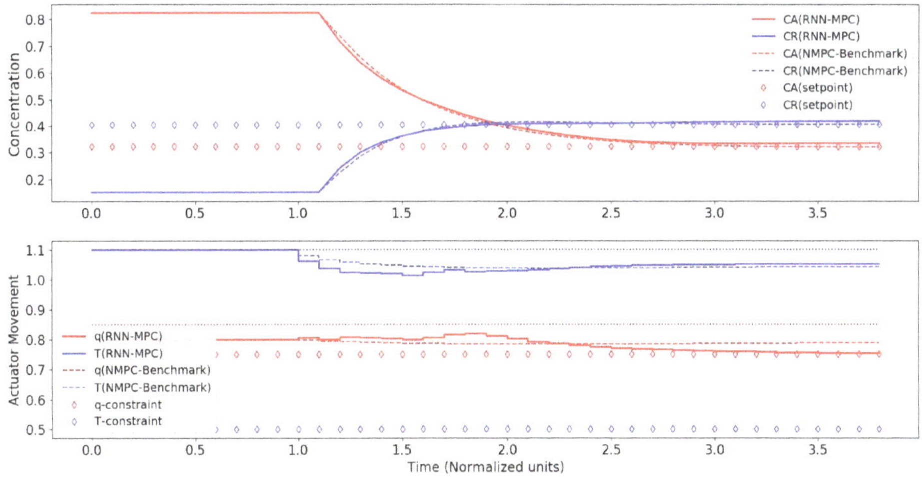

(**b**)

Figure 10. Control performance of RNN-MPC with 2000 hidden nodes and two RNN layers. (**a**) Control performance for case I (start-up). (**b**) Control performance for case II (upset-recovery).

5. Conclusions and Future Research

In this study, a multiple-input multiple-output CSTR that houses a reaction exhibiting output dynamics was studied. The methods for generating the control-oriented RNNs and integrating their use into MPC controllers were articulated, and the performance of the RNN-MPC controllers was benchmarked against NMPC controllers that used the true plant model directly as its internal model.

Two control scenarios were simulated, one involving reactor start-up and the other involving upset-recovery, and the results showed that even if the RNNs do not perfectly describe the underlying plant dynamics, RNN-MPC controllers can give stable closed-loop control performance for these scenarios as long as the global dynamics of the system are captured. These RNNs can be refined through hyperparameter optimisation to obtain configurations that are capable of capturing the subtler dynamics of the system while avoiding overfitting on the training set, further augmenting their effectiveness as control-oriented models for MPC.

It is finally noted again that linear controllers with integral action perform poorly for systems exhibiting input multiplicities, and that rigorous physics-based models like the model used for the benchmark NMPC controller can be difficult to obtain in practice and may be too expensive to evaluate to be useful for online MPC control. The approach presented in this study exploits the ability of RNNs to capture the temporal dynamics of the system, and it was shown that RNN-based system identification methods can give control-oriented models for MPC that offer satisfactory performance for two important control scenarios. As the creation of these RNNs is a data-driven process which yields progressively more effective models with larger amounts of process and sensor data, these models serve indeed as an alternative source of control-oriented models that capture the dynamics of highly non-linear systems sufficiently well while being relatively easy to obtain and evaluate online. This opens promising avenues for the application of RNNs or other ML models in enabling continuous pharmaceutical manufacturing practices.

A potential way to bring this study forward is to compare different ML or RNN architectures for MPC control-oriented system identification. Measurement noise can be incorporated to simulate realistic sampling scenarios and evaluate the robustness to noise of the ML system identification and the ML-MPC control techniques. Cross-validated optimisation of more hyperparameters characterising the ML model, which can potentially include the regular dropout rates between hidden layers, the learning rates and gradient clipping values among others, as well as a more detailed study of the effects of the experimental sequences on the quality of the ML model trained, can also give future practitioners useful heuristics for creating and tuning MLs for their own purposes.

Future research can also involve extending the methodology to multiple CSTRs, and to reaction kinetics of increasing complexity and of more immediate relevance to the pharmaceutical industry. These can be done in view of an eventual real-world implementation of these RNN-MPC or ML-MPC control strategies, from which further comparisons of these controllers to conventional PID controllers, and other MPC strategies like linear MPC and offset-free MPC, can also be made.

Author Contributions: Wee Chin Wong and Xiaonan Wang conceived and designed the methodology; Wee Chin Wong and Ewan Chee performed the simulations; Wee Chin Wong, Ewan Chee, Jiali Li and Xiaonan Wang analyzed the data and wrote the paper.

Funding: The authors thank the MOE AcRF Grant in Singapore for financial support of the projects on Precision Healthcare Development, Manufacturing and Supply Chain Optimization (R-279-000-513-133) and Advanced Process Control and Machine Learning Methods for Enhanced Continuous Manufacturing of Pharmaceutical Products (R-279-000-541-114).

Acknowledgments: We would like to acknowledge S. A. Khan from the Department of Chemical and Biomolecular Engineering, National University of Singapore for his input on reaction kinetics and multiple fruitful discussions. We also thank the Pharma Innovation Programme Singapore (PIPS), which provided the source of motivation for this work.

Conflicts of Interest: The authors declare no conflict of interest.

Abbreviations

The following abbreviations are used in this manuscript:

Adam	Adaptive Moment Estimation
ANN	Artificial Neural Network
API	Active Pharmaceutical Ingredient
BPTT	Back-Propagation Through Time
CPP	Critical Process Parameter
CQA	Critical Quality Attribute
CSTR	Continuous-Stirred Tank Reactor
FBU	Feeding Blending Unit
LSTM	Long Short-Term Memory
MPC	Model Predictive Control
MV	Manipulated Vector
NLP	Non-Linear Programming
NMPC	Non-Linear MPC
PID	Proportional-Integral-Derivative (Control)
QbD	Quality by Design
QDMC	Quadratic Dynamic Matrix Control
RMSE	Root-Mean-Square Error
RNN	Recurrent Neural Network
RNN-MPC	RNN-based MPC
SLSQP	Sequential Least Squares Quadratic Programming

Appendix A. Kinetic Parameters for Plant Model

For this study, C_{A0} was assigned a value of 0.8. The vector of Arrhenius pre-exponentials, k_0, was defined as $[1.0, 0.7, 0.1, 0.006]'$, and the vector of normalized activation energies, $\frac{E}{RT_0}$, as $[8.33, 10.0, 50.0, 83.3]'$.

Appendix B. Long Short-Term Memory Cells

In conventional RNNs, the BPTT algorithm results in the partial derivative of the error function being multiplied numerous times to the weights corresponding to the various connections, or edges, of the RNN. Depending on the spectral radius of the recurrent weight matrix, this leads to either the vanishing or exploding gradients problem, which severely limits the learning quality of the network.

LSTM cells have been proposed to mitigate this issue. These cells use several gating functions, like the 'forget', 'input' and 'output' gating functions, that serve to modulate the propagation of signals between cells. This cell structure avoids the gradient vanishing or exploding problem.

The basic LSTM cell structure is mathematically expressed as follows in Equations (A1)–(A6) below:

$$h_k = o_k * \tanh(C_k) \tag{A1}$$

$$C_k = f_k * C_{k-1} + i_k * \tilde{C}_k \tag{A2}$$

$$\tilde{C}_k = \tanh(W_C[h_{k-1}, u_k]' + b_c) \tag{A3}$$

$$i_k = \sigma(W_i.[h_{k-1}, u_k]' + b_i) \tag{A4}$$

$$f_k = \sigma(W_f.[h_{k-1}, u_k]' + b_f) \tag{A5}$$

$$o_k = \sigma(W_o.[h_{k-1}, u_k]' + b_o) \tag{A6}$$

where $k \in \mathbb{Z}_+$ is the time index, $h_k \in \mathbb{R}^N$ the hidden state variable, and $u_k \in \mathbb{R}^2$ the input variable. $f_k \in [0,1]$, $i_k \in [0,1]$ and $o_k \in [0,1]$ are the 'forgetting', 'input' and 'output' gates respectively, and are characterised by their respective weight matrices and bias vectors, with $\sigma : \mathbb{R}^N \rightarrow [0,1]^N$ an activation function. The input gate controls the degree to which the cell state, represented by Equation (A2) and distinct from the hidden state variable, is affected by candidate information, and the output gate controls how this cell state affects other cells. The forget gate modulates the self-recurrent connection of the cell, allowing it thus to partially remember the previous cell state in a fashion similar to traditional RNNs. $*$ refers to a point-wise multiplication.

References

1. Lakerveld, R.; Benyahia, B.; Heider, P.L.; Zhang, H.; Wolfe, A.; Testa, C.J.; Ogden, S.; Hersey, D.R.; Mascia, S.; Evans, J.M.; et al. The application of an automated control strategy for an integrated continuous pharmaceutical pilot plant. *Org. Process Res. Dev.* **2015**, *19*, 1088–1100. [CrossRef]
2. Schaber, S.D.; Gerogiorgis, D.I.; Ramachandran, R.; Evans, J.M.B.; Barton, P.I.; Trout, B.L. Economic analysis of integrated continuous and batch pharmaceutical manufacturing: A case study. *Ind. Eng. Chem. Res.* **2011**, *50*, 10083–10092. [CrossRef]
3. Glasnov, T. *Continuous-Flow Chemistry in the Research Laboratory: Modern Organic Chemistry in Dedicated Reactors at the Dawn of the 21st Century*; Springer International Publishing: Basel, Switzerland, 2016; p. 119.
4. Gutmann, B.; Cantillo, D.; Kappe, C.O. Continuous-flow technology—A tool for the safe manufacturing of active pharmaceutical ingredients. *Angew. Chem. Int. Ed.* **2015**, *54*, 6688–6728. [CrossRef] [PubMed]
5. Poechlauer, P.; Colberg, J.; Fisher, E.; Jansen, M.; Johnson, M.D.; Koenig, S.G.; Lawler, M.; Laporte, T.; Manley, J.; Martin, B.; et al. Pharmaceutical roundtable study demonstrates the value of continuous manufacturing in the design of greener processes. *Org. Process Res. Dev.* **2013**, *17*, 1472–1478. [CrossRef]
6. Benyahia, B.; Lakerveld, R.; Barton, P.I. A plant-wide dynamic model of a continuous pharmaceutical process. *Ind. Eng. Chem. Res.* **2012**, *51*, 15393–15412. [CrossRef]
7. Susanne, F.; Martin, B.; Aubry, M.; Sedelmeier, J.; Lima, F.; Sevinc, S.; Piccioni, L.; Haber, J.; Schenkel, B.; Venturoni, F. Match-making reactors to chemistry: A continuous manufacturing-enabled sequence to a key benzoxazole pharmaceutical intermediate. *Org. Process Res. Dev.* **2017**, *21*, 1779–1793. [CrossRef]
8. Mascia, S.; Heider, P.L.; Zhang, H.; Lakerveld, R.; Benyahia, B.; Barton, P.I.; Braatz, R.D.; Cooney, C.L.; Evans, J.M.B.; Jamison, T.F.; et al. End-to-end continuous manufacturing of pharmaceuticals: Integrated synthesis, purification, and final dosage formation. *Angew. Chem. Int. Ed.* **2013**, *52*, 12359–12363. [CrossRef] [PubMed]
9. Brueggemeier, S.B.; Reiff, E.A.; Lyngberg, O.K.; Hobson, L.A.; Tabora, J.E. Modeling-based approach towards quality by design for the ibipinabant API step modeling-based approach towards quality by design for the ibipinabant API step. *Org. Process Res. Dev.* **2012**, *16*, 567–576. [CrossRef]
10. Mesbah, A.; Paulson, J.A.; Lakerveld, R.; Braatz, R.D. Model predictive control of an integrated continuous pharmaceutical manufacturing pilot plant. *Org. Process Res. Dev.* **2017**, *21*, 844–854. [CrossRef]
11. Rasoulian, S.; Ricardez-Sandoval, L.A. Stochastic nonlinear model predictive control applied to a thin film deposition process under uncertainty. *Chem. Eng. Sci.* **2016**, *140*, 90–103. [CrossRef]
12. Rasoulian, S.; Ricardez-Sandoval, L.A. A robust nonlinear model predictive controller for a multiscale thin film deposition process. *Chem. Eng. Sci.* **2015**, *136*, 38–49. [CrossRef]
13. Hussain, M.A. Review of the applications of neural networks in chemical process control simulation and online implementation. *Artif. Intell. Eng.* **1999**, *13*, 55–68. [CrossRef]
14. Cheng, L.; Liu, W.; Hou, Z.G.; Yu, J.; Tan, M. Neural-network-based nonlinear model predictive control for piezoelectric actuators. *IEEE Trans. Ind. Electron.* **2015**, *62*, 7717–7727. [CrossRef]
15. Xiong, Z.; Zhang, J. A batch-to-batch iterative optimal control strategy based on recurrent neural network models. *J. Process Control* **2005**, *15*, 11–21. [CrossRef]
16. Tian, Y.; Zhang, J.; Morris, J. Modeling and optimal control of a batch polymerization reactor using a hybrid stacked recurrent neural network model. *Ind. Eng. Chem. Res.* **2001**, *40*, 4525–4535. [CrossRef]
17. Mujtaba, I.; Hussain, M. *Applications of Neural Networks and Other Learning Technologies in Process Engineering*; Imperial College Press: London, UK, 2001.
18. Nagy, Z.K. Model based control of a yeast fermentation bioreactor using optimally designed artificial neural networks. *Chem. Eng. J.* **2007**, *127*, 95–109. [CrossRef]
19. Alanqar, A.; Durand, H.; Christofides, P.D. On identification of well-conditioned nonlinear systems: Application to economic model predictive control of nonlinear processes. *AIChE J.* **2015**, *61*, 3353–3373. [CrossRef]

20. Wang, X.; El-Farra, N.H.; Palazoglu, A. Proactive Reconfiguration of Heat-Exchanger Supernetworks. *Ind. Eng. Chem. Res.* **2015**, *54*, 9178–9190. [CrossRef]
21. Byeon, W.; Breuel, T.M.; Raue, F.; Liwicki, M. Scene labeling with LSTM recurrent neural networks. In Proceedings of the 2015 IEEE Conference on Computer Vision and Pattern Recognition (CVPR), Boston, MA, USA, 7–12 June 2015; pp. 3547–3555.
22. Cho, K.; van Merrienboer, B.; Gülçehre, Ç.; Bougares, F.; Schwenk, H.; Bengio, Y. Learning phrase representations using RNN encoder-decoder for statistical machine translation. *arXiv* **2014**, arXiv:1406.1078.
23. Lee, J.H.; Shin, J.; Realff, M.J. Machine learning: Overview of the recent progresses and implications for the process systems engineering field. *Comput. Chem. Eng.* **2018**, *114*, 111–121. [CrossRef]
24. Rehrl, J.; Kruisz, J.; Sacher, S.; Khinast, J.; Horn, M. Optimized continuous pharmaceutical manufacturing via model-predictive control. *Int. J. Pharm.* **2016**, *510*, 100–115. [CrossRef] [PubMed]
25. Rawlings, J.B.; Mayne, D.Q. *Model Predictive Control: Theory and Design*; Nob Hill: Madison, WI, USA, 2009.
26. Tatjewski, P. *Advanced Control of Industrial Processes, Structures and Algorithms*; Springer: London, UK, 2007.
27. Garcia, C.E.; Morshedi, A. Quadratic programming solution of dynamic matrix control (QDMC). *Chem. Eng. Commun.* **1986**, *46*, 73–87. [CrossRef]
28. Pan, Y.; Wang, J. Model predictive control of unknown nonlinear dynamical systems based on recurrent neural networks. *IEEE Trans. Ind. Electron.* **2012**, *59*, 3089–3101. [CrossRef]
29. Seyab, R.A. Differential recurrent neural network based predictive control. *Comput. Chem. Eng.* **2008**, *32*, 1533–1545. [CrossRef]
30. Koppel, L.B. Input multiplicities in nonlinear, multivariable control systems. *AIChE J.* **1982**, *28*, 935–945. [CrossRef]
31. Seki, H.; Ooyama, S.; Ogawa, M. Nonlinear model predictive control using successive linearization—Application to chemical reactors. *Trans. Soc. Instrum. Control Eng.* **2004**, *E-3*, 66–72.
32. Bequette, B.W. Non-linear model predictive control : A personal retrospective. *Can. J. Chem. Eng.* **2007**, *85*, 408–415. [CrossRef]
33. Kingma, D.P.; Ba, J. Adam: A Method for stochastic optimization. *arXiv* **2014**, arXiv:1412.6980.
34. Pascanu, R.; Gulcehre, C.; Cho, K.; Bengio, Y. How to construct deep recurrent neural networks. *arXiv* **2013**, arXiv:1312.6026.

MDPI
St. Alban-Anlage 66
4052 Basel
Switzerland
Tel. +41 61 683 77 34
Fax +41 61 302 89 18
www.mdpi.com

Mathematics Editorial Office
E-mail: mathematics@mdpi.com
www.mdpi.com/journal/mathematics

www.ingramcontent.com/pod-product-compliance
Lightning Source LLC
Chambersburg PA
CBHW051839210326
41597CB00033B/5708
* 9 7 8 3 0 3 8 9 7 4 2 0 8 *